Mass spectroscopy

Second edition

H. E. Duckworth
Emeritus Professor of Physics, Universities of Manitoba and Winnipeg, Canada

R. C. Barber
Professor of Physics, University of Manitoba, Canada

V. S. Venkatasubramanian
Formerly Professor of Physics, Indian Institute of Science, Bangalore

CAMBRIDGE UNIVERSITY PRESS
Cambridge
New York Port Chester
Melbourne Sydney

CAMBRIDGE UNIVERSITY PRESS
Cambridge, New York, Melbourne, Madrid, Cape Town, Singapore,
São Paulo, Delhi, Dubai, Tokyo, Mexico City

Cambridge University Press
The Edinburgh Building, Cambridge CB2 8RU, UK

Published in the United States of America by Cambridge University Press, New York

www.cambridge.org
Information on this title: www.cambridge.org/9780521386890

First published 1958
Second edition 1986
First paperback edition 1990
Re-issued 2010

A catalogue record for this publication is available from the British Library

Library of Congress catalogue card number: 84-23066

British Library Congress catalogueing in publication data

Duckworth. H.E.
Mass spectroscopy - (Cambridge monographs
on physics)
1. Mass spectrometry
I. Title II. Barber, R. C.
III. Venkatasubramanian, V.S.
543'.0873 QC454.M3

ISBN 978-0-521-23294-5 Hardback
ISBN 978-0-521-38689-6 Paperback

CAMBRIDGE MONOGRAPHS ON PHYSICS

Mass spectroscopy

Contents

Preface to first edition

In this monograph I have undertaken to give a concise description of two aspects of mass spectroscopy. The first of these is concerned with the principles which govern the operation of mass spectroscopes. This includes not only the schemes which have been employed to obtain mass resolution and dispersion, but also the methods which have been devised for the production and detection of positive ion beams. The second aspect is concerned with the applications of mass spectroscopy, which, it will be realized, are extraordinarily diverse. The many applications in physics are described in some detail, as are those in geophysics and in certain branches of chemical physics. However, the part played by the mass spectrometer in the study of the mechanism and rate of chemical reactions is no more than mentioned, in the belief that these experiments are primarily of interest to the chemist. A third aspect of mass spectroscopy, which might be termed the engineering aspect, concerned as it is with electronic and high vacuum techniques, is not felt to be a proper subject for inclusion in a monograph of this size. The monograph concludes with tables of isotopic abundances and atomic masses, plus an extensive bibliography. It should be useful, both as a self-contained compendium and as a generous source of further references, to all those whose work involves mass spectroscopy in any of its multifarious forms.

I have made use of several of the large-scale discussions of mass spectroscopy which have been published in book or hand-book form. These include F. W. Aston, *Mass Spectra and Isotopes* (Edward Arnold and Co., 1942); M. G. Inghram, *Modern Mass Spectroscopy* (in *Advances in Electronics*, vol. 1, Academic Press Inc., 1948); A. Guthrie and R. K. Wakerling, eds., *Characteristics of Electrical Discharges in Magnetic Fields* (McGraw-Hill, 1949); K. T. Bainbridge, 'Charged Particle Dynamics and Optics, Relative Isotopic Abundances of the Elements, Atomic Masses' (in *Experimental Nuclear Physics*, vol. 1, Wiley, 1953); *Mass Spectroscopy in Physics Research* (National Bureau of Standards Circular 522, 1953); G. P. Barnard, *Modern Mass Spectrometry* (the Institute of Physics, London, 1953); H. Ewald and H. Hintenberger, *Methoden und Anwendun-*

gen der Massenspektroskopie (Verlag Chemie, GMBH, Weinheim, 1953); *Applied Mass Spectrometry* (The Institute of Petroleum, London, 1954); A. J. B. Robertson, *Mass Spectrometry* (Methuen, 1954); M G. Inghram and R. J. Hayden, *A Handbook on Mass Spectroscopy* (Nuclear Energy Series, Report No. 14, 1954); C. E. Berry and J. K. Walker, 'Industrial Applications–Mass Spectrometry' (*Annual Review of Nuclear Science*, vol. v, Annual Reviews, Inc., 1955); M. L. Smith ed., *Electromagnetically Enriched Isotopes and Mass Spectrometry* (Butterworths, 1956); and J. T. Wilson, R. D. Russell and R. M. Farquhar, 'Radioactivity and Age of Minerals' (*Handbuch der Physik*, 1956, part x, chapter 14).

I acknowledge much help in the preparation of the monograph, especially that of Professor H. G. Thode, F. R. S., who has encouraged me in many ways and has read chapters 9 and 11, that of Dr G. R. Bainbridge, who has read the entire manuscript and has prepared several of the figures, that of Dr R. H. Tomlinson, who has read portions of chapter 9, and that of Mr C. F. Eve, who has read chapter 4. In addition, Professor A. O. C. Nier, Dr K. S. Quisenberry, Dr W. H. Johnson, Dr R. M. Farquhar, Dr Harmon Craig, Dr C. E. Berry, Dr J. R. Sites and Dr T. J. Kennett have kindly made available to me pre-publication information; Professor J. Mattauch, Professor A. O. C. Nier and Dr K. S. Quisenberry have provided photographs which appear in the frontispiece; many persons and publishers have given permission to reproduce figures in the text (these are individually acknowledged *in situ*); Mrs J. R. Dingwall and Mrs Marion Tweedley have competently performed the secretarial work associated with the manuscript; Mrs N. R. Isenor has assisted greatly in the reading of proof and in the preparation of the index. I am particularly indebted to Professor N. Feather, F. R. S., one of the General Editors of this series, who has corrected in the manuscript a few errors in fact and has suggested very many improvements in form.

Much of the manuscript was written in England in the autumn of 1955 during the tenure of a much appreciated Travelling Fellowship of the Nuffield Foundation.

Hamilton, Ontario H.E.D.
August 1957

Preface to second edition

The suggestion to prepare a second edition came from the late Professor V. S. Venkatasubramanian, whose permanent affiliation was with the Indian Institute of Science in Bangalore, but who spent three extended periods as a colleague of ours in Canada. The senior author and he enlisted the collaboration of R. C. Barber. It is tragic that Professor Venkatasubramanian's premature death in July, 1984 prevented him from seeing the completion of the project to which he had contributed so materially.

Users of the first edition appear to have appreciated its two-fold purpose: (a) to outline the principles underlying the design and performance of mass spectrometers and (b) to describe the principal uses to which mass spectrometers have been put. Accordingly, the second edition has been written with the same general purpose in mind. Because of the increased complexity of the subject, however, many of the uses are touched on lightly, in the expectation that readers will pursue special interests through some of the numerous references.

Developments in mass spectroscopy are recorded in the regular literature, in specialized periodicals (e.g., *International Journal of Mass Spectrometry and Ion Physics, Mass Spectrometry Specialist Periodic Reports, Journal of Biomedical Mass Spectrometry*), in specialized abstracts (e.g. *Chemical Abstracts Selects – Mass Spectrometry, Mass Spectrometry Bulletin*), in treatises on specific topics (as referred to in the text), and in *Proceedings* of periodic international conferences (e.g., *Mass Spectrometry ('IMSC')* – tenth held in Swansea, 1985; *Atomic Masses and Fundamental Constants ('AMCO')* – seventh held in Darmstadt, 1984; *Nuclei Far from Stability* – sixth held in Helsingor, 1981; and *Electromagnetic Isotope Separators and Techniques Related to Their Applications ('EMIS')* – eleventh held in Zinal, 1984). We have drawn heavily on these principal sources in preparing our general account of the subject.

We acknowledge much help in the preparation of the manuscript, including that of: Professor John B. Westmore, who read Chapter 11; Dr Ronald J. Ellis, who prepared the Index and assisted in other aspects;

Mr James N. Lanfaer, who drew most of the figures; Miss Ingrid Riesen, who typed most of the manuscript and together with Mr Gary Dyck and Mr Robert Beach assisted with the references; Mrs Norma Gwizon, who typed a portion of the manuscript; Mr Clifford A. Lander who assisted with the proofreading; and the many persons and publishers who gave permission to reproduce figures (as acknowledged in the text). Finally, we acknowledge the support of the Universities of Manitoba and Winnipeg and the cooperation of the several members of the Cambridge University Press who had responsibility for approval, editing and production of the monograph.

Winnipeg, Canada H. E. D.
June, 1985 R. C. B.

1

The development of mass spectroscopy

1.1 Introduction

Positive rays were discovered by Goldstein (1886) in a low-pressure electrical discharge tube. He observed, while using a perforated cathode, luminous streamers passing through the perforations or 'canals' into the space behind the cathode. He attributed this luminosity to rays of some sort which he called 'Kanalstrahlen'. These canal rays were subsequently deflected by Wien (1898, 1902) in magnetic and electric fields, and were shown thereby to be positively charged particles, with values of specific charge very much smaller than that of the electron.

J. J. Thomson (1897) had meanwhile demonstrated the existence of electrons, and had shown that cathode rays consist of streams of these particles. His attention was then drawn toward the counterpart of these cathode rays and, under his probing, positive rays began to reveal their remarkably complex character.

1.2 J. J. Thomson's positive-ray parabolae

Thomson (1913) conducted his investigations using the positive-ray parabola apparatus shown schematically in fig. 1.1. This type of analyser was first employed by Kaufman (1901) in his study of cathode rays.

The positive rays, formed in the discharge bulb A, enter the space behind the cathode, or analysing region, as a collimated beam, after traversing a long narrow tube. There they are acted upon by parallel and coterminous electric and magnetic fields, which analyse the beam into its various charge-to-mass (q/M) components.

In the uniform electric field E, of length x, the charged particle experiences an acceleration in the z direction of Eq/M. If v be the velocity of the particle, the time spent by it in the field is x/v, and it emerges from the field with vertical displacement

$$z = \tfrac{1}{2}(Eq/M)(x/v)^2, \tag{1.1}$$

where this value is in metres if mks units are employed. Simultaneously, in the uniform magnetic field B, the particle experiences an acceleration in the y direction of Bqv/M and emerges with lateral displacement given, for small deflections, by

$$y = \tfrac{1}{2}Bqx^2/Mv. \tag{1.2}$$

From equations (1.1) and (1.2)

$$y^2 = kz, \tag{1.3}$$

where the constant k depends upon the specific charge of the ion, the values of the applied deflecting fields and the geometry of the apparatus.

Thus, particles with the same value of q/M, regardless of their velocity, all strike the photographic plate G in a parabola whose vertex lies at the undeflected position of the particles. There is a different parabola for each value of specific charge represented in the positive-ion beam.

Thomson obtained different positive-ray parabolae by introducing different gases into the electrical discharge, and established that the positive-ray particles are the massive fragments remaining when one or more electrons are removed from the neutral atom or molecule. In this work strong evidence was found (Thomson, 1912) that neon exists in two isotopic forms, one of atomic weight 20 and the other, a much rarer variety, of atomic weight 22. This was the first indication that isotopes exist among the stable elements, although the concept of isotopy had been advanced by Soddy two years earlier in order to explain identical chemical properties among certain radioactive atoms, and had rapidly gained acceptance. F. W. Aston, then a research student at the University of

Fig. 1.1. Thomson's positive-ray parabola apparatus. The letters indicate the following: *A*, discharge tube; *B*, cathode; *C*, water jacket for cooling cathode; *D*, anode; *E*, gas inlet; *F*, pump lead; *G*, photographic plate; *I*, magnetic shield; *M*, *M'*, magnetic poles; *N*, *N'*, mica for electrical insulation of *P*, *P'*, which are pieces of soft iron to serve both as condenser plates and to define the magnetic field. (From Aston, 1942.)

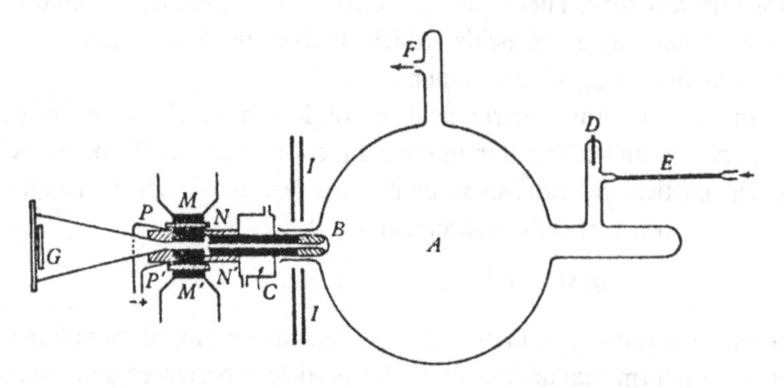

Cambridge, was encouraged by Thomson to investigate the matter of stable isotopes by other methods.

1.3 Aston's first mass spectrograph

Aston first attempted to enrich neon in one of its suspected isotopes by a series of fractionating operations, but the enrichment achieved was slight and barely conclusive. He then devised an instrument, which he termed the 'mass spectrograph', which possessed the property of focusing as well as analysing positive rays. This provided him with a much higher resolving power than was available with the parabola method.

Aston's mass spectrograph, which was completed in the Cavendish Laboratory in 1919 (Aston, 1919), and improved versions of which were completed in 1925 and 1937, is shown schematically in fig. 1.2. The positive rays, usually originating in a discharge tube, and, therefore, possessing a wide range of energies, are collimated by two narrow slits, S_1 and S_2, spaced some distance apart. The beam is first deflected by a uniform electric field, between P_1 and P_2, through an angle θ. Subsequently, it is deflected in the opposite direction, through more than twice this angle, by a circular uniform magnetic field centred at O. The foci lie along the line ZB which determines the position of the photographic plate GF.

This arrangement possesses a 'velocity focusing' property. Thus, despite

Fig. 1.2. The scheme of Aston's mass spectrograph. S_1 and S_2 are collimating slits, P_1 and P_2 are condenser plates, D is a diaphragm for selecting a portion of the beam emerging from the condenser, Z is the virtual source for the rays emerging from the condenser, O is the centre of the uniform magnetic field, GF is the photographic plate, θ and ϕ are the deflections in the electric and magnetic fields, respectively. (From Aston, 1942.)

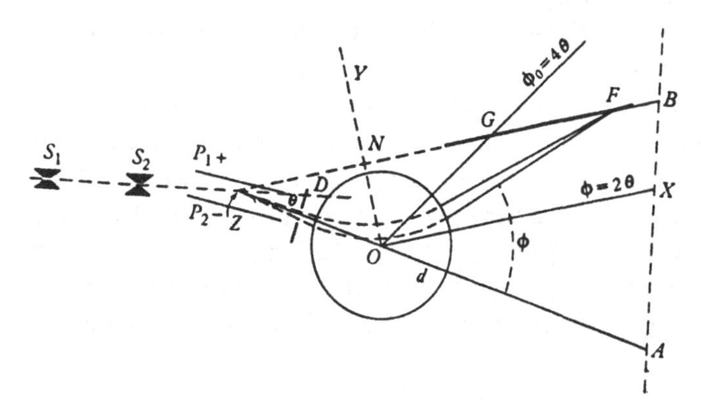

a velocity spread, ions with the same values of q/M are focused at one point. There is no 'direction focusing' – rather, the broadening of the image caused by an angular spread in the initial positive-ion beam is reduced to manageable proportions by the highly selective collimating system. In this first mass spectrograph the resolution achieved was approximately 1 part in 130, representing an improvement of a factor of 10 over the parabola method. The mean dispersion was 1.1 mm for 1% mass difference.

The first experiments with the mass spectrograph provided unequivocal proof (Aston, 1920) that neon consists of two isotopes, each of whole number mass. As the work rapidly progressed to chlorine, mercury, nitrogen and the noble gases, it became clear that the masses of all atoms are, to a close approximation, integrally related. This relationship was expressed by Aston as the 'whole number rule', oxygen and carbon being taken as standard simple elements.

The whole number rule represents the modern renascence of the unitary theory of matter. A similar idea, namely, that all atoms are aggregations of hydrogen atoms, had been advanced by Prout in 1815, but had fallen through disrepute into oblivion as it became established that many elements possessed atomic weights which are far from integral. Chlorine, with atomic weight 35.46, had been particularly recalcitrant with respect to Prout's rule. Aston's discovery that chlorine possesses the two isotopes ^{35}Cl and ^{37}Cl (the former being about three times as abundant as the latter) provided a natural explanation for fractional atomic weights and eventually led to an independent method of calculating them.

1.4 Dempster's first mass spectrograph

Meanwhile, a different type of apparatus for analysing positive rays had been constructed by A. J. Dempster (1918), a Canadian, at the University of Chicago. His method, shown in fig. 1.3, applied to the study of positive rays the scheme which Classen (1907) had used to determine e/m for electrons.

The charged particles from a monoenergetic source fall, in the region G, through a potential difference V, acquiring kinetic energy $\frac{1}{2}Mv^2 = Vq$. A portion of this beam, selected by the slit S_1, is then deflected through 180° in a circle of radius r by a uniform magnetic field B and emerges from the field through S_2 to strike the ion collector. During the ion's passage through the magnetic field the force equation is $Mv^2/r = Bqv$, so that its specific charge may be calculated from the relationship

$$q/M = 2V/B^2r^2. \quad (1.4)$$

The separate ion groups are in turn brought to the collector for detection by appropriately varying the accelerating potential.

This semicircular uniform magnetic field arrangement possesses direction focusing; that is, a beam of ions diverging from a point at the entrance boundary of the magnetic field is focused after completing the semicircle. The monoenergetic source of ions needed in the absence of velocity focusing came from a heated salt of the element to be studied. With salts of the alkali metals the metallic atoms evaporated in the ionized state, but, with the other elements, the evaporated material was bombarded with electrons in order to produce the ions. The resolution of this instrument was 1 part in 100.

With this apparatus, Dempster (1920) discovered and made an accurate abundance determination of the isotopes of magnesium, and shortly afterward made similar studies (Dempster, 1921, 1922) of lithium, potassium, calcium and zinc.

1.5 Early determinations of atomic mass

By 1923 Aston had already observed (Aston, 1923) small divergences from the whole number rule, but the first refined study of these

Fig. 1.3. Dempster's first mass spectrograph. *A* indicates the analysing chamber, *B* and *C* are brass walls of the vacuum chamber, *D* is a diaphragm to prevent transmission of reflected rays, *E* is the ion collector unit, *G* is the ion source region, S_1 and S_2 are slits. (From Dempster, 1918.)

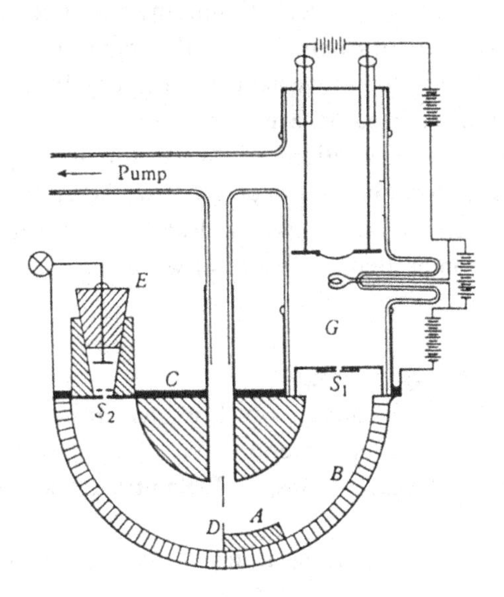

divergences was made by Costa (1925) in Paris, using a modification of Aston's mass spectrograph. In this work Costa obtained an accurate value for the mass of ^{1}H, and made useful mass comparisons involving ^{4}He, ^{6}Li, ^{7}Li, ^{12}C and ^{14}N. The accuracy of these determinations was 1 part in 3000, three times that available with Aston's first instrument.

The realization that the study of these divergences could be highly rewarding in the study of nuclear structure impelled Aston to construct his second mass spectrograph. This was completed (Aston, 1927) in 1925, possessing a resolution of 1 part in 600, a dispersion of 2.2 mm for 1% mass difference, and its accuracy of mass determination approached 1 part in 10 000.

With this instrument Aston systematically investigated the whole number divergences for a score of elements. He it was who chose ^{16}O to be the standard of isotopic mass, and invented the term 'packing fraction' to express the extent of the divergences:

$$\text{packing fraction} = \frac{\text{atomic mass} - \text{mass number}}{\text{mass number}} = \frac{M - A}{A}. \tag{1.5}$$

Thus, greatest stability is associated with those atoms whose packing fractions are algebraically the smallest. In 1927 Aston published his celebrated packing fraction curve in which was depicted, for the first time, the general manner in which nuclear stability varies as a function of mass number. This original curve correctly started with a large positive value for hydrogen, then dropped swiftly through lithium and boron, crossed the zero line at ^{16}O, descended to a minimum in the region of nickel, from which it rose slowly to a small positive value for mercury. In Aston's own words (Aston, 1942), it depicted 'in a general way the changes of energy to be expected from transmutations of nuclei, not only by the aggregation of light atoms to form heavier ones, but also the prodigious release of energy to be expected from the fission of uranium by neutron bombardment, a phenomenon entirely undreamt of when the curve was first drawn'.

1.6 Bainbridge's first mass spectrograph

Notwithstanding the great interest which existed in his atomic mass determinations, Aston enjoyed for several years a monopoly in this type of investigation. Then in the early 1930s, K. T. Bainbridge began his outstanding career.

Bainbridge first used a high resolution Dempster-type semicircular instrument which had been constructed at the Bartol Research Institute under the direction of Swann (1930). To remove the necessity for a monoenergetic ion source, which severely limited the number of elements which could be studied at that time, Bainbridge (1933*c*) added to the apparatus a Wien velocity filter, as shown in fig. 1.4.

Here the ions, formed in an ordinary discharge tube, traverse the velocity filter, bounded by S_2 and S_3, before entering the magnetic analyser. The velocity filter consists of crossed and coterminous electric and magnetic fields which exert forces upon the ions in opposite directions. For an ion to be undeflected, and so enter the magnetic analyser, it is required that these forces balance; that is, $Eq = Bqv$, so that the transmitted ion possesses a unique velocity given by $v = E/B$. The radius of curvature of the path of such an ion in the subsequent magnetic deflection is inversely proportional to its q/M value. The mass scale is, therefore, linear – a convenient dispersion law, particularly for use with photographic detection.

With this apparatus Bainbridge (1933*a*) made a number of valuable

Fig. 1.4. Bainbridge's first mass spectrograph. S_1, S_2 and S_3 are slits, P_1 and P_2 are condenser plates providing the electric field of the Wien velocity filter. As indicated, the radius of curvature in the magnetic analyser is linear with ion mass for constant degree of ionization. (From Aston, 1942.)

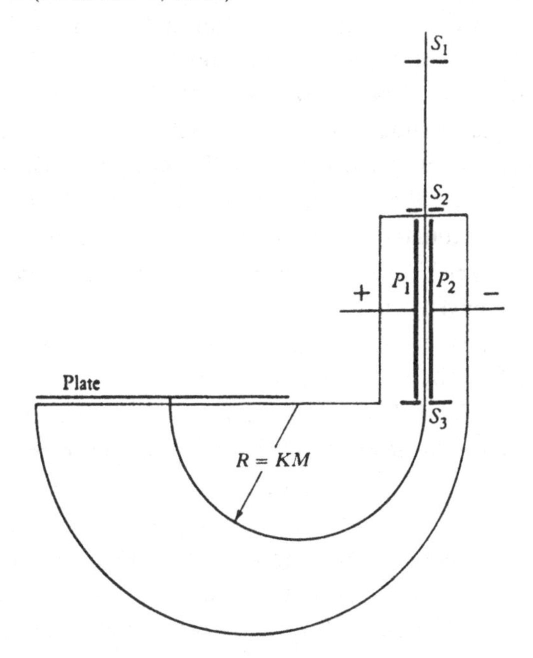

isotopic mass determinations for the lighter atoms. These included the newly discovered heavy hydrogen atom and the atoms involved in the nuclear reaction $^{1}H + {}^{7}Li \rightarrow 2{}^{4}He$. This work provided (Bainbridge, 1933*d*) the first experimental proof of the Einstein mass–energy relationship. In addition, important isotopic analyses were made by him (Bainbridge, 1932*a*, *b*, 1933*b*) of the elements zinc, germanium and tellurium.

1.7 Introduction of modern techniques

The pioneer work of Thomson, Aston, Dempster and Bainbridge was done on the basis of primitive knowledge of ion optics and in spite of grave limitations in experimental technique. Beginning in the 1930s there was a massive change in these conditions.

To begin with, several studies of the focusing properties of electric and magnetic fields culminated in 1934 in the definitive work of Herzog and Mattauch (Herzog & Mattauch, 1934; Herzog, 1934), in which general focusing equations were derived for radial electric and/or homogeneous magnetic fields. This made possible the design of high resolution, *double focusing*, mass spectrographs which accommodated ion beams possessing both angle and energy spread, and which were used for the precise determination of atomic masses (Chapter 5). The details of these and subsequent advances in ion optics are outlined in Chapter 2.

On the technical side, the crucial improvements were in vacuum technology, in ion sources and in electrical detector systems. In 1935 Nier brought together state-of-the-art components from these three fields to demonstrate the sensitivity and accuracy that could be achieved in the determination of isotopic abundances. This seminal work established a standard which subsequent workers have steadily improved. Developments in ion sources are described in Chapter 3, in electrical detectors in Chapter 4, and in overall abundance determinations in Chapter 7.

Finally, the uses to which mass spectrometers were put expanded dramatically with the advent, beginning about 1950, of commercial instruments. Prior to that time, many temperamental pieces of equipment were persuaded to perform by their sympathetic builders but, in general, non-physicists avoided the field of mass spectrometry. Commercial instruments provided many chemists, biologists and geologists with means to study *phenomena* of interest, with little concern for the basic instrumentation. At the present time most major developments in instrumentation become commercially available soon thereafter. Although we do not describe commercial equipment *per se*, many of the applications of mass spectrometry described in Chapters 10–14 would not have taken place without it.

2

Positive-ion optics

In this chapter we describe the ion-focusing properties of certain magnetic and electric fields which are used in deflection-type mass spectrometers. These properties may be appreciated by examining the results of first order theory in which it is assumed that the fields possess well defined boundaries. Such an assumption would be unrealistic, were it not feasible to approximate effective boundaries by methods described in §§2.6 and 2.7. The more precise, but frequently unnecessary, results of higher order focusing theory are then described.

2.1 Direction focusing in homogeneous magnetic fields

The homogeneous magnetic field, discriminating as it does between ions of different momenta, provides the most convenient arrangement for mass analysis of an ion beam. The special focusing property of the semicircular field, first applied to positive rays by Dempster (1918), formed the basis of many mass spectrometers, particularly before the focusing theory was developed for the so-called 'sector' fields. Although Aston (1919) considered the focusing properties of such fields, he published no detailed analysis. The later analytical treatment by Barber (1933), and independently by several others (Brüche & Scherzer, 1934; Herzog, 1934; Stephens, 1934), led to the widespread use of sector magnets, which can be built at a much lower cost than can those of the semicircular type.

The focusing properties of homogeneous magnetic fields will be described, employing the terminology of Herzog. This approach possesses the advantage that the focusing properties of radial electrostatic fields and of Wien velocity filters may also be expressed in the same form, a form which is similar to that used to describe the geometric optics of thick lenses.

Referring to fig. 2.1, let us suppose that ions of mass M_0 and velocity v_0 emerge from the object point O with a half-angular spread in the plane of the paper α ($\alpha \ll 1$). An ion with median direction, after traversing the distance l'_m, enters the magnetic field normally, where it is constrained to follow a circular path of radius r_m. After deflection through the angle Φ_m, it emerges from the field normally and proceeds to the image point I, at which there is a convergence of the ion beam which diverged from O.

Herzog likened this arrangement to the optical combination of prism plus cylindrical lens, and demonstrated that its focal length is given by

$$f_m = r_m/\sin\Phi_m. \tag{2.1}$$

The object and image distances, l'_m and l''_m, are related through the equation

$$(l'_m - g_m)(l''_m - g_m) = f^2_m, \tag{2.2}$$

where $g_m = f_m \cos\Phi_m$ is the distance from the boundary of the field to the principal focus. The relationship given in equation (2.2) was conveniently expressed by Barber in the statement that the object, the centre of curvature and the image lie on a straight line, as shown in fig. 2.1.

The displacement of the image, b''_m, corresponding to a displacement of the object, b'_m, a change of the ionic mass to $M = M_0(1+\gamma)$ and of the ionic velocity to $v_0(1+\beta)$, where both γ and β are small quantities, is given by

$$b''_m = r_m(\beta+\gamma)\left[1+\left(\frac{f_m}{l'_m - g_m}\right)\right] - b'_m\frac{f_m}{l'_m - g_m}. \tag{2.3}$$

From equation (2.3), considering the original ion group (M_0, v_0), one finds the magnification to be

$$\frac{b''_m}{b'_m} = \frac{-f_m}{l'_m - g_m}, \tag{2.4}$$

which, as can readily be seen from equation (2.2), has the value unity for all symmetrical arrangements, that is, for those for which the image and object

Fig. 2.1. Focusing properties of a homogeneous magnetic field. O, I and C are the positions of the object, image and centre of curvature, respectively, l'_m and l''_m are the distances of object and image, respectively, from the boundaries of the magnetic field, b'_m and b''_m are used to describe the object or image widths, respectively, or to measure the displacement of object or image in a direction normal to the central path, r_m is the radius of curvature in the magnetic field, Φ_m is the angle of deflection and α is the half-angular direction spread of the ions.

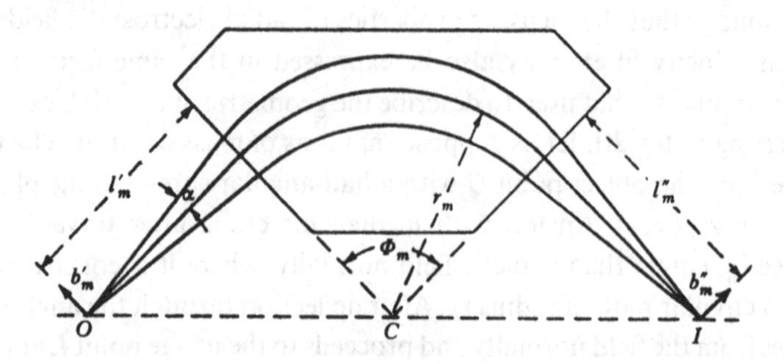

distances are equal. The negative sign preceding the last term in equation (2.3) implies an image inversion as required by Barber's rule.

If $S_0 = 2b'_m$ is the width of the object slit, located at O, the images corresponding to two mass groups, M_0 and $M = M_0(1+\gamma)$, are just resolved when their centres are separated by the distance $2b''_m = S_0 f_m/(l'_m - g_m)$. For the common case of a monoenergetic ion beam, where $\gamma + 2\beta = 0$, one computes the resolution from (2.3) to be

$$\frac{\Delta M}{M} = \gamma = \frac{2S_0 f_m}{r_m(l'_m - g_m)\left[1 + \left(\dfrac{f_m}{l'_m - g_m}\right)\right]}. \tag{2.5}$$

For a symmetrical arrangement, equation (2.5) simplifies to $\Delta M/M = S_0/r_m$. If the beam be other than monoenergetic, the resolution may be computed in the light of the existing circumstances. Thus, for example, the resolution is improved by a factor of two if the monoenergetic ion beam be replaced by one in which all the ions possess the same velocity.

The assumption has so far been made that the ion detector, located at I, is of such a type that the image width, as indicated by it, is not broadened. Such is the case with a photographic plate. The normal arrangement, however, is to place at I a slit of width S_i, followed by a sensitive electrical detector. In this case the resolution for a symmetrical arrangement becomes $(S_0 + S_i)/r_m$. For maximum sensitivity this image slit should be no smaller than the actual width of the image. In the important case where the optimum intensity and resolution are required simultaneously, the width of the collector slit is made equal to the image width and the resolving power is decreased by a factor of two from the corresponding photographic case.

The dispersion is found from (2.3) by computing the displacement of the image corresponding to a 1% mass change. For a monoenergetic ion beam and symmetrical spectrometer, the dispersion is

$$d = \frac{r_m}{100} \quad \text{for 1\% mass difference,} \tag{2.6}$$

measured in the plane normal to the central beam.

Herzog (1934) and Cartan (1937) also considered field configurations in which the boundaries are not normal to the ion trajectory. The position of the focus is given by equation (2.2) where we now have

$$g'_m = \frac{r_m \cos\varepsilon'_m \cos(\Phi_m - \varepsilon''_m)}{\sin(\Phi_m - \varepsilon'_m - \varepsilon''_m)}, \tag{2.7}$$

$$g''_m = \frac{r_m \cos\varepsilon''_m \cos(\Phi_m - \varepsilon'_m)}{\sin(\Phi_m - \varepsilon'_m - \varepsilon''_m)}, \tag{2.8}$$

$$f_m = \frac{r_m \cos \varepsilon'_m \cos \varepsilon''_m}{\sin(\Phi_m - \varepsilon'_m - \varepsilon''_m)}, \tag{2.9}$$

and the magnification is given by

$$\frac{b''_m}{b'_m} = \frac{\cos(\Phi_m - \varepsilon'_m)}{\cos \varepsilon'_m} - \left(\frac{l''_m}{f_m}\right). \tag{2.10}$$

In this connection Cartan has given a useful graphical method, shown in fig. 2.2, for locating the image position. See p. 22 for sign convention for ε's.

It should be emphasized that Herzog's theory, as outlined in this section, is based upon the assumption that α and β (describing the angular spread and the velocity spread, respectively) are small quantities whose squares may be neglected. It is, therefore, a first order theory and, in certain special instances, its results are inadequate. The question of higher order focusing will be discussed in §§2.9 and 2.10.

2.2 Direction focusing in radial electrostatic fields

The term 'radial electrostatic field' refers to the field which exists between the plates of a cylindrical condenser. It was shown by Hughes & Rojansky (1929) and Hughes & McMillan (1929) that such a field of angular extent $\pi/\sqrt{2}$ corresponds to the field of a semicircular magnet, in that a beam of ions diverging from a point at the entrance boundary is brought to a focus at the exit boundary. Smythe (1934) and Herzog & Mattauch (1934) demonstrated that, with appropriate angular deflection, either the image or object could be moved back from the edge of the field, and then Herzog (1934) developed the general focusing equations. These equations take the

Fig. 2.2. Cartan's graphical method for locating the image, I, of an object, O, for the general case of a sector magnetic field. In addition to the symbols defined in fig. 2.1, ε' and ε'' are the angles which the incoming and emerging ions, respectively, make with the normals n' and n'' to the magnetic field boundaries. (From Bainbridge, 1953.)

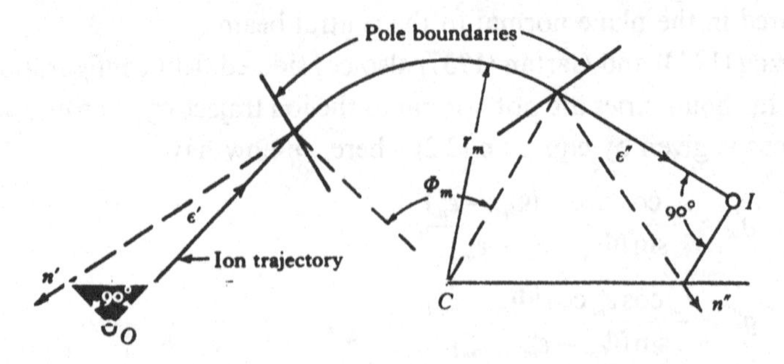

same form as those given for homogeneous magnetic fields, and the symbols used have the same significance.

The object and image distances are related through

$$(l'_e - g_e)(l''_e - g_e) = f_e^2, \tag{2.11}$$

where

$$f_e = r_e/\sqrt{2}\sin\sqrt{2}\Phi_e \tag{2.12}$$

and

$$g_e = f_e \cos\sqrt{2}\Phi_e. \tag{2.13}$$

By analogy with the magnetic case (fig. 2.1) it can be seen that a diagram can be constructed in which Barber's rule holds, if the angle in the diagram is $\sqrt{2}\Phi_e$ and the radius of curvature is $r_e/\sqrt{2}$. The image displacement b'' is given by

$$b''_e = r_e(\beta + \tfrac{1}{2}\gamma)\left[1 + \left(\frac{f_e}{l'_e - g_e}\right)\right] - b'_e\left(\frac{f_e}{l'_e - g_e}\right), \tag{2.14}$$

from which one may compute the magnification, resolution and dispersion, as in §2.1.

In the case of a monoenergetic ion beam, for which $\beta = -\frac{1}{2}\gamma$, all ions, regardless of their mass, are focused at the same point. The radial electrostatic analyser is, thus, an energy selector, and has been frequently used as such to determine the energies of charged particles involved in nuclear reactions. Its chief use in mass spectroscopy is in linear combination with a magnetic analyser to form a double focusing arrangement (§2.4).

The radial electrostatic analyser is a special case of the toroidal condenser, whose focusing properties, which are two-directional, are given by Ewald & Leibl (1955).

$$(l'_r - g_r)(l''_r - g_r) = f_r^2 \quad \text{and} \quad (l'_z - g_z)(l''_z - g_z) = f_z^2, \tag{2.15}$$

where

$$\begin{aligned} f_r &= r_r/\kappa \sin\kappa\Phi_e, \quad f_z = \sqrt{(r_r r_z)}/\sin\sqrt{(r_r/r_z)}\Phi_e, \\ g_r &= f_r \cos\kappa\Phi_e, \quad g_z = f_z \cos\sqrt{(r_r/r_z)}\Phi_e, \\ \kappa &= \sqrt{\{2 - (r_r/r_z)\}}. \end{aligned} \tag{2.16}$$

Here the subscripts r and z refer to the radial and axial directions, respectively. The shape of the central equipotential surface between the two deflecting plates is described in terms of r_r and r_z which are, respectively, its radial and axial radii. The path of the median ion within the condenser lies in this equipotential surface.

The image displacement in the radial direction b_r'', resulting from a similar type of displacement of the object b_r' and the presence of ions of mass $M = M_0(1+\gamma)$ and velocity $v = v_0(1+\beta)$ is

$$b_r'' = \frac{r_r}{\kappa^2}(2\beta+\gamma)\left[1+\left(\frac{f_r}{l_r'-g_r}\right)\right] - b_r'\left(\frac{f_r}{l_r'-g_r}\right). \tag{2.17}$$

As expected, equations (2.15), (2.16) and (2.17) reduce to equations (2.11), (2.12), (2.13) and (2.14) when $r_z = \infty$, as in the radial (cylindrical) electrostatic analyser. The condition $r_r = r_z$ corresponds to the spherical condenser (Aston, 1919; Brüche & Henneberg, 1935; Purcell, 1938; Hachenberg, 1948; Browne, Craig & Williamson, 1951; and Herzog, 1953*a*). In this case the radial and axial focusing properties are identical, thus making possible an electrostatic analyser with two-dimensional direction focusing. Here the object and image lie on a line through the centre of curvature, exactly as in Barber's rule (§2.1) for the magnetic case. The spherical condenser has not been used extensively, but has been incorporated in the very high resolution instrument of Stevens *et al.* (1960, 1963, §5.11) and the radio-frequency (rf) mass spectrometer of Smith (1967; §6.9).

Whereas the cylindrical condenser possesses radial but no axial focusing, and the spherical condenser possesses both radial and axial focusing with identical focal lengths, the work of Ewald & Liebl makes possible other combinations of radial and axial focusing properties. For example, Ewald and his colleagues (Ewald & Sauermann, 1956; Ewald, Liebl & Sauermann, 1957) have employed a toroidal condenser from which the ion beam emerges parallel in the radial direction, while at the same time converging in the axial direction, an example of astigmatic focusing. Similarly, Matsuda and his colleagues (Matsuda, Fukumoto & Kuroda, 1966; Matsuda & Fujita, 1975; Taya, Tsuyama, Kanomata, Noda & Matsuda, 1978) have exploited the properties of toroidal analysers. Further, Matsuda (1961) has used an analyser having cylindrical deflection plates and an additional pair of flat electrodes ('Matsuda plates') as shown in fig. 2.3. A potential V_M, applied to both plates, may be used to curve the equipotential surfaces in the central region to approximate the field of a toroidal analyser, thus modifying the focal lengths.

Clearly, where it is intended that the field shape correspond to the shape of the deflection plates, either the height of the plates must be sufficiently large, relative to the analyser gap, or shims (Nier, 1960; Dietz, 1961) must be added.

2.3 Wien velocity filter

The velocity filter consisting of crossed electric and magnetic fields, which had been used by Wien (1902) to determine the specific charge of positive-ray particles and which subsequently served as an important element in Bainbridge's first mass spectrograph (§1.6), was found by Herzog (1934) to possess a direction focusing property. The focusing condition, expressed in the standard form, is

$$(l'_w - g_w)(l''_w - g_w) = f_w^2, \tag{2.18}$$

with

$$f_w = r_m/\sin\left(\frac{L}{r_m}\right) \quad \text{and} \quad g_w = f_w \cos\left(\frac{L}{r_m}\right), \tag{2.19}$$

where r_m is the radius of curvature that would result if the magnetic field of the filter were acting alone, and L is the length of filter. The image displacement is given by

$$b''_w = -r_m\beta\left[1 + \left(\frac{f_w}{l'_w - g_w}\right)\right] - b'_w\left(\frac{f_w}{l'_w - g_w}\right) \tag{2.20}$$

from which it will be seen that the Wien filter is fundamentally a device for producing a velocity dispersion.

The special case ($l'_w = l''_w = 0$) in which the object and image are located at the entrance and exit boundaries, respectively, requires

$$\cos\left(\frac{L}{r_m}\right) = \pm 1, \quad \text{where} \quad L = n\pi r_m, \quad n = 1,2,\ldots. \tag{2.21}$$

Although not used widely, the Wien filter may be combined with an electrostatic field to form a double focusing (§2.4) arrangement (Ioanoviciu & Cuna, 1974; Taya, Takiguchi, Kanomata & Matsuda, 1978). The

Fig. 2.3. The electrode structure for an electrostatic analyser having 'Matsuda plates'. A potential V_M applied to these plates affects the focal lengths for both radial and axial focusing.

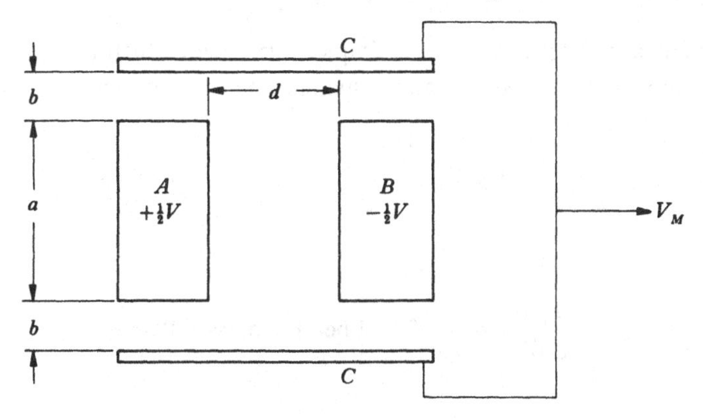

focusing properties of crossed electric and magnetic fields, with special reference to the Wien filter, are given in greater detail by Holmlid (1975) and by Ioanoviciu (1974) in terms of transfer matrices (see §2.10). Wien filters with inhomogeneous fields are discussed by Ioanoviciu (1973).

2.4 Double focusing in consecutive field combinations

In each of the three preceding sections, the expression for b'', which gives the position of the image along an axis perpendicular to the central ion trajectory, contains a term in β. This term gives the velocity dispersion associated with that particular field, and makes it possible to plan a two-field arrangement in which the velocity dispersion produced in one will be counter-balanced by that produced in the other. If measures are simultaneously taken to ensure that direction focusing is achieved by the combination, the final image will be independent of both velocity and direction spread among the ions emerging from the object slit. The combination will, therefore, be double focusing.

The most commonly employed double focusing arrangement is one in which the image formed by a radial electrostatic field serves as the object for a following magnetic field. Thus, from (2.3) and (2.14), the final image displacement is given by

$$b''_m = r_m(\beta + \gamma)\left(1 + \frac{f_m}{l'_m - g_m}\right) - \left(\frac{f_m}{l'_m - g_m}\right) \times \left[r_e(\beta + \tfrac{1}{2}\gamma)\left(1 + \frac{f_e}{l'_e - g_e}\right) - b'_e\left(\frac{f_e}{l'_e - g_e}\right)\right]. \tag{2.22}$$

Velocity focusing will occur when the coefficient of β vanishes, that is, when

$$r_m\left[\frac{l'_m - g_m}{f_m} + 1\right] - r_e\left[1 + \frac{f_e}{l'_e - g_e}\right] = 0, \tag{2.23}$$

in which case (2.22) reduces to a simpler form from which the resolution (for photographic detection) and dispersion may be computed to be

$$\frac{\Delta M}{M} = \frac{2S_0}{r_e}\frac{\left(\frac{f_e}{l'_e - g_e}\right)}{\left[1 + \left(\frac{f_e}{l'_e - g_e}\right)\right]} \tag{2.24}$$

$$\text{and } d = \frac{r_m}{200}\left[1 + \frac{f_m}{(l'_m - g_m)}\right] \text{ per 1\% mass difference.} \tag{2.25}$$

If the ions strike the photographic plate at an angle θ the dispersion is increased by the factor $1/\sin\theta$.

It will be noted in (2.24) and (2.25) that the resolution depends only on the constants of the electrostatic analyser, while the dispersion is a function only of the constants of the magnetic analyser. The constants are, of course, related to each other through equation (2.23).

The velocity focusing condition (2.23) implies that for a given electrostatic analyser (that is, fixed r_e and Φ_e) and magnetic sector (Φ_m) velocity focusing occurs at one value of r_m. If electrical detection is used, such an instrument is a double focusing mass spectro*meter*. If, however, the principal slit is located at the principal focus of the electrostatic analyser, $l'_e = g_e$ and equation (2.23) becomes indeterminate. It can be rewritten, after appropriate substitution and division, in the form

$$\frac{r_m}{l''_e}[1-\cos\Phi_m] + \left(\frac{l'_m}{l''_e}\right)\sin\Phi_m - \left(\frac{r_e}{l''_e}\right)(1-\cos\sqrt{2}\Phi_e) - \sqrt{2}\sin\sqrt{2}\Phi_e = 0. \tag{2.26}$$

Writing $l'_m = \Delta - l''_e$, where Δ is the separation between the electrostatic and magnetic analyser, and proceeding to the limit $l''_e \to \infty$, one obtains (Mattauch & Herzog, 1934) the double focusing condition in the form

$$\sin\Phi_m = -\sqrt{2}\sin\sqrt{2}\Phi_e, \tag{2.27}$$

which is independent of r_m, as desired. In this case Φ_m and Φ_e are of opposite signs as the deflections in the two analysing fields are in opposite directions. The condition $l''_e = \infty$ implies a parallel beam entering the magnetic analyser, whence $l''_m = g_m$, as remarked above. Instruments which satisfy (2.27) for all values of r_m may be used as mass spectro*graphs*. The most notable of these is the Mattauch–Herzog instrument (§5.7).

Double focusing has also been achieved with a Wien filter-magnetic field combination, and it is likewise possible with a Wien filter–radial electrostatic field combination, although no instrument based on this latter principle has yet been constructed.

The manner in which double focusing has been achieved in actual instruments will be discussed in Chapter 5.

2.5 Focusing in crossed electric and magnetic fields

Although the consecutive field arrangement has been widely used, the first double focusing scheme (Bartky & Dempster, 1929) was not of this sort, but consisted of a deflection through $\pi/\sqrt{2}$ radians in crossed magnetic and radial electrostatic fields. The object and image were located at the entrance and exit boundaries respectively.

Some years later, Bleakney & Hipple (1938) showed that the combination of crossed uniform magnetic and electric field possessed perfect double focusing in the plane normal to the magnetic field. In this plane the ions follow trochoidal paths, the nature of which depends on the initial conditions. An instrument based on this principle was constructed by Hipple & Sommer (1953). Although such a device has a linear mass scale, and offers in theory the possibility of high resolution, problems with space and surface charges have discouraged its further development.

The velocity and two-directional focusing properties of crossed electric and magnetic fields which possess cylindrical symmetry have been studied by Svartholm (1950) and Fischer (1952).

2.6 Effect of fringing magnetic fields

The lateral displacement (that is, in the plane of the optic axis) of ions caused by the fringing field of a magnet has been studied by Dempster (1918) and Herzog (1934) and, in greater detail, by Coggeshall (1947), Bainbridge (1949, 1953), Ploch & Walcher (1950), Reuterswärd (1951, 1952), Barnard (1953, 1955), Paul (1953), König & Hintenberger (1955), Herzog (1955), Kerwin (1963), Enge (1964), Baril & Kerwin (1965) and Ezoe (1970).

If the object slit be taken as the origin of coordinates, as in fig. 2.4, with the magnetic field in the z-direction and the ions following the positive x-axis, the fringing field in the median plane will normally be a function of x only, that is, $B(x) = h(x)B_0$, where $h(x)$ is the ratio of the fringing field to the field B_0 in the uniform region between the poles. The effective field

Fig. 2.4. The coordinate system for studying fringing magnetic fields. δ is the distance which the effective field extends beyond the physical boundary and t is the pole thickness.

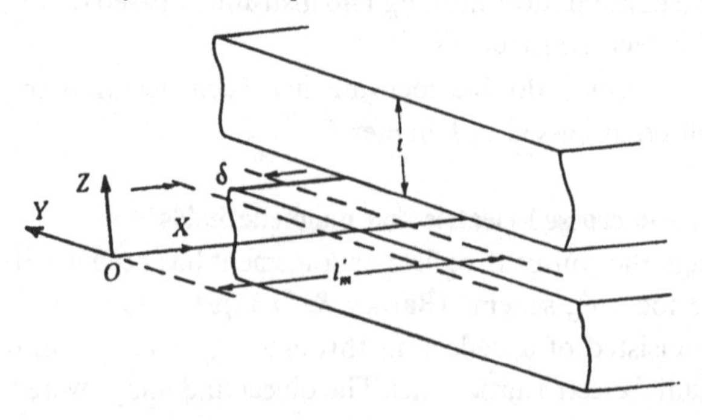

extends beyond the physical boundary a distance

$$\delta = \int_0^{l'_m + b} h(x)\, dx - b, \tag{2.28}$$

where the upper limit recognizes that fringing field effects are felt to a distance b inside the physical boundary. The distance δ gives the position of the virtual boundary, where the uniform field may be considered as suddenly commencing. The angular deflection computed on this assumption will be, for all values of x greater than $l'_m + b$, identical with that actually experienced by an ion leaving the source with an initial direction normal to the real boundary.

Coggeshall (1947) has used the Schwarz–Christoffel transformation to obtain $h(x)$ for the type of pole faces (assuming infinite permeability) shown in fig. 2.4, that is, poles of thickness t extending to infinity in the $-y$, $+y$, and $+x$ directions. From (2.28) he has then obtained the values of δ shown in table 2.1, which indicate, among other things, that the thickness of the poles has a surprisingly small effect upon the shape of the fringing field. Barnard (1953) has experimentally confirmed the correctness of some of Coggeshall's calculated values of $h(x)$.

The results in table 2.1 should be generally applicable as long as those dimensions which are assumed to be infinite in the idealized case are kept large compared to the gap width. If there is doubt as to the validity of this approximation for any particular magnet, $h(x)$ can be determined experimentally with a small search coil. It will be noted from table 2.1. that, for thick pole faces, the effective field extends ~ 1.15 gap widths beyond the physical boundary. This figure is similar to the values 1.05 (calculated) and 1.10 (observed) given by König & Hintenberger (1955) and the one gap width allowance used by Nier (1940) for magnets not unlike that of fig. 2.4.

A common method of taking the fringing field into account is to apply Barber's rule using the point of intersection of the two virtual boundaries

Table 2.1. *The separation, δ, between the physical and effective boundaries of the magnetic field as a function of object distance, l'_m, and pole thickness, t. All distances are given in gap widths (from Coggeshall, 1947).*

$l'_m \backslash t$	0.5	1	2.5	5	∞
5	0.69	0.84	0.92	0.95	0.98
10	0.81	0.97	1.07	1.14	1.20

as the centre of curvature, rather than the ideal one. In this approximation δ is frequently taken to be equal to one gap width. This, however, ignores the fact that the ions are both deflected and displaced before reaching the virtual entrance boundary, in accordance with the equation of their trajectory (Coggeshall & Muskat, 1944)

$$\frac{dy}{dx}=\frac{\int_0^x h(x)dx}{\left(r_m^2-\left[\int_0^x h(x)dx\right]^2\right)^{1/2}}, \tag{2.29}$$

so that the actual centre of curvature is not at the apex of the virtual boundaries. Here $r_m = mv_0/eB_0$ is the radius of curvature of the ion path in the uniform field B_0.

Bainbridge has suggested that the significant effect of the fringing field is that it causes the ions to cross the physical entrance and exit boundaries obliquely. He recommends, therefore, that the object and image distance be calculated by means of Herzog's or Cartan's theory for oblique incidence. The actual object and image positions will then be l'_m and l''_m from the physical boundaries, and will be displaced from the normals to the points of actual entrance and exit by an amount y_0, found by integrating (2.29) between the limits $x=0$ and $x=l'_m$ (or $x=l''_m$).

Kerwin (1963) has suggested an iterative procedure for the symmetric sector case. The location of the effective boundary is initially assumed to be at one gap width and the nominal value of l'_m plus one gap width is taken to give an initial source position relative to the physical boundary of the magnet. Integrating along the path between this initial position and a distance one gap width outside the physical boundary, one calculates the angle ε at which the ion beam enters the effective field. With the design sector angle and ε, one calculates from Herzog's equations the position of the image, l''_m. From symmetry, this gives the position of a virtual object $l'_m = l''_m$, and an improved calculation of the position of the effective boundary is made on the basis of equation (2.28). This leads to a more precise limit of integration and a new value of δ is calculated and subsequently an improved value of $l = l'_m = l''_m$. Finally the lateral displacement of the ion beam is calculated with the improved values of l and the position of the effective boundary. The comparison of the various methods, given in fig. 2.5, is from Kerwin (1963).

Enge (1964, 1967) has considered the differences in the focusing produced by a magnet which has an extended fringe field and one for which the field is sharply terminated. For straight boundaries he shows that, if the ion paths for the two cases coincide in the main field, the path for the extended fringe field will be displaced laterally toward the outside, but first order focusing in the median plane is unaffected. When curved boundaries are considered, additional first order effects must be considered.

König & Hintenberger (1955), Herzog (1955), Wollnik & Ewald (1965), Hübner & Wollnik (1970) and Czok *et al.* (1971) have discussed the use of an iron diaphragm (fig. 2.6) to terminate arbitrarily the fringing field,

Fig. 2.5. Calculated positions of the image formed by a magnetic analyser taking the fringing field into account. (After Kerwin, 1963.) Here r_m is 10.2 cm = 3.33 gap widths.

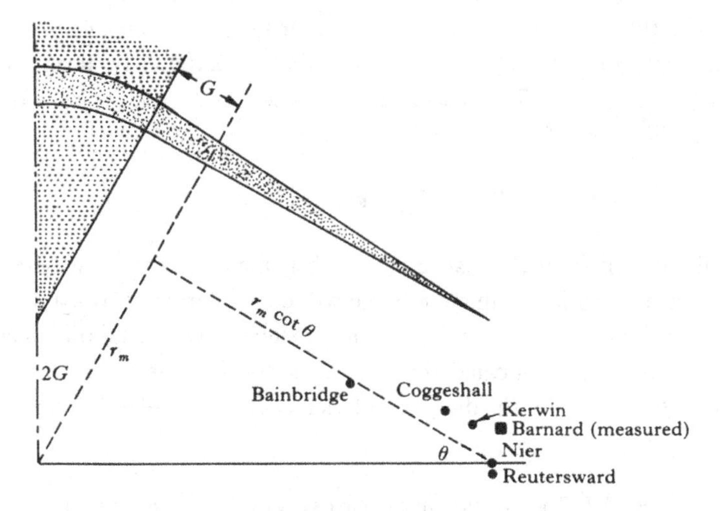

Fig. 2.6. The use of an iron diaphragm to terminate the fringing magnetic field. $2b$ is the diaphragm opening, d is the distance between diaphragm and magnet poles and $2k$ is the pole gap.

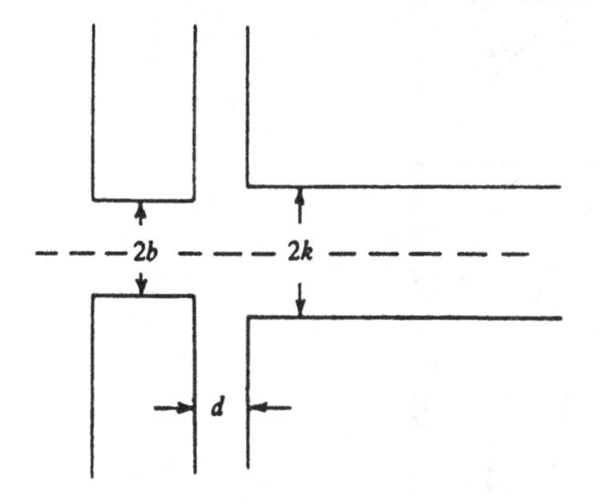

in a manner similar to that proposed by Herzog for an electrostatic analyser (see §2.7). The general effect of such a diaphragm is shown in fig. 2.7. Here δ, as in equation (2.28), gives the position, relative to the plane of the pole pieces, of the effective boundary of the magnetic field when infinite permeability is assumed. The effective screening action of this diaphragm removes the dependence of the position of the virtual boundary upon the object distance, l'_m, explicit in equation (2.28). Hübner & Wollnik (1970) consider the realistic case where the permeability is not only finite but also dependent on the value of the magnetic field.

A vertical ('axial') acceleration is also experienced in the fringing field (Cotte, 1938; Lavatelli, 1946; Herzog, 1951; Camac, 1951; Cross, 1951; Herzog, 1953*b*) by ions travelling above or below the median plane and approaching or leaving the field boundary obliquely. In terms of the coordinate system of fig. 2.4, when $\mathbf{B} = \mathbf{i}B_x + \mathbf{k}B_z$ and $\mathbf{v} = \mathbf{i}v_x + \mathbf{j}v_y + \mathbf{k}v_z$, the acceleration, $e\mathbf{v} \times \mathbf{B}/M$, is

$$\frac{e}{M}[\mathbf{i}v_yB_z + \mathbf{j}(v_zB_x - v_xB_z) - \mathbf{k}v_yB_x]. \tag{2.30}$$

It will be seen from the last term, which is the one of interest here, that ions not moving in the median plane will be accelerated toward it if they approach or leave the boundary from the side of the normal on which the centre of curvature is located, that is, ε is positive. There is, thus, a z-focusing effect associated with the direction of incidence and a z-defocusing effect if

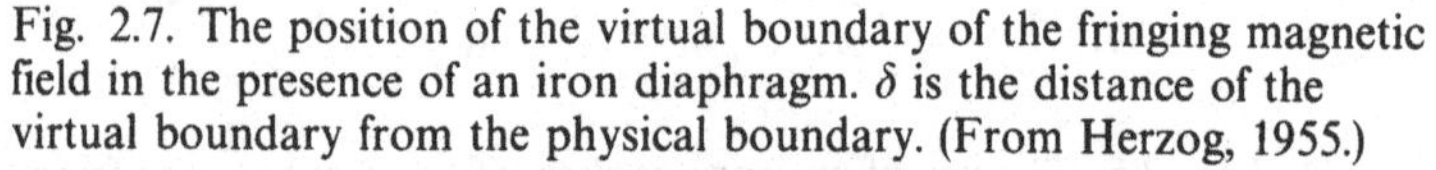

Fig. 2.7. The position of the virtual boundary of the fringing magnetic field in the presence of an iron diaphragm. δ is the distance of the virtual boundary from the physical boundary. (From Herzog, 1955.)

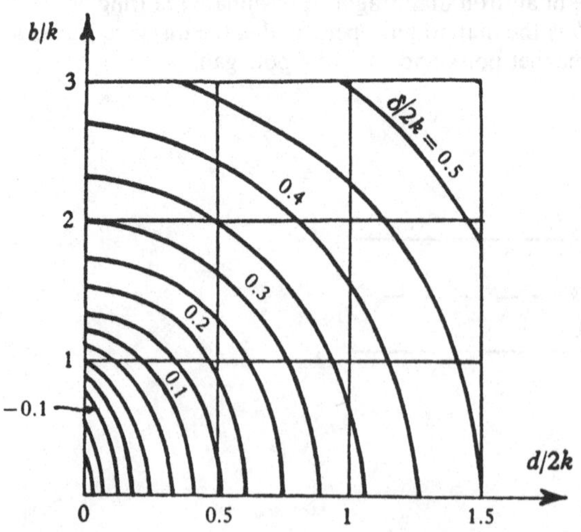

the approach is made from the side of the normal away from the centre of curvature.

Herzog (1953*b*) has shown that, in the z-direction, the effect of each fringing field is that of a thin cylindrical lens of focal length

$$f_z = r_m/\tan \varepsilon. \tag{2.31}$$

Using the equations developed by Herzog (1934) and by Cotte, Cross showed that the position of the axial focus, l''_z, was given by

$$\left(\frac{r}{l''_z}\right) = \tan \varepsilon'' - \left(\frac{1}{\Phi_m - \cot \eta}\right), \tag{2.32}$$

where Φ_m is in radians,

$$\tan \eta = \tan \varepsilon' - \left(\frac{r_m}{l'_z}\right), \tag{2.33}$$

and l'_z, l''_z are object and image distances for the focusing in the z-direction only. These results may also be derived by analogy to the optical case of two thin lenses having focal lengths given by (2.31) and separated by a distance measured along the optic axis between the effective field boundaries.

Both Camac and Cross have described arrangements possessing simultaneous focusing in the radial and axial directions,† with the latter extending his calculation to eliminate second order radial aberrations. It should be noted that, in general, the effective field boundaries for focusing in the radial and axial directions will not be the same (Ezoe, 1967). Moreover, in the case where the fringe field extends over an appreciable distance, the thin lens approximation must be replaced by a thick lens treatment in which a decrease in the focusing action occurs (Enge, 1967).

2.7 Fringing field of a cylindrical condenser

Herzog (1935) has shown that convenient effective boundaries may be established for cylindrical condensers by the use of grounded diaphragms. The geometrical arrangement is similar to the magnetic case, shown in fig. 2.6, with the magnet poles replaced by condenser plates whose potentials are $+\frac{1}{2}V$ and $-\frac{1}{2}V$, with respect to the grounded diaphragm.

When the diaphragm opening, $2b$, is small compared to the distance between condenser plates, $2k$, the effective field extends a distance δ beyond

† This 'two directional' focusing is, regrettably, sometimes referred to as 'double focusing', but should not be confused with double focusing as described in §2.4. The prior introduction of 'double focusing' in the latter sense (Mattauch & Herzog, 1934; Mattauch, 1936) prompts the usage adopted here.

the physical boundary of the condenser, where δ is given by

$$\delta = \frac{2d}{\pi}\tan^{-1}\left(\frac{k}{d}\right) - \left(\frac{k}{\pi}\right)\ln\left(\frac{4k^2}{d^2+k^2}\right). \tag{2.34}$$

The quantity δ is positive when $d > 0.52k$ and negative when $d < 0.52k$. When the diaphragm opening is not negligibly small, δ is dependent upon the diaphragm thickness. The values of the variables b, d and k for which $\delta = 0$ are shown in fig. 2.8 for the cases of both thick and thin diaphragms.

These results of Herzog's are actually for a parallel plate condenser, but they will hold for a cylindrical condenser provided $r_e \gg k$. Herb, Snowdon & Sala (1949) have considered the effect of rounded condenser plates on the effective field boundary, while Rogers (1940) has given a method of calculating the lateral displacement of the ion beam caused by the normal fringing field without a diaphragm.

Fig. 2.8. Values of the diaphragm opening $2b$, the condenser spacing $2k$, and the distance between condenser and diaphragm, d, which cause the effective boundary of the condenser to coincide with its physical boundary. Curve A refers to a thin diaphragm and curve B to a thick diaphragm. (From Herzog, 1935.)

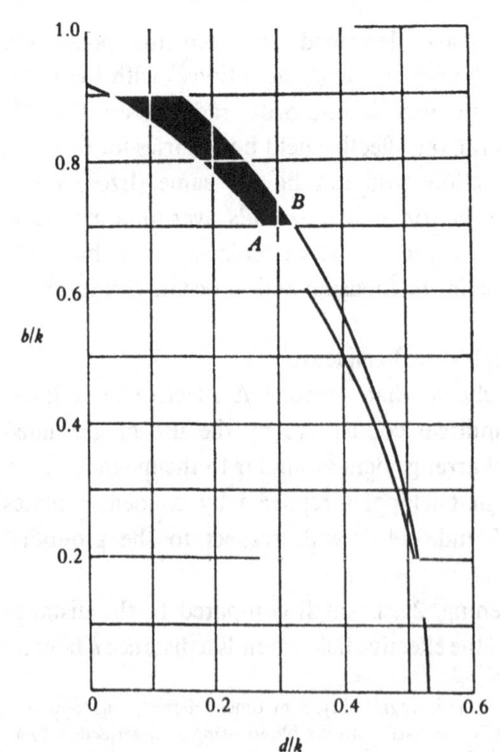

2.8 Focusing in non-uniform fields

It was shown by Siegbahn & Svartholm (1946) that, with proper pole shaping, a magnetic spectrometer may provide coincident axial and radial focusing. This matter has been subsequently studied by Shull & Dennison (1947*a*, *b*), Kurie, Osoba & Slack (1948), Judd (1949), Rosenblum (1949) and Lee-Whiting & Taylor (1956).

The focusing principle is based on facts, first established by Kerst & Serber (1941), concerning the free oscillations of a charged particle moving in a magnetic field. In fields which are cylindrically symmetrical, the frequency of these oscillations depends upon the radial variation of the field. In particular, for a field of the form

$$B(r) = B_0\left(\frac{r_0}{r}\right)^n, \quad 0 < n < 1, \tag{2.35}$$

where B and r are values in the median plane, and B_0 and r_0 refer to the optic axis, the angular frequencies of the radial and axial oscillations are given by

$$\omega_r = \sqrt{(1-n)}\omega_0 \tag{2.36}$$

and

$$\omega_z = \sqrt{n}\omega_0, \tag{2.37}$$

respectively, where $\omega_0 = B_0 e/M$ is the angular cyclotron frequency. It is seen from equations (2.36) and (2.37) that, for $n = \frac{1}{2}$, the radial and axial frequencies are equal. Thus, particles diverging from a point source in such a magnetic field will perform their oscillations and meet again on the equilibrium orbit after half a cycle. Since the cyclotron frequency is $\sqrt{2}$ times as large as ω_r and ω_z, this image point will occur after an angular deflection $\Phi_m = \sqrt{2}\pi$ radians or 254.6°. Such was the scheme proposed by Siegbahn & Svartholm (1946), which has been embodied in a number of β ray spectrometers.

In general, the field in the median plane may be expressed, according to Taylor's expansion, as

$$B(r) = B(r_0) + (r - r_0)\left(\frac{dB}{dr}\right)_{r=r_0} + \frac{(r-r_0)^2}{2!}\left(\frac{d^2B}{dr^2}\right)_{r=r_0} + \frac{(r-r_0)^3}{3!}\left(\frac{d^3B}{dr^3}\right)_{r=r_0} + \ldots. \tag{2.38}$$

For fields of the general form given by equation (2.35), this becomes

$$B(r) = B(r_0)\left[1 - n\left(\frac{r-r_0}{r_0}\right) + \frac{n(n+1)}{2!}\left(\frac{r-r_0}{r_0}\right)^2 - \frac{n(n+1)(n+2)}{3!}\left(\frac{r-r_0}{r_0}\right)^3 + \cdots\right] \tag{2.39}$$

$$= B(r_0)\left[1 - A_1\left(\frac{r-r_0}{r_0}\right) + A_2\left(\frac{r-r_0}{r_0}\right)^2 - \cdots\right]. \tag{2.40}$$

For a field which follows equation (2.35) strictly and for which $n = \frac{1}{2}$, the coefficient $A_2 = \frac{3}{8}$ and second order *axial* defocusing is eliminated (Shull & Dennison, 1947 *a*, *b*). However, from equation (2.38), it is clear that the size of $(dB/dr)_{r=r_0}$ determines the slope of the plot of B vs r about the optic axis, $r = r_0$, while $(d^2B/dr^2)_{r=r_0}$ determines the curvature of the plot. Hence $(d^2B/dr^2)_{r=r_0}$ may be treated as a somewhat arbitrary constant. Thus for $A_1 = \frac{1}{2}$ and $A_2 = \frac{1}{8}$, first order two-directional focusing with second order *radial* focusing is achieved.

In the more general case (Judd, 1949; Rosenblum, 1949) the object and image points may be located outside the field, and n may take on any value between zero and unity. The first order focusing properties for both radial and axial directions are given by the modified diagrams for Barber's rule (fig. 2.9). The particle tracing the equilibrium path leaves the source

Fig. 2.9. Focusing properties of an inhomogeneous magnetic field of the form given by equation (2.35). These are modified diagrams for Barber's rule (Judd, 1949) for radial (*a*) and axial (*b*) focusing.

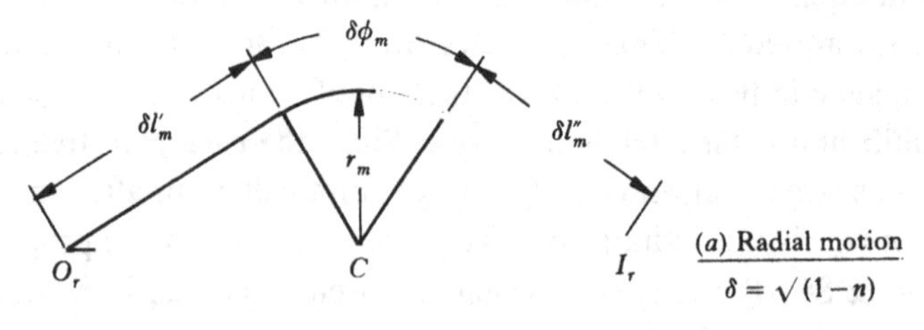

travelling in the $z = 0$ plane, and enters and leaves the magnetic field normally, suffering an angular deflection Φ_m. With object distance l'_m, particles diverging in the $z = 0$ plane are focused a distance l''_m beyond the end of the field, while the image for those following trajectories whose projections in the $z = 0$ plane lie along the equilibrium path is located at the distance l''_{mz}.

By comparing equations (2.38) and (2.39), one sees that the index, n, of the field is

$$n = -\frac{r_0}{B_0}\left(\frac{\partial B}{\partial r}\right)_{\substack{r=r_0\\z=0}}. \tag{2.41}$$

A second parameter, X, is used to described the size of the second derivative and is defined by

$$X(1-n) - 2n = -\frac{r_0}{B_0}\left(\frac{\partial^2 B}{\partial r^2}\right)_{\substack{r=r_0\\z=0}}. \tag{2.42}$$

Equations giving the focusing properties to second order in terms of these parameters have been derived by Tasman (1959).

Axial focusing is important in instruments for which high transmission is required, for example isotope separators, and in the special case of the ion microprobe. It is, however, of special interest to look at the advantages that inhomogeneous fields offer in the radial direction. Judd has shown that the lateral magnification is unity for all symmetrical arrangements, that is, where $l'_m = l''_m$. Further, from his work, assuming monoenergetic ions, the dispersion and resolution in the radial direction are found to be

$$d = \frac{r_0}{100(1-n)} \quad \text{for 1\% mass difference,} \tag{2.43}$$

$$\frac{\Delta M}{M} = \frac{(S_0 + S_i)(1-n)}{r_0}. \tag{2.44}$$

For $n = 0$ these expressions reduce to the standard homogeneous field values given in §2.1. Also, it will be noted that the dispersion and resolution of the Siegbahn & Svartholm arrangement ($n = \frac{1}{2}$) are both better by a factor of 2 than in the homogeneous field case. Furthermore, if n takes on values greater than 0.5, the dispersion and resolution will improve further, and at a very rapid rate.

This potential improvement is to some extent misleading since equations (2.43) and (2.44) have not reckoned with second order image aberrations. However, the fact that much is to be gained by increasing n from 0.5 towards 1.0 is demonstrated by the enhanced dispersion and resolution values achieved by Alekseevskii, Prudkovskii, Kosourov & Filimonov (1955). These investigators have used the values $n = 0.87 - 0.89$ and $\Phi_m = \pi$ radians to increase the resolution by factors of ~ 8 over homogeneous field spectrometers of similar radius of curvature. The enhancement is in approximate agreement with that predicted by equation (2.40). In the report of this work the image and object distances for symmetrical arrangements are given as

$$l'_m = l''_m = \frac{r_0 \cot[\sqrt{(1-n)}\frac{1}{2}\Phi_m]}{\sqrt{(1-n)}}. \tag{2.45}$$

Thus, for values of n approaching 1.0, the image and object distances become quite long, resulting in a serious loss of ions in the z direction.

Matsuda, Fukumoto & Kuroda (1966) have investigated the special case of a field where $n = 1$. As can be seen from fig. 2.9, such a field has no direction focusing in the radial direction but focuses strongly in the axial direction. However, the property of special interest is the strong velocity and mass dispersion which is produced in the radial direction. Matsuda *et al.* have shown that the first order ion orbit is

$$\frac{d^2\rho}{d\phi^2} = \gamma + \beta, \tag{2.46}$$

where a cylindrical coordinate system (r, ϕ, z) is used with $r = r_0(1 + \rho)$ and $\gamma = dM/M$. Such a field has been incorporated in the remarkable instrument at Osaka University (§5.13).

Certain fields which vary in ways other than as r^{-n} have also played a role in mass spectroscopy. Bock (1933) and Beiduk & Konopinski (1948) have shown that the focusing properties of a 180° magnetic spectrometer will be greatly improved by shaping the field in the radial direction according to the relation

$$B(r) = B_0[1 - \tfrac{3}{4}\eta^2 + \tfrac{7}{8}\eta^3 - \tfrac{9}{16}\eta^4 + \tfrac{51}{320}\eta^5 \ldots], \tag{2.47}$$

where B_0 is the field strength along the median path of radius r_0, $B(r)$ is its value at radius r, and $\eta = (r - r_0)/r_0$. This shaped field (Langer & Cook, 1948; Kistemaker & Zilvershoon, 1953) permits the use in the source of values of α, the half-angular divergence, which would be intolerably large in a uniform field spectrometer because of the α^2 broadening of the image.

Finally, it should be mentioned that several studies have been made of ion trajectories in non-uniform electric and magnetic fields which have not yet found applications in mass spectroscopy (for example, Coggeshall & Muskat, 1944; Coggeshall, 1946).

2.9 Higher order direction focusing in homogeneous magnetic fields

It has been mentioned that Herzog's theory assumes α, the half-angular divergence of the ions emerging from the source, to be small, and neglects terms containing it to the second power or higher. As a result, in mass spectrometers designed on the basis of Herzog's theory, the expression for the actual image width contains an image broadening of amount $\sim \alpha^2 r_m$. Detailed studies of the higher order direction focusing properties of homogeneous magnetic fields have been made by Smythe, Rumbaugh & West (1934), Bainbridge (1947), Hintenberger (1948*a*, *b*, 1949), Kerwin (1949) and Kerwin & Geoffrion (1949). These have shown that the $\alpha^2 r_m$ term may be made equal to zero (second order focusing) by properly curving the magnetic field boundaries or, in the case of a straight-line field boundary, by appropriately inclining the ion path to the boundary.

The shape of the boundaries of a symmetrically shaped uniform magnetic field which produces perfect focusing of an ion beam may be readily deduced (Kerwin, 1949) by reference to fig. 2.10.

Here, the ions leaving O and focused at I must cross the axis of symmetry AY normally and, therefore, the centre of curvature of the ion's path in the magnetic field will lie on this axis. Also

$$y/(x_0 - x) = \tan\theta = x/\sqrt{(r_m^2 - x^2)},$$

whence

$$y = \frac{x(x_0 - x)}{\sqrt{(r_m^2 - x^2)}}. \tag{2.48}$$

This curve when plotted gives the symmetrical field, shown shaded in fig. 2.10, whose shape is a complicated one to generate. However, it may be approximated by either a circle or a straight line in such a way that second

Fig. 2.10. Perfect focusing in a homogeneous magnetic field. (From Kerwin, 1949.)

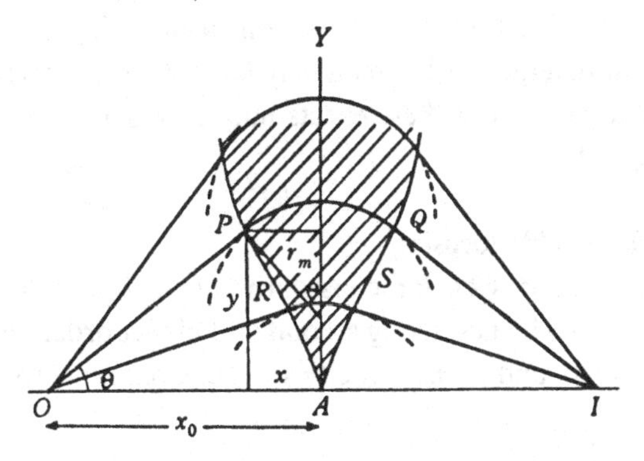

order rather than perfect focusing is obtained. Furthermore, this second order focusing and Herzog's first order focusing occur at the same point I. These arrangements will now be described.

If the median ion beam crosses the boundaries normally, as at P and Q, second and first order focusing will be obtained simultaneously if the exact pole shape given by (2.48) be approximated by a circle whose radius is given by

$$R = r_m \cot^3 \theta = r_m \cot^3 (\tfrac{1}{2}\Phi_m). \tag{2.49}$$

Thus, for a total deflection of $\frac{1}{2}\pi$ radians, the radii of curvature of the exit and entrance field boundaries are simply each made equal to the radius of curvature of the ion beam. For the widely used $\frac{1}{3}\pi$ sector field $R = 5.2r_m$.

The straight-line approximation takes advantage of the points of inflection (R and S in fig. 2.10) in the ideal field boundaries. It is found that field boundaries which are tangent (equation (2.48)) at these inflection points to the ideal boundaries result in second order focusing. This is the so-called 'inflection' case and obviously requires the ions to enter and leave the magnetic field obliquely. For simultaneous first and second order focusing the angle ε between the ion beam and the normal to the boundary is given by

$$\tan \varepsilon = -\tfrac{1}{2} \tan (\tfrac{1}{2}\Phi_m), \tag{2.50}$$

and the distance, l, from the object (or image) to the point of inflection is

$$l = \tfrac{2}{3} r_m \cot (\tfrac{1}{2}\Phi_m). \tag{2.51}$$

With this arrangement there is a vertical defocusing as described in §2.6. A positive value for ε indicates that the ion beam approaches the normal to the field boundary from the side on which the centre of curvature is located.

Simultaneous first and second order focusing may also be obtained (Hintenberger, 1949) for a great variety of non-symmetrical arrangements.

It is emphasized that while such magnetic fields produce higher order direction focusing, they also possess a first order velocity dispersion. Hence they are appropriate to situations where the ion beam is highly homogeneous in energy, but diverges with a relatively large angular spread. An assessment of the relative sizes of these effects may be made through the equations given in §2.10.

2.10 Second order double focusing

The results given in §2.4 for consecutive field double focusing combinations were derived by neglecting second and higher order terms in α, the half-angular spread, and β, the velocity spread. Consequently, there are image aberrations in these cases which hamper the achievement of very high resolution.

It is usually found that the magnitude of α is much larger than that of β, so that the dominating second order aberration is the term in α^2. Accordingly, Nier & Roberts (1951) and Johnson & Nier (1953) described a double focusing combination (§5.8) possessing, at a given point, direction focusing to second order together with velocity focusing to first order. Similarly, several other instruments have been constructed, in which the α^2 aberration has been corrected. For example, the Mattauch–Herzog instrument (§5.7) produces a second order angle (α^2) focus at the position given by $(r_e/r_m) = 1.683$ (Hintenberger & König, 1959). Similarly, the 2.54m radius instrument at the Argonne National Laboratory (Stevens *et al.*, 1960, 1963; Stevens & Moreland, 1967; §5.11) corrects for the α^2 aberration.

An extensive investigation of the second order focusing properties of cyclindrical electrostatic analysers and uniform magnetic fields was done by Hintenberger & König (1959). The investigation was restricted to the median plane of such devices and the fields in question were assumed to have well defined, abrupt boundaries. The ion path after deflection in each field is described as a power series in α and β. Thus, for the cylindrical condenser,

$$y_e = r_e\{K_1\alpha_e + K_2\beta + K_{11}\alpha_e^2 + K_{12}\alpha_e\beta + K_{22}\beta^2\} + x_e\{L_1\alpha_e + L_2\beta + L_{11}\alpha_e^2 + L_{12}\alpha_e\beta + L_{22}\beta^2\}, \qquad (2.52)$$

Fig. 2.11. Coordinate system and parameters used to describe the focusing properties of the field produced by a cylindrical electrostatic analyser (this is, by a radial electric field).

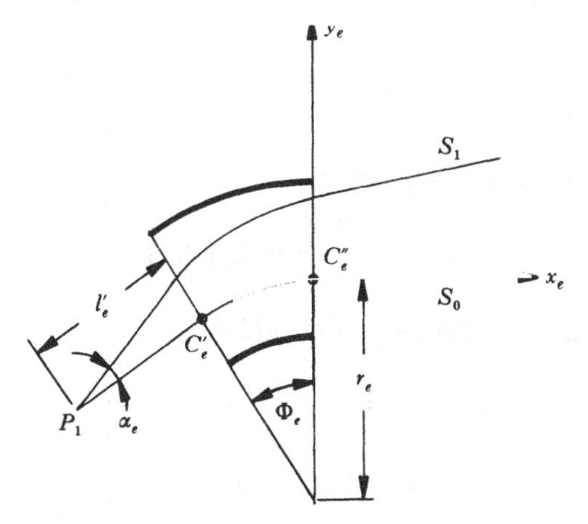

where α_e, r_e, l'_e, Φ_e and the coordinates are shown in fig. 2.11. The coefficients in equation (2.52) are given by

$$\left.\begin{aligned} K_1 &= \kappa_{1a} + \kappa_{1b}\frac{l'_e}{r_e} \\ K_2 &= \kappa_{2a} \\ K_{11} &= \kappa_{11a} + \kappa_{11b}\frac{l'_e}{r_e} + \kappa_{11c}\left(\frac{l'_e}{r_e}\right)^2 \\ K_{12} &= \kappa_{12a} + \kappa_{12b}\frac{l'_e}{r_e} \\ K_{22} &= \kappa_{22a} \end{aligned}\right\}, \tag{2.53}$$

$$\left.\begin{aligned} L_1 &= \lambda_{1a} + \lambda_{1b}\frac{l'_e}{r_e} \\ L_2 &= \lambda_{2a} \\ L_{11} &= \lambda_{11a} + \lambda_{11b}\frac{l'_e}{r_e} + \lambda_{11c}\left(\frac{l'_e}{r_e}\right)^2 \\ L_{12} &= \lambda_{12a} + \lambda_{12b}\frac{l'_e}{r_e} \\ L_{22} &= \lambda_{22a} \end{aligned}\right\}, \tag{2.54}$$

where the quantities κ and λ depend on the geometry of the electric fields and are given by

$$\left.\begin{aligned} \kappa_{1a} &= \frac{1}{\sqrt{(2)}}\sin(\sqrt{(2)}\phi_e) \\ \kappa_{1b} &= \cos(\sqrt{(2)}\phi_e) \\ \kappa_{2a} &= 1 - \cos(\sqrt{(2)}\phi_e) \\ \kappa_{11a} &= -\tfrac{3}{8} + \tfrac{2}{3}(\sqrt{(2)}\phi_e) - \tfrac{7}{24}\cos(2\sqrt{(2)}\phi_e) \\ \kappa_{11b} &= -\frac{4}{3\sqrt{(2)}}\sin(\sqrt{(2)}\phi_e) + \frac{7}{6\sqrt{(2)}}\sin(2\sqrt{(2)}\phi_e) \\ \kappa_{11c} &= -\tfrac{3}{4} + \tfrac{1}{6}\cos(\sqrt{(2)}\phi_e) + \tfrac{7}{12}\cos(2\sqrt{(2)}\phi_e) \\ \kappa_{12a} &= \frac{7}{3\sqrt{(2)}}(\sin(\sqrt{(2)}\phi_e) - \tfrac{1}{2}\sin(2\sqrt{(2)}\phi_e)) \\ \kappa_{12b} &= \tfrac{3}{2} - \tfrac{1}{3}\cos(\sqrt{(2)}\phi_e) - \tfrac{7}{6}\cos(2\sqrt{(2)}\phi_e) \\ \kappa_{22a} &= \tfrac{1}{4} - \tfrac{5}{6}\cos(\sqrt{(2)}\phi_e) + \tfrac{7}{12}\cos(2\sqrt{(2)}\phi_e) \end{aligned}\right\} \tag{2.55}$$

$$\left.\begin{aligned}
\lambda_{1a} &= \cos(\sqrt{(2)}\phi_e) \\
\lambda_{1b} &= -\sqrt{(2)}\sin(\sqrt{(2)}\phi_e) \\
\lambda_{2a} &= \sqrt{(2)}\sin(\sqrt{(2)}\phi_e) \\
\lambda_{11a} &= -\frac{\sqrt{(2)}}{3}(2\sin(\sqrt{(2)}\phi_e) - \sin(2\sqrt{(2)}\phi_e)) \\
\lambda_{11b} &= -\tfrac{4}{3}(\cos(\sqrt{(2)}\phi_e) - \cos(2\sqrt{(2)}\phi_e)) \\
\lambda_{11c} &= -\frac{\sqrt{(2)}}{6}(\sin(\sqrt{(2)}\phi_e) + 4\sin(2\sqrt{(2)}\phi_e)) \\
\lambda_{12a} &= \tfrac{4}{3}(\cos(\sqrt{(2)}\phi_e) - \cos(2\sqrt{(2)}\phi_e)) \\
\lambda_{12b} &= \tfrac{4}{3}\sqrt{(2)}(\sin(\sqrt{(2)}\phi_e) + \sin(2\sqrt{(2)}\phi_e)) \\
\lambda_{22a} &= -\frac{\sqrt{(2)}}{6}(\sin(\sqrt{(2)}\phi_e) + 4\sin(2\sqrt{(2)}\phi_e))
\end{aligned}\right\}. \tag{2.56}$$

Similarly for a uniform magnetic field (fig. 2.12) the ion path after deflection is given by

$$\begin{aligned} y_m = r_m\{M_1\alpha_m + M_2\beta + M_{11}\alpha_m^2 + M_{12}\alpha_m\beta + M_{22}\beta^2\} \\ + x_m\{N_1\alpha_m + N_2\beta + N_{11}\alpha_m^2 + N_{12}\alpha_m\beta + N_{22}\beta^2\}. \end{aligned} \tag{2.57}$$

Here the coefficients are given by

$$\left.\begin{aligned}
M_1 &= \mu_{1a} + \mu_{1b}\frac{l'_m}{r_m} \\
M_2 &= \mu_{2a}
\end{aligned}\right|$$

Fig. 2.12. Coordinate system and parameters used to describe the focusing properties of a sector magnetic field.

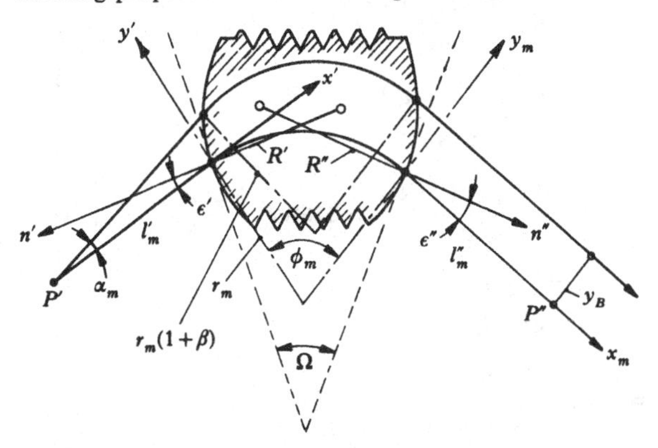

$$\left.\begin{aligned}M_{11}&=\mu_{11a}+\mu_{11b}\frac{l'_m}{r_m}+\mu_{11c}\left(\frac{l'_m}{r_m}\right)^2\\M_{12}&=\mu_{12a}+\mu_{12b}\frac{l'_m}{r_m}\\M_{22}&=\mu_{22a}\end{aligned}\right\}, \quad (2.58)$$

$$\left.\begin{aligned}N_1&=\nu_{1a}+\nu_{1b}\frac{l'_m}{r_m}\\N_2&=\nu_{2a}\\N_{11}&=\nu_{11a}+\nu_{11b}\frac{l'_m}{r_m}+\nu_{11c}\left(\frac{l'_m}{r_m}\right)^2\\N_{12}&=\nu_{12a}+\nu_{12b}\frac{l'_m}{r_m}\\N_{22}&=\nu_{22a}\end{aligned}\right\}, \quad (2.59)$$

and the quantities μ and ν depend on the geometry of the magnetic field, namely,

$$\left.\begin{aligned}\mu_{1a}&=\sin\phi_m\\\mu_{1b}&=\frac{\cos(\phi_m-\varepsilon')}{\cos\varepsilon'}\\\mu_{2a}&=1-\cos\phi_m\\\mu_{11a}&=\tfrac{1}{2}[\cos\phi_m(1-\cos\phi_m)+\tan^2\varepsilon''\sin^2\phi_m]\\\mu_{11b}&=\frac{\cos(\phi_m-\varepsilon')}{\cos\varepsilon'}\left[\tan\varepsilon'+\tan\varepsilon''\frac{\cos(\phi_m-\varepsilon'')}{\cos\varepsilon''}\right]\\&\quad+\cos\phi_m\frac{\sin(\phi_m-\varepsilon'-\varepsilon'')}{\cos\varepsilon'\cos\varepsilon''}\\\mu_{11c}&=-\frac{1}{2}\frac{\sin(\phi_m-\varepsilon'-\varepsilon'')}{\cos\varepsilon'\cos\varepsilon''}\\&\quad\times\left[\frac{\sin(\phi_m-\varepsilon'-\varepsilon'')}{\cos\varepsilon'\cos\varepsilon''}+2\tan\varepsilon''\frac{\cos(\phi_m-\varepsilon')}{\cos\varepsilon'}\right]\\&\quad+\frac{\sin\phi_m}{2\cos^3\varepsilon'}\frac{r_m}{R'}\\\mu_{12a}&=\sin\phi_m\frac{1-\cos\phi_m}{\cos^2\varepsilon''},\end{aligned}\right\}, \quad (2.60)$$

$$\left.\begin{aligned}
\mu_{12b} &= \tan\varepsilon''\left[\tan\varepsilon'' + \frac{\sin(\phi_m - \varepsilon'')}{\cos\varepsilon''}\right]\frac{\cos(\phi_m - \varepsilon')}{\cos\varepsilon'} \\
&\quad + \sin\phi_m \frac{\sin(\phi_m - \varepsilon'' - \varepsilon')}{\cos\varepsilon'\cos\varepsilon''} \\
\mu_{22a} &= \tan\varepsilon''\left[\tan\varepsilon'' + \frac{\sin(\phi_m - \varepsilon'')}{\cos\varepsilon''}\right](1 - \cos\phi_m) \\
&\quad - \frac{1}{2}\left[\tan\varepsilon'' + \frac{\sin(\phi_m - \varepsilon'')}{\cos\varepsilon''}\right]^2
\end{aligned}\right\}$$

and

$$\left.\begin{aligned}
v_{1a} &= \frac{\cos(\phi_m - \varepsilon'')}{\cos\varepsilon''} \\
v_{1b} &= -\frac{\sin(\phi_m - \varepsilon' - \varepsilon'')}{\cos\varepsilon'\cos\varepsilon''} \\
v_{2a} &= \tan\varepsilon'' + \frac{\sin(\phi_m - \varepsilon'')}{\cos\varepsilon''} \\
v_{11a} &= -\frac{1}{2}\left[\frac{\sin(\phi_m - \varepsilon'')}{\cos\varepsilon''} + \tan\varepsilon''\frac{\cos^2(\phi_m - \varepsilon'')}{\cos^2\varepsilon''}\right] \\
&\quad + \frac{\sin^2\phi_m}{2\cos^3\varepsilon''}\frac{r_m}{R''} \\
v_{11b} &= -\frac{\sin(\phi_m - \varepsilon' - \varepsilon'')}{\cos\varepsilon'\cos\varepsilon''}\left[\tan\varepsilon' - \tan\varepsilon''\frac{\cos(\phi_m - \varepsilon'')}{\cos\varepsilon''}\right] \\
&\quad + \frac{\sin\phi_m}{\cos^3\varepsilon''}\frac{\cos(\phi_m - \varepsilon')}{\cos\varepsilon'}\frac{r_m}{R''} \\
v_{11c} &= -\tfrac{1}{2}\tan\varepsilon''\frac{\sin^2(\phi_m - \varepsilon' - \varepsilon'')}{\cos^2\varepsilon'\cos^2\varepsilon''} + \frac{\cos^2(\phi_m - \varepsilon')}{2\cos^2\varepsilon'\cos^3\varepsilon''}\frac{r_m}{R''} \\
&\quad + \frac{\cos(\phi_m - \varepsilon'')}{2\cos^3\varepsilon'\cos\varepsilon''}\frac{r_m}{R'} \\
v_{12a} &= -\frac{\cos(\phi_m - \varepsilon'')}{\cos\varepsilon''}\left[\tan\varepsilon'' + \frac{\sin(\phi_m - \varepsilon'')}{\cos\varepsilon''}\right]\tan\varepsilon'' \\
&\quad + \frac{\sin\phi_m(1 - \cos\phi_m)}{\cos^3\varepsilon''}\frac{r_m}{R''} \\
v_{12b} &= \frac{\sin(\phi_m - \varepsilon' - \varepsilon'')}{\cos\varepsilon'\cos\varepsilon''}
\end{aligned}\right\} \cdot (2.61)$$

$$
\left.
\begin{aligned}
&\times\left\{1+\tan\varepsilon''\left[\tan\varepsilon''+\frac{\sin(\phi_m-\varepsilon'')}{\cos\varepsilon''}\right]\right\}\\
&+\frac{\cos(\phi_m-\varepsilon')}{\cos\varepsilon'}\,\frac{1-\cos\phi_m}{\cos^3\varepsilon''}\,\frac{r_m}{R''}\\
v_{22a}&=-\tfrac{1}{2}\tan\varepsilon''\left[\tan\varepsilon''+\frac{\sin(\phi_m-\varepsilon'')}{\cos\varepsilon''}\right]^2\\
&-\left[\tan\varepsilon''+\frac{\sin(\phi_m-\varepsilon'')}{\cos\varepsilon''}\right]+\frac{(1-\cos\phi_m)^2}{2\cos^3\varepsilon''}\,\frac{r_m}{R''}
\end{aligned}
\right\}
$$

For a combination of such electric and magnetic fields in tandem (fig. 2.13) the displacement, y_B, of the ion path from the optic axis at a distance $x = l''_m$ is given by

$$y_B = r_m\{B_1\alpha_e + B_2\beta + B_{11}\alpha_e^2 + B_{12}\alpha_e\beta + B_{22}\beta^2\}. \tag{2.62}$$

Here the coefficients are

$$
\left.
\begin{aligned}
B_1 &= \pm S_{1a}L_1 \pm S_{1b}T_1\\
B_2 &= \pm S_{1a}L_2 \pm S_{1b}T_2 + S_{2a}\\
B_{11} &= \pm S_{1a}L_{11} \pm S_{1b}T_{11} + S_{11a}L_1^2 + S_{11b}L_1T_1 + S_{11c}T_1^2\\
B_{12} &= \pm S_{1a}L_{12} \pm S_{1b}T_{12} + 2S_{11a}L_1L_2 + S_{11b}(L_1T_2 + L_2T_1)\\
&\quad + 2S_{11c}T_1T_2 \pm S_{12a}L_1 \pm S_{12b}T_1\\
B_{22} &= \pm S_{1a}L_{22} \pm S_{1b}T_{22} + S_{11a}L_2^2 + S_{11b}L_2T_2 + S_{11c}T_2^2\\
&\quad \pm S_{12a}L_2 \pm S_{12b}T_2 + S_{22a}.
\end{aligned}
\right\} \tag{2.63}
$$

Fig. 2.13. Focusing properties of a radial electrostatic field and magnetic field in tandem. Deflection in the two fields is in the same sense.

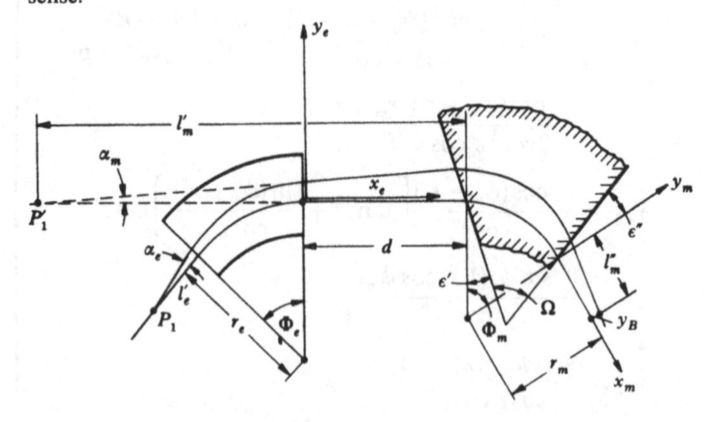

and

$$S_i = \mu_i + \nu_i \frac{l''_m}{r_m}$$

$$(i = 1a,\ 1b,\ 2a,\ 11a,\ 11b,\ 11c,\ 12a,\ 12b,\ 22a), \qquad (2.64)$$

$$T_k = K_k \frac{r_e}{r_m} + L_k \frac{d}{r_m},$$

$$(k = 1,\ 2,\ 11,\ 12,\ 22) \qquad (2.65)$$

with the coefficients as given in (2.60), (2.61), (2.53), and (2.54). The choice of sign in (2.63) is (+) for deflection in the same sense in the two fields and (−) for deflection in the opposite sense.

Clearly when $B_1 = B_2 = 0$ in (2.62), first order direction and velocity focusing are attained. Hintenberger & König's purpose in studying second order effects was to seek solutions to the system of five simultaneous equations, namely,

$$B_1 = B_2 = B_{11} = B_{12} = B_{22} = 0. \qquad (2.66)$$

In the case of mass spectrometers having straight magnetic field boundaries, the coefficients in (2.62) are functions of eight geometric parameters: Φ_m, ε', ε'', Φ_e, (d/r_m), (r_e/r_m), (l'_e/r_m), (l''_m/r_m). Thus three parameters may be chosen arbitrarily, and a unique solution found for the remaining five parameters. Hintenberger & König presented a number of geometric arrangements (table 2.2) which provide complete second order focusing (that is, equation (2.66) is satisfied). One of these arrangements (number 8) is the basis for a mass spectrometer which has been constructed at the University of Manitoba (Barber *et al.*, 1971; §5.12). Solutions to equation (2.66) were also presented for certain cases with curved magnetic field boundaries.

In the case of instruments intended for use with photographic detectors, solutions were sought in which first order double focusing occurs along a straight line and second order focusing occurs at one point. Such solutions were found only for $\sqrt{2}\Phi_e > \pi$. However, where the deflections in the two fields are in the same sense, the ion paths tend to be very long. The solutions for deflections in the opposite sense appear to be more practical, but involve inclined magnetic field boundaries which produce strong z defocusing. Solutions were also found for instruments in which the second order coefficients were small (although not zero) over extended regions of a photographic plate.

Hintenberger & König also investigated numerically the tolerances on the design values and have shown that, when adjustments are made to achieve a first order focus, the second order coefficients are small even when angles and lengths differ from nominal values by $\sim 1°$ and 1% of r_m respectively. Further, Barber (1965) has shown, for one case given by Hintenberger & König, that departures by 1% from uniformity in the magnetic field also have little effect on the second order coefficients.

As was described earlier, toroidal electrostatic analysers (§2.2) and non-uniform magnetic fields (§2.8) have focusing properties in both the radial and axial directions. Accordingly, second order expressions analogous to (2.52), (2.53), (2.54), (2.55) and (2.56) have been given for the toroidal condenser by Ewald & Liebl (1957*a*, *b*, *c*) for both radial and axial directions. Similar expressions for the non-uniform magnetic field have been given by Tasman (1959), Tasman & Boerboom (1959), Wachsmuth, Boerboom & Tasman (1959) and Tasman, Boerboom & Wachsmuth (1959). As in the calculations of Hintenberger & König, the

Table 2.2. *Parameters given by Hintenberger & König (1959) for second order double focusing mass spectrometers ($B_{11} = B_{12} = B_{22} = 0$) with deflection in the same sense in the electric and magnetic field. Magnetic field boundaries are straight*

No.	ϕ_m	ε'	ε''	$\sqrt{(2)}\phi_e$	ϕ_e	d/r_m	r_e/r_m	l'_e/r_m	l''_m/r_m
1	55	0	0	115.272	81.510	3.4364	1.2762	0.4768	1.4239
2	60	0	−3	128.692	90.999	2.9930	1.2203	0.2835	1.2145
3	60	0	−15	80.000	56.569	4.5200	1.8143	1.8204	0.8377
4	60	−15	−15	138.380	97.849	2.8459	1.0081	0.0879	0.6637
5	70	15	−15	70.761	50.036	4.9392	2.6663	4.1875	0.9116
6	72	15	−15	83.582	59.101	4.0507	2.2773	2.5560	0.9082
7	80	15	−15	129.381	91.486	2.3494	1.5454	0.5406	0.8768
8	90	27	−15	133.855	94.650	1.5955	1.5938	0.7084	0.9477
9	105	32	−35	65.392	46.239	3.8878	4.5927	45.8907	0.4426
10	110	31	−35	154.540	109.276	1.3432	2.0185	0.7614	0.4424
11	111	31	−35	167.058	118.128	1.2136	1.8983	0.4473	0.4392
12	116	35.9	−45	182.770	129.238	1.1190	2.5988	0.7416	0.3174
13	60	15	−15	183.220	129.556	0.9270	3.8756	5.8805	0.9434
14	60	23	−15	227.282	160.713	0.9449	4.5896	2.8751	0.9870
15	70	15	−15	140.724	99.507	0.6908	3.6209	22.1367	0.9092
16	72	15	−15	133.451	94.364	0.6499	3.6208	45.2327	0.0917
17	80	33	−15	183.292	129.607	0.7162	3.3174	2.9046	0.9528
18	90	33	−15	128.778	91.060	0.7056	2.6338	6.3650	0.9789
19	90	33	−30	250.548	177.164	0.7458	3.8714	0.1440	0.5230
20	105	32	−35	144.110	101.901	1.0577	3.1492	3.8777	0.4463
21	110	33	−45	158.727	112.237	1.2991	3.6810	3.1829	0.3210
22	115	34	−40	184.544	130.492	1.0553	2.0840	0.3143	0.3714

assumption was made that the field terminates abruptly at the boundaries.

The matrix representation for calculating ion trajectories in ion optical systems to second and third order has been much exploited because it is convenient in computer calculations (Penner, 1961; Brown, Belbeoch & Bounin, 1965; Wollnik, 1967*a*; Enge 1967). The way in which the matrix expressions are related to the second order equations of Hintenberger & König is discussed by Takeshita (1966).

A transfer matrix of the desired order may be derived for each section of the instrument, namely, the drift spaces, fringe fields and deflection fields. The ion beam is transferred through the instrument by multiplying the separate matrices together in the reverse order corresponding to the instrumental arrangement. There is one such equation which describes the ion beam in the radial direction and a second equation for the axial direction.

Transfer matrices to second order have been given for the magnetic and electric deflection fields by Wollnik (1967*b*) and, from these, expressions for second order image aberrations have been calculated (Wollnik 1967*c*, 1968). Similarly, third order trajectories have now been determined for cylindrical electric fields (Matsuo, 1975), toroidal electric fields (Matsuo, Matsuda & Wollnik, 1972) and, finally, the general case which describes any electric or magnetic sector (Fujita, Matsuda & Matsuo, 1977).

The effects of the fringe fields for homogeneous magnetic fields and cylindrical electrostatic fields have been studied to second order by Wollnik & Ewald (1965), Wollnik (1967*b*) and Enge (1964, 1967). More recently the general case of the fringe field of an inhomogeneous magnetic field has been described by Matsuda & Wollnik (1970*a*, *b*), to third order. Similarly, the general case of the fringe field of a toroidal electrostatic analyser has been described to third order by Matsuda (1971).

The use of multipole elements for the correction of image aberrations has been considered by Wollnik (1972) and the higher order focusing properties, including fringe effects, have been examined by Fujita & Matsuda (1975) and by Matsuda & Wollnik (1972).

Finally, Matsuo, Matsuda, Fujita & Wollnik (1975) have given in detail a computer program third (order) ion optics ('TRIO') for the calculation of ion trajectories through any combination of drift spaces, cylindrical or toroidal electric sector fields, homogeneous or inhomogeneous magnetic sector fields and magnetic or electrostatic quadrupole lenses. The effect of the fringe fields is included and the calculation is carried out to third order for the radial direction and to second order for the axial direction.

3

Ion sources for mass spectroscopy

Ion sources take many forms depending upon the element to be studied and upon the nature of the study. Ideally, the source should supply a monoenergetic ion beam which is rich in the desired ions and free of undesired ones. In practice it is frequently necessary to relax one or other of these requirements in order to improve the performance of the source in a particular desired respect.

3.1 Gas discharge

The positive ions which are created in a discharge in a gas at low pressures were first observed by Goldstein (1886). Thereafter, for four decades, the gas discharge was the principal source of positive ions and, in some laboratories, in modified forms, it is again in use. The classic form of this source is described in detail by Ewald & Hintenberger (1953).

The ions represent either gases present in the discharge tube, or electrode material which has been volatilized (or sputtered) under bombardment of the cathode or anode rays. Some workers have exploited this latter aspect by introducing foreign material into one or other of the electrodes to be ionized on volatilization (Gehrcke & Reichenheim, 1906, 1907, 1908; Aston & Thomson, 1921; Aston, 1923; Bainbridge, 1932; Mattauch, 1937; Aston, 1942; Coburn, Taglauer & Kay, 1974; Mattson, Bentz & Harrison, 1976). After formation, the ions are accelerated toward a perforated cathode, through which they pass, en route to the analyser, with a wide range of energy and degree of ionization. The wide energy spread amongst the ions is a disadvantage, as it limits the use of this type of source to velocity or double focusing instruments. The presence of multiply charged ions is normally also a disadvantage, although in the 1920's and 1930's it provided an important means of obtaining doublets needed in atomic mass determinations, prior to the development of the modern electron impact source (§3.3.). It is necessary, of course, to maintain different pressures between the discharge and analyser regions. The voltage necessary to excite the discharge is determined by the pressure and the electrode geometry. In early sources, which were operated at relatively low pressures, the voltage was

20 000–50 000 V and conditions were notoriously unsteady.[†]

The gas discharge source has recently been revived for possible use in the chemical analysis of solids. In this modern reincarnation the discharge takes place in a specially provided discharge gas (usually neon or argon) whose pressure is controlled, the sample to be analysed forms all or part of one of the discharge electrodes and is sputtered therefrom, and strong differential pumping is provided between the discharge region and the first stage of the mass spectrometer. Thus, Harrison & Magee (1974) and Mattson *et al.* (1976) have shown that the gas discharge source may be a useful alternative to the vacuum spark (§3.6). In their work both hollow and coaxial cathodes have been run in the pressure range $(1–7) \times 10^2$ Pa (1–5 Torr) with voltages 400–1000 V and currents 0.5–5 mA; sensitivities for trace element detection (few ppm) have been achieved that are comparable to those found with the vacuum spark source. In a somewhat related work Coburn, Taglauer & Kay (1974) have used a low voltage rf glow discharge between plane electrodes at a pressure of 1–10 Pa (10^{-2}–10^{-1} Torr). They have found that elements forming the electrodes are ionized with roughly equal efficiency and that ions can be extracted from the glow discharge with a sufficiently narrow energy spread that a quadrupole mass spectrometer (§6.11) can be used for their analysis.

The microwave glow discharge has also shown promise as an intense source with a large cross section beam, and has been used (Sakudo, Togikuchi, Koike & Kanomata, 1978) for a high current ion implantation unit. The discharge occurs in a ridged waveguide coupled to a magnetron and yields resolved ion currents of P^+ of a few milliamperes with PH_3 gas.

3.2 Surface ionization

The emission of positive ions from a salt coated on a hot filament was first observed by Gehrcke & Reichenheim (1906, 1907), and was first put to use in mass spectroscopy by Dempster (1918). Kunsman (1925, 1926) did important pioneer work concerned with the mechanism of this positive-ion emission and with methods of catalysing it.

The efficiency of this ionization process, which also includes the re-emission as ions of neutral atoms striking the hot filament, has been shown theoretically (Langmuir & Kingdon, 1925) to depend exponentially upon the difference between the work function of the filament and the ionization

† For a detailed and personalized account of gas discharge sources, the reader is referred to Aston (1942).

potential of the evaporated atom, according to the relationship

$$N_+ = N_0 \exp\left[\frac{e(\omega - \phi)}{kT}\right]. \tag{3.1}$$

In table 3.1, ω, the work function, and ϕ, the ionization potential, are given for a number of elements. Of the pure metals, platinum has the highest work function; partly because of this, but mostly because of its availability in ribbon form, it was the first widely used filament material. Subsequently tungsten, tantalum and rhenium came into use, in that order. Although these last-named materials have lower work functions than platinum, they are able to withstand higher temperatures. Zone refined rhenium, which offers an attractive compromise between work function and melting point, is the filament material now commonly used.

Fig. 3.1 shows, in two views, the source geometry developed by Inghram (Inghram & Hayden, 1954). The use of a new 'hat' assembly for each sample eliminated cross contamination effects, while the tertiary suppressor inhibited surface ionization by the filament of negative ions produced by bombardment of the first collimating slit (Hess, Wetherill & Inghram, 1951).

Table 3.1. *Work functions, ionization potentials and melting points of some elements*

Element	Ionization potential (eV)	Work function (eV)	Melting point (°C)
Li	5.39	—[a]	—
Na	5.14	—	—
K	4.34	—	—
Rb	4.18	—	—
Cs	3.89	—	—
Be	9.32	—	—
Mg	7.64	—	—
Ca	6.11	—	—
Sr	5.69	—	—
Ba	5.21	—	—
Ni	—	5.03	1453
Rh	—	4.80	1966
Pd	—	4.99	1552
Ta	—	4.19	2996
W	—	4.52	3410
Re	—	5.1	3180
Pt	—	5.32	1772

[a] – Indicates information not included (no relevance to present discussion).

For a given element, the yield of positive ions from a heated filament depends strongly upon the chemical compound used. For example, Inghram & Hayden (1954) found that when 10 μg/mm^2 of CsCl were placed on a tungsten filament, the ratio N_+/N_0 for cesium was approximately 10^{-4}, whereas for a similar coating of Cs_2SO_4, N_+/N_0 was ~ 1. Apparently most of the CsCl evaporated before the ionization temperature was reached. It was to overcome this difficulty of volatilization taking place before ionization temperatures were attained which led Inghram & Chupka (1953) to construct a two filament source. Here the volatile salt was evaporated from one filament at a relatively low temperature (thus increasing the sample life), whilst the neutral vapour was ionized by a second filament at a much higher temperature (thus achieving good ionization efficiency). The present form of this important innovation is the *triple filament* source (Craig, 1959; Shields, 1966), in which two filaments for holding the sample are arranged symmetrically with respect to the ionizing filament, as shown in fig. 3.2.

As the surface ionization source produces ions which are practically monoenergetic (~ 0.2 eV energy spread), it is well suited for use in instruments which possess direction focusing only. A further important virtue is the small amount of material needed for isotopic analysis. Ordinarily 10^{-9}–10^{-6} g of sample are sufficient to provide an adequate ion beam (Hintenberger, 1966). Further, this sensitivity can be greatly increased by using isotopic dilution techniques (§7.9); thus, 2×10^{-13} g of uranium (5×10^8 atoms) have been detected with a precision of 1–2% (Hess *et al.*, 1951).

An unfavourable feature of the surface ionization source is the isotopic fractionation which inevitably takes place during evaporation of the material. If it were simply Rayleigh distillation, in which the rate of

Fig. 3.1. Surface ionization source. *H* is the filament 'hat', *F* is the ion filament, *S* is the tertiary ion suppressor, *C* are collimating slits and *B* represents the beam-centring plates. (After Inghram & Hayden, 1954.)

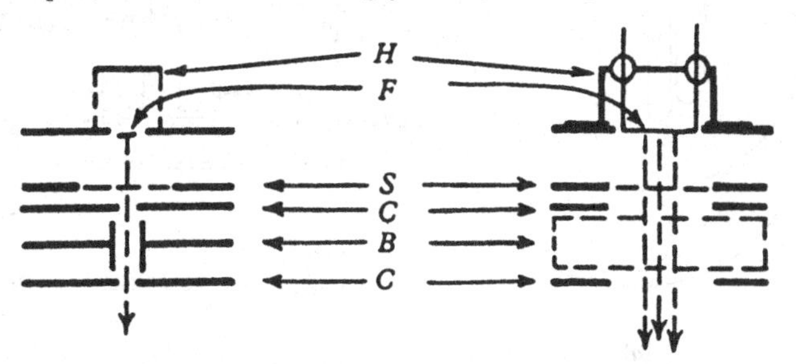

evaporation is inversely proportional to the mass, the effect could be corrected for. However, as Kanno (1971), de Bièvre (1978) and (especially) Moore & Heald (1978) have emphasized, there are other important considerations. For one thing, the sample becomes depleted in the lighter isotopes as a result of the fractionation and is therefore no longer representative. In addition, both molecules and atoms may be involved in the evaporation. These and other factors are discussed further in §§7.4–7.7.

The surface ionization source is suitable for a large number of elements, in particular for the alkalis, the alkaline earths, the rare earths and transuranium elements (see tabulation by Inghram & Hayden, 1954). Many elements are used in the form of the oxide or of a salt which reduces to the oxide on heating (for example, uranyl nitrate to uranium oxide) and frequently the monoxide ion rather than the atomic ion is studied. The source is especially useful for studying isotopic ratios in samples (Sr, Pb, etc.) of geochemical significance (see §§13.1, 13.2). In this work, where samples are small, special techniques have been developed for coating filaments with the sample material, for example, using borax, P_2O_5 or silica gel for lead samples (Cameron, Smith & Walker, 1969). Use of the surface ionization source for certain other elements is described in Rees (1969) – Mo; Rossman & de Laeter (1975) – Cd; Terra, Burnett & Wasserburg (1970) – Sm, Nd; N'Guyen & de Saint Simon (1972) – rare earths; Schwegler & White (1968) and Nief & Roth (1974) – U.

Fig. 3.2. Triple filament source: (*a*) shows the triple filament assembly whilst (*b*) shows a section through the source and lens assembly. (After Palmer, 1959.)

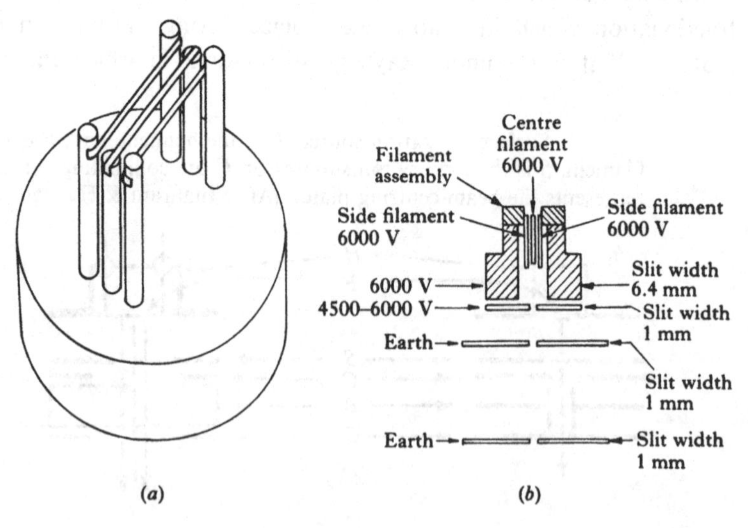

In the early 1970s the ionization of lanthanide elements in high temperature *cavities* was investigated on the assumption that such cavities would behave as high temperature surface ionizers. Observed efficiencies were much higher than predicted by the Langmuir equation (3.1), a behaviour attributed to the creation of a plasma in the cavity. The advantages of this arrangement over the conventional thermionic source – not only in efficiency, but also in the number of elements for which it is effective – are described by Kirchner (1981).

3.3 Electron impact source

This type of source, also called 'electron bombardment' source, was devised by Dempster (1918), but Bleakney (1929) and Nier (1940, 1947) are largely responsible for its present form. Two versions of this source, representing differing degrees of sophistication, are shown in fig. 3.3. Reference will be made first to the earlier and simpler version shown on the left.

Here the sample material, in the form of a gas or vapour, passes through the opening, shown at the top of the diagram, to the ionization chamber, where it is bombarded by a beam of electrons whose energy may be varied

Fig. 3.3. Two versions of the electron impact source: (*a*) simple, and (*b*) with focusing plates. *S* is the opening for sample introduction, *R* is the repeller, *E* is the bombarding electron beam, *I* is the region in which ionization takes place, *A* is the slit at the lower boundary of the ionization region, *B* is the drawing-out plate, *C* is a focusing plate, *D* and *G* are grounded collimating slits, *F* is the filament, *T* is the electron trap and *J*, *K* are beam-centring plates. (After Inghram & Hayden, 1954.)

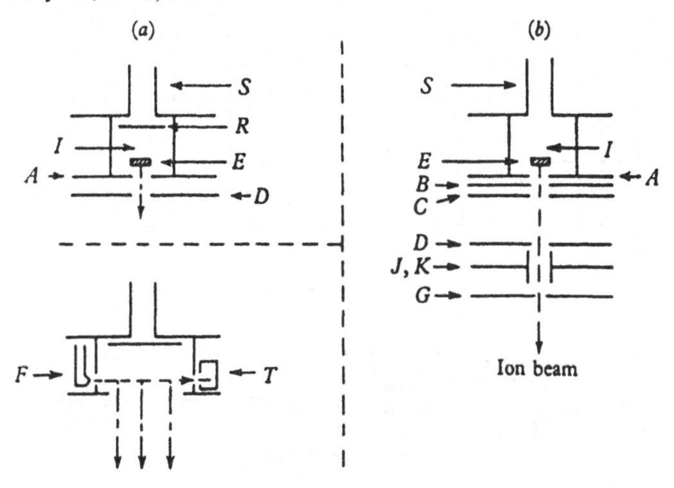

to secure maximum ionization efficiency. For most singly charged ions the ionization cross section reaches a maximum in the range 70–90 V, and falls off slowly thereafter. The behaviour of argon (Bleakney, 1930) is shown in fig. 3.4. Although the general shape of the ionization curve is the same for most gases, the numerical values of the cross section, σ, differ greatly: for Ar^+, $\sigma_{max} = 4 \times 10^{-16}\,cm^2$. The optimum electron energy increases with the degree of ionization, as can be seen also from fig. 3.4.

The positive ions so formed are urged downward by a weak drawing-out field between the repeller and the opposite wall of the ionization chamber, and emerge from the latter through a slit which is parallel to the direction of the electron beam. They are then accelerated by the strong field between *A* and the grounded diaphragm, *D*, and pass through a slit in *D* to the analyser region. The intensity of the bombarding electron beam is measured by the current to the electron trap, and is usually of the order of 10 μA.

The later arrangement, shown on the right of fig. 3.3, incorporates some additional features. The field between *A* and *B* extends through the slit in *A* into the ionization region and provides the drawing-out field; *C* is a focusing plate, and the acceleration takes place between *C* and *D*; *J* and *K* are beam-centring plates; while *G* provides a final collimating slit. Both the intensity and collimation of the ion beam are improved by these means, and the beam-centring plates provide a useful control for the lateral displacement of the beam.

The ions arising from the electron impact source are nearly homogeneous in energy. This is primarily due to the fact that, since the ionizing

Fig. 3.4. Number o argon ions, *N*, formed per electron per cm per mm Hg pressure at 0 °C. (From Bleakney, 1930.)

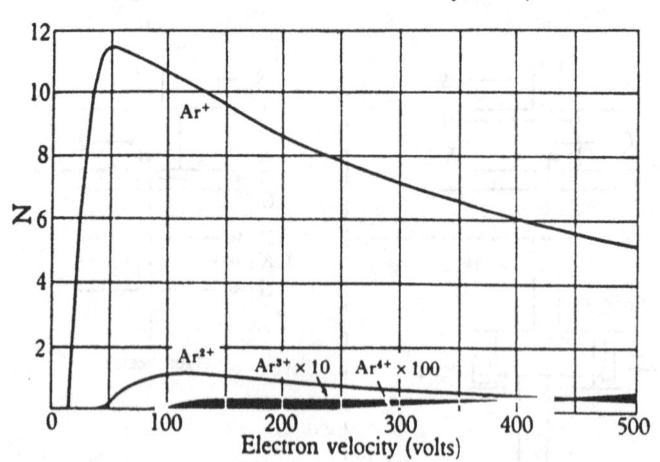

electron beam is narrow, the ions are created more or less along an equipotential surface and experience similar drawing-out accelerations. Usually the width of the electron beam is reduced even further by the use of an externally applied magnetic field parallel to the direction of the electron path. Electrons which would otherwise diverge from the specified path are in this way constrained to describe close helices about it. This generally results in an increase in the intensity of the positive-ion beam, concomitant with the improvement in monochromaticity which was sought. A homogeneity in energy of 0.05 eV can be achieved (Inghram & Hayden, 1954) provided the ions do not possess significant dissociation energy.

The most convenient elements to use with the electron bombardment source are those which at room temperature are either gases or possess an appreciable vapour pressure in the elemental or in a compound state. In these cases the sample may be led to the ionization chamber through an appropriate leak which controls the pressure in the chamber. The design of the leak and the sample line is of great importance and will be discussed in §7.5. In the case of non-volatile solids a furnace or oven may be located just above the ionization chamber.

Reference has been made to the narrow energy spread in the ion beam and to the versatility of this type of source in dealing with materials ranging from gases to low-vapour-pressure solids. Both these features represent outstanding advantages. There is the further advantage, in the case of gaseous materials, that changes may be made in the constitution of the gas under analysis without opening the vacuum system. Minute quantities of a gas are sufficient for an analysis (Thode & Graham, 1947; Macnamara & Thode, 1950; Inghram & Reynolds, 1950; Fleming & Thode, 1953*a*, *b*; Wetherill, 1953); for example 2×10^{-9} cc (5×10^{10} atoms) of a single isotope is adequate in the case of xenon (Kennett & Thode, 1956). Much smaller samples can be detected (xenon – 5×10^5 atoms, any isotope; ^{36}Ar – 1.4×10^7 atoms) if all pumping of the mass spectrometer is suspended after the entire gas sample has been admitted (Reynolds, 1956). Finally, with certain modifications (§10.1), the electron impact source is much used to obtain fundamental physico–chemical data, such as ionization and appearance potentials.

The chief disadvantage of the electron impact source is that ions are created of all gases present in the ionization chamber. This can give rise to a background which interferes with the analysis of the sample itself. In addition, with organic samples, the large amount of vibrational and electronic excitation which arises from electron impact results in the dissociation of the molecular ions, thus making mass and structure determination difficult. This last point has led to the development of other sources in which the integrity of the sample material is preserved (§§3.4, 3.5, 3.10, 3.13, 3.14).

Notwithstanding its shortcomings for certain analytic applications, the standard electron impact source is the most widely used in mass spectroscopy. In addition, it has been modified for certain specific purposes, for example Fox, Hickam, Kjeldaas & Grove (1951) – to provide a monochromatic source; Čermák & Herman (1962) – to study ion–molecule reactions (by addition of extraction voltage to the electron collector); Melton (1968) – to analyse primaries and reaction products (using a dual electron beam); Swingler (1970) – to serve as a source for a quadrupole filter; Hayden & Nier (1974) – to serve as an open ion source in space flight; Isler (1974) – to provide a particularly stable source (by addition of axial field coils plus feedback).

3.4 Chemical ionization source

In this source a primary or reagent ion is formed by electron impact and then reacts with a neutral molecule to produce a molecular ion, usually by proton transfer. The process was first recognized as such when CH_5^+ was observed in the electron bombardment of CH_4 (Tal'rose & Lyubimova, 1952; Field, Lampe & Franklin, 1957). The two-stage process in this case is

$$CH_4 + e^- \rightarrow CH_4^+ + 2e^-, \tag{3.2}$$

$$CH_4 + CH_4^+ \rightarrow CH_5^+ + CH_3. \tag{3.3}$$

It was then realized that H_3^+ ions which had been observed by Thomson, Dempster and others must have been formed in a gas discharge by an analogous *ion–molecule reaction*, namely,

$$H_2 + e^- \rightarrow H_2^+ + 2e^-, \tag{3.4}$$

$$H_2 + H_2^+ \rightarrow H_3^+ + H. \tag{3.5}$$

In the *chemical ionization* source, which has been developed to utilize this effect (Munson & Field, 1962; Munson, 1971), two gases are introduced into the source – the reagent gas (say CH_4) and the sample gas (M). After the primary ionization of the reagent gas by electron impact, ion–molecule reactions within this gas itself produce reactive species (for example, CH_5^+, as above) which react further with the sample gas to produce, say, $(M + H)^+$ ions. In this way, the sample gas gives rise to a relatively simple mass spectrum, characterized by little dissociation and a prominent $(M \pm 1)$ peak. Chemical ionization can also be used to generate negative ions from a sample gas and, in so doing, often achieves much greater sensitivity (Hunt & Sethi, 1978).

In the chemical ionization source, shown schematically in fig. 3.5, a conventional electron impact source is modified to accommodate the reagent-sample gas mixture at high pressure (Munson, 1971; Beggs, Vestal, Fales & Milne, 1971; Field, 1972; Gierlich, Heindricks & Beckey, 1974; Mather & Todd, 1979). The ionization chamber is made gas tight except for small apertures for the electron beam and for the ion beam slit. Thus, with an electron entrance aperture ~ 0.5 mm in diameter and an ion exit slit $\sim 5 \times 0.05$ mm, a pressure differential of $\sim 10^4$ can be maintained between the ion chamber box and the region outside. The reagent and sample gases are admitted into the ion chamber at partial pressures of 10^2 Pa and 10^{-1} Pa (1 Torr and 10^{-3} Torr), respectively. The electron energy is 100–500 eV, that is, considerably higher than normal for the electron impact source.

Variations of the chemical ionization source include the following: Hoegger & Bommer (1974)–employing an rf discharge; Hunt, McEwen & Harvey (1975) – employing a Townsend discharge between mesh electrodes; Kambara & Kanomata (1977) – a needle source; Cohen & Karasek (1970) – using beta rays from ^{63}Ni to produce the initiating ions in

Fig. 3.5. Chemical ionization source. (After Beggs *et al.*, 1971.)

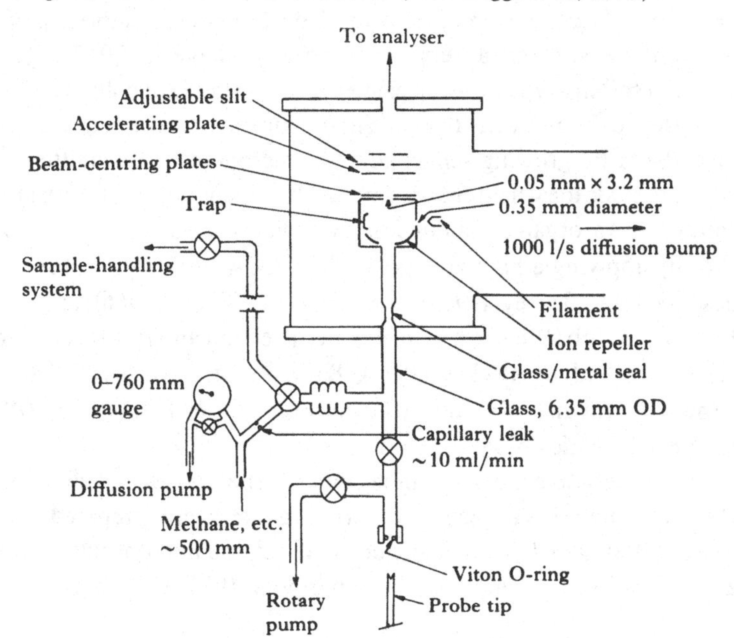

a carrier gas such He or Ne ('atmospheric pressure ionization source'); Boehme, Goodings & Ng (1977) – using the *in situ* chemical ionization in a premixed CH_4–O_2 flame.

3.5 Field ionization (and field desorption) source

The desorption of positive ions from surfaces under the action of strong electrostatic fields was first observed by Müller (1953), who made use of the effect in a *field emission ion microscope.* The strong field (of $\sim 10^8$ V/cm) was established between the tip of a sharpened tungsten wire and a coaxial ring. The screen of the microscope was located beyond the ring. Inghram & Gomer (1954, 1955) and Gomer & Inghram (1955) investigated the ions so formed by making a small hole in the screen through which a portion of the ion beam could pass into the mass spectrometer. In this ion beam, *parent* peaks were found to predominate, indicating that quantum mechanical tunnelling of an electron from an adsorbed gas molecule to the metal surface (that is, Fermi level) had produced a positive ion of the molecule. Furthermore, this molecular ion was stable, with little electronic or vibrational excitation. In addition, it was found that molecular ions were similarly formed of gas molecules near the metal surface, but not adsorbed on it. These ions are said to be formed by *field ionization.*

In its early form the field ionization source consisted of a sharp metal point or surface (for example a razor blade) as anode, a cathode perforated with an exit slit and a very high applied voltage (5–20 kV). This was followed by lens-type focusing elements and a standard collimating system which led to the analyser. The current practice is to activate the metal points or surfaces by growing semiconductor microneedles, or emitter needles, which provide local field enhancement. This is generally done by heating the emitter in an organic vapour (for example, benzonitrile) whilst simultaneously applying a high voltage. Early work with this type of source was described by Beckey (1963), Robertson & Viney (1966) and Beynon, Fontaine & Job (1966), whilst the modern version and its preparation have been described by Beckey (1971), Barofsky & Barofsky (1974), Cross, Brown & Anbar (1976) and Matsuo, Matsuda & Katakuse (1979). It is shown schematically in fig. 3.6.

Field desorption of organic molecules is a special case of field ionization. Here, the analysis sample is transferred to a specially prepared anode – a fine tungsten wire (10 μm diameter) on which the microneedles mentioned above have been grown (Barofsky & Barofsky, 1974; Rollgen, Giessmann &

Reddy, 1976; Reddy *et al.*, 1977) – by dipping the latter in a solution of the sample. After careful evaporation of the solvent the field anode is introduced into the field ionization source where positive ions are ejected from the molecules of the adsorbed layer under the action of the strong field. This is in contrast to field ionization where the compound reaches the vicinity of the field tip as a vapour.

The efficiency with which field ionization can produce stable molecular ions of most compounds is being exploited in organic structure analysis. Small samples can be analysed and high molecular ion intensities achieved even for polar molecules. The method shows particular promise in biochemistry (Beckey, 1978). The components of very fast reactions (on the time scale $\sim 10^{-11}$ s) can also be studied in the high field gradients of the field ionization source (§11.9). Finally, the technique is useful for the study of surface phenomena themselves. For example, in a modification of the source (*atom probe field ion microscope*) Müller, Panitz & McLane (1968) have been able to inject ions from selected lattice locations, as located in the field emission microscope, into a mass spectrometer for analysis.

Colby & Evans (1976) have described a source in which a sharply pointed liquid meniscus is produced in strong electric field. At sufficiently high fields, ions are formed at the tip of the meniscus by field evaporation. This so-called 'electrohydrodynamic ionization' has the advantage that the liquid tip is self-forming and self-replacing.

Reviews of field ionization mass spectrometry are given by Beckey (1971), Block (1976) and Reynolds (1979).

Fig. 3.6. Field ionization source including slit lens and principal slit of the mass spectrometer. (After Matsuo, Matsuda & Katakuse, 1979.)

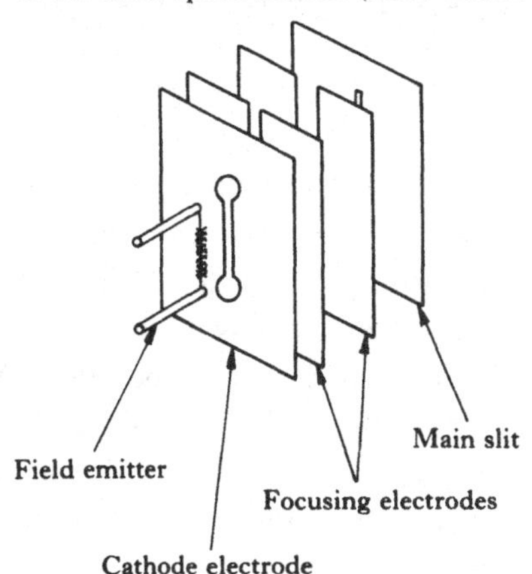

3.6 Vacuum spark source

The vacuum spark was developed as a source of positive ions by Dempster (1935, 1936). Three types of spark were tried which were known to be effective sources for the study of atomic spectra: the *trembleur dans le vide* or *vacuum vibrator* (Fabry & Perot, 1900*a*, *b*), the *hot sparks* from a large condenser discharge (Millikan, Sawyer & Bowen, 1921) and, finally, a spark coupled inductively to a high frequency oscillating circuit, that is, a *Tesla spark*. The high frequency spark proved highly successful, yielding an abundance of ions of the elements comprising the spark electrodes, and permitting for the first time isotopic analysis of the four elements Pd, Ir, Pt and Au.

This vacuum spark is a high voltage discharge ($\sim$ 50 000 V) with a high instantaneous current followed by a relatively long 'off' period. High local temperatures at the time of the discharge result in copious evaporation of the electrode materials. This vapour is ionized in the discharge, frequently to a high degree, while the long 'off' period prevents melting of the electrodes. The detailed behaviour of this pulsed rf spark has been described by Franzen (1963, 1972).

A variation of the vacuum spark is the low voltage triggered discharge where the arc phase of the high frequency spark is extended by the addition of a low voltage supply (Honig, 1966; Franzen & Schuy, 1967).

The principal merit of the vacuum spark is its ability to produce ions of *all* elements present in the electrodes, plus any others present in the spark gap in the form of a vapour or gas. While the spark exercises some discrimination between elements, it does so to a vastly lesser degree than does a filament or an oven. This attribute led Dempster (1946) and others (Gorman, Jones & Hipple, 1951; Hannay, 1954; Hannay & Ahearn, 1954; Brown, Craig & Elliott, 1956) to explore its use in the mass spectroscopic determination of the composition of solids, a use which is now well established.

The relative sensitivity of this source for various elements, as well as the relative abundance of different charge states, have been studied in many laboratories (Chakravarty, Venkatasubramanian & Duckworth, 1962; Schuy & Hintenberger, 1963; Ahearn, 1966; Davis & Miller, 1969; Farrar, 1972; Honig, 1966; Billon, 1976; Stüwer, 1976; Berthod, 1976; Venkatasubramanian, Swaminathan & Rajagopalan, 1977). It is found that the sensitivity for two-thirds of the elements is the same to within a factor of $\sim$ 3, whilst almost all elements fall within a range of $\sim$ 10. As the spark takes place between conducting electrodes, metals to be analysed are used as one, or both, of the electrodes. Non-conductors may be mixed with conducting powders or may be packed into conducting tubes to form electrodes. Under favourable circumstances, this source may be used to detect a few ppb (parts per billion, that is, 10^{-9}) of trace elements.

It is also found that in addition to producing singly charged ions of all elements present, this spark source produces a copious supply of multiply charged ions, certainly up to six charges for many metallic elements, and probably higher. This is an advantage or disadvantage depending upon the application. For early atomic mass determinations it was an advantage, as it enabled mass comparisons to be made between atoms from widely separated parts of the mass table, for example $^{65}Cu^{+}$ and $^{195}Pt^{3+}$. For analytic work, on the other hand, the presence of multiply charged ions can interfere with the detection of singly charged ions of lighter elements (for example, a doubly charged ion of mass M falls on the mass spectrum at mass $\frac{1}{2}M$).

The two chief demerits of the spark source are its unsteadiness and the very wide energy spread among the ions. Because of the energy spread, which may exceed 1000eV (see, for example, Woolston & Honig, 1964), its use has been limited to double focusing instruments. Because of its unsteadiness, it has usually been employed with a photographic detection system, which integrates the current. Although electrical detection has been used (Straus, 1941; Gorman, Jones & Hipple, 1951; Chakravarty *et al.*, 1962) by determining the ratio of the desired isotope current to that of another portion of the total ion current, it shows little sign of competing with photographic detection for this type of source (see, for example, Conzemius & Svec, 1972).

One of the two other types of vacuum spark originally investigated by Dempster as a source of positive ions – the vacuum vibrator – has been re-examined (Venkatasubramanian & Duckworth, 1963; Schuy & Hintenberger, 1963; Wilson & Jamba, 1967; Stroud, 1969). In this modification an arc is struck between a vibrating electrode and a fixed one. Ions from this source have a much narrower energy spread ($\sim$200eV) than for the high frequency spark source (Honig, 1966; Venkatasubramanian & Swaminathan, 1966; Davis & Miller, 1969), but there is a greater spread in the sensitivity for different elements and a less uniform distribution amongst various charge states (especially the doubly charged species) (Venkatasubramanian, Swaminathan & Rajagopalan, 1977).

3.7 Arc source

The so-called *arc* source is the first of three types of sources to be described in which *plasmas* are initiated by bombarding a neutral gas with electrons and superimposing appropriate electric and magnetic fields. Positive ions are extracted from the plasma, primarily for use in isotope separators, ion implantation, particle accelerators and other high ion current applications.

Low voltage arc ion sources were originally developed (Tuve, Dahl & Hafstad, 1935; Lamar, Samson & Compton, 1935) to provide intense ion beams for particle accelerators. Their application to mass spectroscopy in 1941 by Massey and his coworkers (Burhop, Massey & Page, 1949) and by Koch (1942) represented an important stage in the development of large-scale isotope separators.

The vacuum arc differs from an ordinary discharge in that it is not self-sustaining, but requires a beam of electrons for its operation. It may be operated with or without a magnetic field and the electron beam may be transverse to or along the direction in which the ion beam emerges. An example of a transverse electron beam with magnetic field is the arrangement shown in fig. 3.7, in which the arc or plasma region is constricted by a metal capillary, with a hole in one side for positive-ion extraction (Lamar *et al.*, 1935; Yates, 1938).

To start the discharge, the metal wall of the capillary is connected to the anode through a resistance of ~ 25 000 ohms. A low current discharge then starts in the cathode enclosure, the walls of the enclosure acting as the anode. Electrons from this discharge ionize the gas in the capillary and strike an arc to the anode, the ion density within the capillary being particularly high. The accelerating field penetrates weakly through the hole

Fig. 3.7. A vacuum arc source, showing the anode *A*, the filament *F*, the collimating slip *C*, and the focusing electrodes *B*. (After Inghram & Hayden, 1954.)

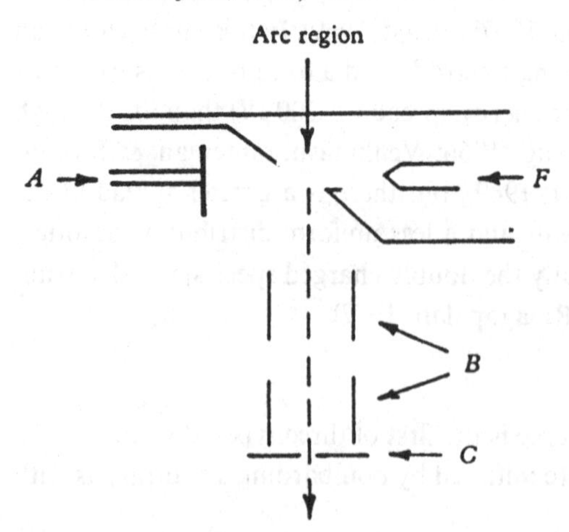

in the capillary and extracts ions from the arc plasma; these ions are then focused and collimated. As most of the ions are created by the primary electrons, these should have an energy of ~ 100 eV, as in the electron impact type of source.

The pressure for stable arc operation is critical for the following reason (Bohm, 1949). Surrounding the filament is a double sheath consisting of ions drifting in from the plasma and electrons emitted by the filament. For stable operation the positive-ion current, i_p, to this sheath must satisfy the Langmuir (1929) condition, $i_p \geqslant \alpha i_e(m_e/M_p)^{\frac{1}{2}}$, where i_e, m_e are the electronic current and electronic mass, respectively, M_p is the mass of the positive ion, and α is a constant of the order of unity. A certain threshold pressure must be maintained in order to ensure the necessary supply of positive ions.

With a source of this type, built by Lyshede (1941), Koch & Bendt-Nielsen (1944) and Koch (1953) have found that, for noble gases, an arc current of 0.2–0.8 A, operating at pressures of 0.3–0.8 Pa ((2–6) × 10^{-3} Torr), can produce an ion current of several hundred microamperes, which is, at the same time, monoenergetic to approximately 0.2 eV. Materials for analysis are introduced into the ion source in the form of gases or vapours.

When a magnetic field is used, it is made parallel to the direction of the primary electrons and effectively collimates them. The arc is, thus, naturally constricted without the use of a capillary. Approximate threshold pressures have been found, for argon and helium, to be 7 × 10^{-2} Pa (5 × 10^{-4} Torr) and 0.3 Pa (2.5 × 10^{-3} Torr), respectively, and these, in general, decrease somewhat with increasing magnetic field (Burhop *et al.*, 1949). The ion current usually emerges from a slot in the side of the arc region, shown in fig. 3.7. In this type of source, which was developed for the large-scale electromagnetic separation of isotopes, as in the calutron, the arc current is usually 2–10A. In favourable cases the positive-ion current may exceed 0.1 A. The element for isotopic separation is introduced into the arc as a vapour at a pressure of 30–130 Pa (0.2–1.0 Torr). This requires the charge material to be heated to the proper temperature and the whole of the arc chamber to a still higher temperature in order to prevent loss by condensation. A source temperature of 650 °C is adequate for more than half of the stable elements. For refractory elements the required source temperatures are much higher – 2800 °C in the case of ruthenium and iridium (Normand *et al.*, 1956). At this temperature vaporization of the graphite source block becomes serious. Other sources based on an arc in a magnetic field have been described by Bernas & Nier (1948), Bergström, Thulin, Svartholm & Siegbahn (1949), Kistemaker, Rol, Shutten & de Vries (1956), Almén & Nielsen (1956), Dawton (1956) and others. A survey of arc sources has been given by Thonemann (1953).

3.8 Oscillating electron plasma source

In this type of source, first developed by Finkelstein (1940), the ionization efficiency is greatly increased by causing the electrons to oscillate in the region occupied by the gas with energies sufficient to ionize the gas. An axial magnetic field is provided to confine the plasma. In the original model 150 mA of H^+ ions were extracted axially from the plasma. Subsequently, the source was modified for use in isotope separators (Nielsen, 1957; Almén & Nielsen, 1957). In this version, often called the 'Nielsen' source, electrons emitted from a filament oscillate between the filament and the other end of a cylindrical discharge chamber (see fig. 3.8.). The chamber lies in a coaxial magnetic field and the wall of the cylinder serves as the anode. Thus, under the combined action of the electric and magnetic fields the electrons move in helical orbits as they oscillate between the filament and the other end of the hollow cylinder. The sample material enters the discharge chamber as a gas or vapour and, for certain materials, the chamber is especially heated to prevent condensation. The whole chamber is made as air-tight as possible. When the gas pressure in the cylinder exceeds a certain threshold value a plasma is formed which fills most of the discharge chamber. Positive ions are extracted from the plasma through a hole in the end plate of the cylinder. As described earlier in connection with the arc source, a double sheath is formed around the filament which limits the electron current to the plasma, in accordance with the Langmuir condition. Despite its name, the 'oscillating' electron source has also been operated with the end plate at anode potential, in which case no oscillations take place.

In recent versions of the Nielsen source discharge chamber pressures are

Fig. 3.8. Nielsen ion source. The end plate may be operated at the same potential as the filament (as shown) or at the anode potential, E_a. (After Nielsen, 1957.)

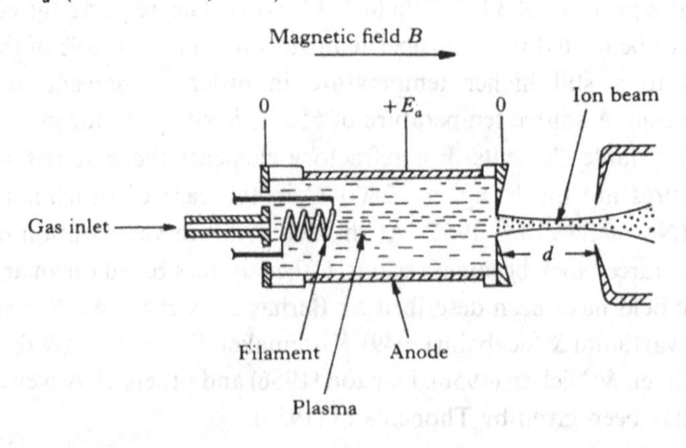

7×10^{-2} Pa (5×10^{-4} Torr) to several 10^{-1} Pa (10^{-3} Torr) and give rise (for xenon) to ion currents of 50–200 μA (Kirchner & Roeckl, 1976). In *on-line* studies of fission products, isotopes of krypton and xenon, produced by bombardment of a UO_2 stearate target, reach the collector of the isotope separator ~ 1 s after they enter the discharge chamber of the source.

The time delay of 1 s and also the fact that the standard Nielsen source will not operate at the low pressures often associated with on-line work led Kirchner & Roeckl (1976) to modify the source by adding a grid in front of the cathode. In this FEBIAD version (Forced Electron Beam Induced Arc Discharge), a stable plasma is formed with pressures ~ 10^{-3} Pa (~ 10^{-5} Torr) and gives rise to ion currents $\leqslant 1\ \mu$A, whilst the time delay in passing through the isotope separator is reduced by a factor of 10. The designers of this version believe that it fills a gap between the Nielsen source (high efficiency, medium ion current) and the electron impact source (low efficiency, low ion current) and, on that account, may have other applications in mass spectroscopy.

Another type of oscillating electron source, developed and extensively used by Barber *et al.* (1971), is shown in fig. 3.9. This source has a filament,

Fig. 3.9. Oscillating electron source: *A*, sample; *B*, oven; *C*, Re filament; *D*, ribbon heater and boron nitride insulation; *E*, electromagnet coil.

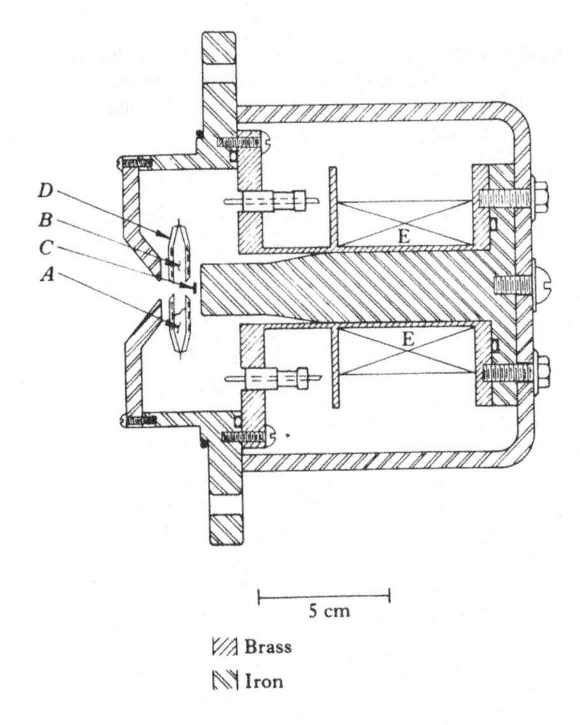

an anode with axial apertures, and a reflector cathode which also serves as source case and as positive ion extractor. The anode is at the same time an oven, in which the substances to be ionized are vaporized by external heating. The filament is at -100 V, and the reflector at -300 V, with respect to the anode. A strong axial field (~ 1 kG) is developed across the gap of the specially shaped yoke by the current in the coil. Thus, the electrons oscillate along the source axis in tight helical trajectories and create a plasma in the oven, from which the ions are extracted. By using chlorides as the sample materials, total ion currents of $\sim 25\,\mu$A have been obtained with energy spreads of less than 2 eV.

In recent years sources which exploit the discharge between a hollow (cylindrical) cathode and an anode have been developed for isotope separators operated on-line with accelerators for nuclear physics ('ISOL' systems; §9.9) and for instruments used for ion implantation experiments (§12.4). These sources are very compact, may be operated at very high temperatures, and have high efficiencies and current densities. Although the original hollow cathode discharges employed cold cathodes, heated cathodes are now invariably used because they permit greater control over the discharge parameters.

Sidenius (1965, 1969) has described two basic versions of the source which are illustrated in fig. 3.10. In the first type (*a*), shown on the left, the

Fig. 3.10. Hollow cathode ion source. In version (*a*) ion extraction is from the anode while in version (*b*) extraction is from the cathode. (After Sidenius, 1965, 1969). Here V_f is the voltage across the heated filament and V_a is the arc voltage.

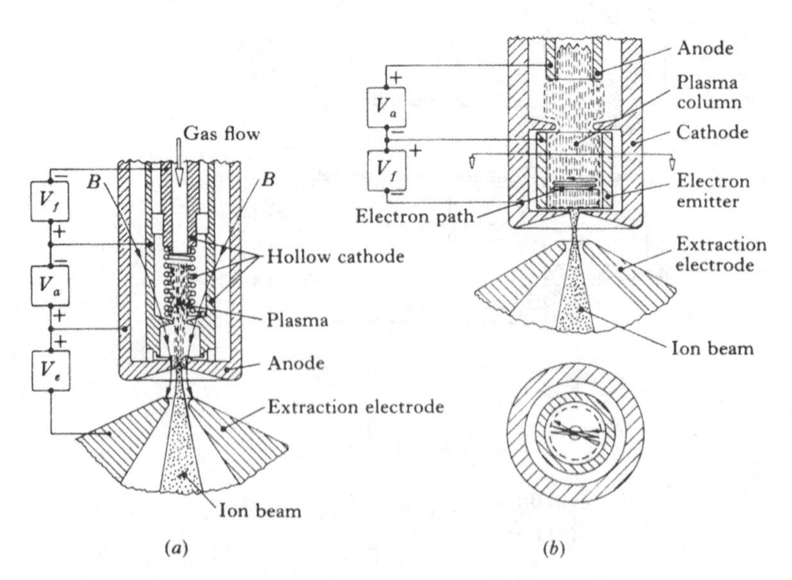

extraction of the ions is through a hole in the *anode*. A strong, inhomogeneous magnetic field is used so that the plasma is concentrated in the region of the extraction hole. The source may be operated at 2000 °C and, with Ar, an efficiency of 40% and a total ion current of 1.3 mA were obtained with a 0.4 mm diameter outlet hole and a 1.5 mm diameter expansion cup.

In the second version of the source (*b*), shown on the right of fig. 3.10, the ion beam is extracted from the *cathode*. In this case the full cathode voltage drop occurs across the plasma sheath so that electrons are emitted and accelerated normally to the cathode. Thus, to a first approximation, the electrons pass through the axis of the hollow cathode and are reflected from the cathode surface on the opposite side. Hence the plasma density is highest in the central region of the discharge. Ions may travel toward the anode as a result of diffusion or of weak fields in the plasma. In this version the source may be operated without an external magnetic field, with a field that just cancels the magnetic field generated by the filament or with a field of ~300 G (Sidenius, 1965, 1969; Äystö *et al.*, 1976; Mazumdar, Wagner, Walcher & Lund, 1976).

Although the total ion current and efficiencies were somewhat lower in the latter geometry (~500 μA for Ar with 25% efficiency), it was found that the stability of operation of this version was superior to that of the former version.

3.9 The duoplasmatron

This source was developed for use with particle accelerators (von Ardenne, 1956) and incorporates two means of increasing the plasma density in the region from which the ions are extracted. In the first stage of development (the 'unoplasmatron') a cone-shaped intermediate electrode was placed between the filament and the anode. In the second stage, the plasma was further compressed by adding a coaxial magnetic field. These features of the design are shown in fig. 3.11. The potential of the conical electrode is about midway between that of the filament and that of the anode, and the electrode itself is of mild steel. Ions which diffuse through the aperture at the tip of the cone – into the space between it and the anode – are either collected by the anode or pass through it and are accelerated. Because of the high current thus drawn from the plasma, the central part of the anode must be heat resistant and is normally made of tungsten, tantalum or molybdenum. Further, the density of the current extracted through the anode may exceed 20 A/cm^2, so that space charge limitation must be taken into account. For this reason an *expansion cup* (shown in fig. 3.11) is provided to allow the ion beam to expand to sufficient diameter. Meanwhile, the shape of the cup provides electrostatic shielding of the beam and does not interfere with its acceleration. General reviews of the duoplasmatron have been given by Morgan, Kelley & Davis (1967). Illgen, Kirchner & Schulte (1972), Green (1974), Lejejune (1974) and Bacon (1978).

Ions are produced from gases specially introduced or from materials vapourized under the action of the discharge in the region between the intermediate electrode and the anode. Thus, with bismuth so placed, bismuth ions provided 60% of the ion current, the auxiliary gas 25% and sputtered ions of electrode materials the rest (Illgen *et al.*, 1972). Also, many multiply charged ions are produced: for example, with xenon at a source pressure of 3.3 Pa (2.5×10^{-2} Torr) and total ion current of ~5 mA the following charge states were found: 1+(7.3%), 2+(21.0%), 3+(26.0%), 4+(20.0%), 5+(13.5%), 6+(7.3%), 7+(3.1%), 8+(1.0%), 9+(0.4%). As we shall see in §3.11, medium charged states are favoured in the duoplasmatron as compared to the Philips ionization gauge (PIG) source.

A modification of the duoplasmatron, intended for higher current densities, is the duoPIGatron (Morgan, Kelley & Davis, 1967; Davis, Morgan, Stewart & Stirling, 1971; Bacon, Bickes & O'Hagan, 1978). Here the cathode and intermediate electrodes are the same but the subsequent ion beam is subject to a Penning-type discharge (§3.11) which further intensifies it.

Fig. 3.11. Duoplasmatron ion source and extraction system (after Rose & Galejs, 1965): *a*, anode aperture; *b*, plasma expansion cone; *c*, extractor electrode; *d*, suppression electrode.

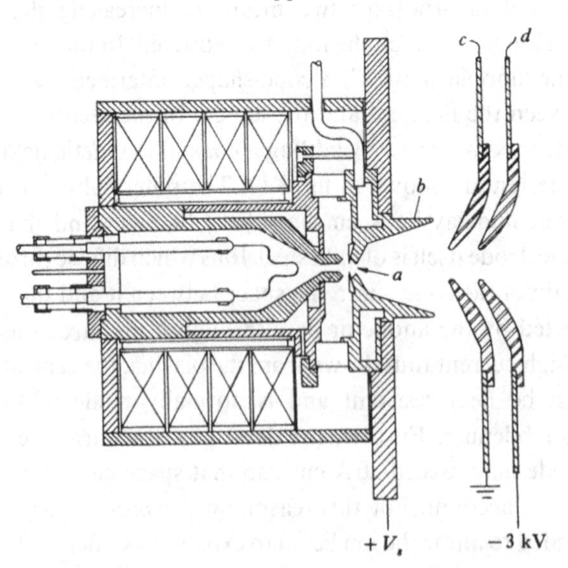

3.10 Secondary ion sources

The development of these sources has led to *secondary ion mass spectrometry* (SIMS) and *ion microprobe spectrometry* for the detailed analysis of solids. Although earlier studies had been made of secondary ion emission (for example, Herzog & Viehböck, 1949), the use of secondary ions as an ion source began with Castaing & Slodzian (1962). In this type of source the material to be analysed is bombarded with primary ions which sputter atoms and molecules from the surface, some of which are ejected as ions. Most sputtered particles come from the top three or four layers of the solid whilst their final ionization takes place outside the solid but within a few angstrom units of it (Blaise, 1978). The primary beam can be focused to achieve a lateral resolution of 1 μm and trace elements can be detected to a sensitivity of 1 ppm. Thus, the secondary ion source sensitivity ($\sim$ 1 ppm) represents a powerful method for surface analysis. It is shown in fig. 3.12, as the first component of an ion microprobe, a use that is described later in §12.2.

Fig. 3.12. Secondary ion source. (After Liebl, 1967.) The primary ions emerge from the source at the upper right, are selected according to mass and are focused on the sample, giving rise to secondary ions. The deflection plates shown prior to the sample are used to scan the primary ions across the sample. The entire diagram shows an ion microprobe (see §12.2) for which the secondary ion source is an essential component.

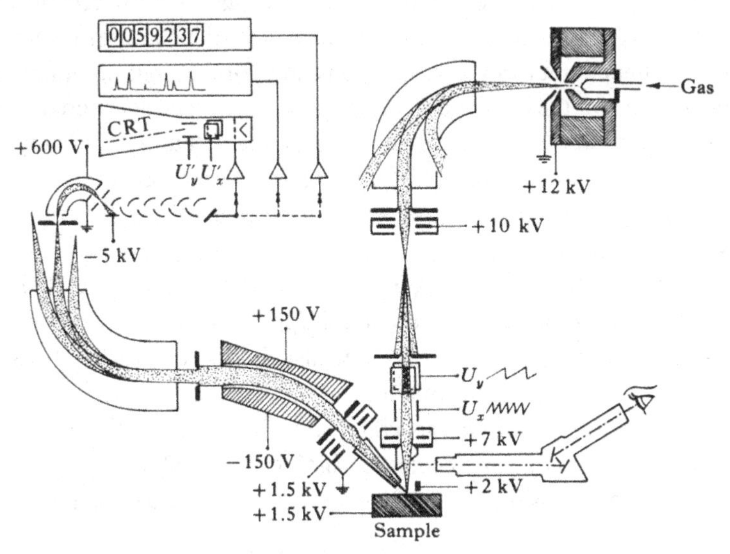

As indicated in the diagram, the primary ions are produced by a duoplasmatron source operating with Ar, N_2 or O_2, are extracted with energies of 1–20 keV, and (in some cases) are selected by a low resolution mass spectrometer. They are then focused by two objective electrostatic lenses and also pass between deflection plates by which the beam ($\sim 1\,\mu$m diameter) may be scanned in both x and y directions over a distance of about 300 μm. The secondary ions are extracted at an angle and focused by a high efficiency lens in the double focusing analyser for mass analysis. Primary beam densities normally range between 10–$10^3\,\mu A/cm^2$. With a beam density of $100\,\mu A/cm^2$ and ion energies of a few keV the surface erosion rate is one atomic layer/s. By reducing the beam density to 10^{-3}–$10^{-1}\,\mu A/cm^2$ the material consumption can be reduced to $\sim 10^{-4}$ monolayer/s.

The production of secondary ions is a complex process, two mechanisms being involved. In the *kinetic* mechanism, the primary ions transfer their energy to the host matrix, breaking lattice bonds and ejecting electrons into the conduction band. Whilst most ions are neutralized before leaving the surface they are in metastable states and undergo Auger and other de-excitation processes, thus escaping as ions. This is the dominant mechanism with argon or other non-reactive primary ions. With reactive gases, such as oxygen, the *chemical* mechanism is dominant. Here the reactive species reduce the number of conduction electrons available for neutralization of sputtered ions and the result is a dramatic increase in secondary ion emission (Anderson, 1969, 1970).

Many ionic species are produced in the secondary ion source. In general, singly charged ions predominate, especially for oxides and with reactive gas bombardment: multiply charged ions are formed but in much lower abundance. On the other hand dimers and other polymers and also inter-element ions are quite abundant (1–20%), the actual values being determined mainly by chemical considerations. The energy spread amongst secondary ions is hundreds (or even thousands) of eV, but the maximum is usually ~ 5 eV.

Morrison & Slodzian (1975), Benninghoven (1975) and Blaise (1978) have summarized the present and potential applications of the secondary ion source in the analysis of surfaces, whilst Smith & Christie (1978) have analysed the source's performance.

Recently there has been a special interest in the production of ions from non-volatile, organic molecules of biological significance by forms of secondary ion emission in which the sample is deposited on a substrate and bombarded by ions or atoms (Sundquist, 1982; Benninghoven, *Proceedings of the International Conference on Ion Formation from Organic Solids* (1982)). The experimental techniques, which are related to each other, have been given the following names: ^{252}Cf plasma desorption mass spectrometry (^{252}Cf–PDMS, Macfarlane & Torgerson, 1976), heavy ion-induced desorption mass spectrometry (HIIDMS, Dück *et al.*, 1980; Håkansson *et al.*, 1981), secondary ion mass spectrometry (SIMS, Benninghoven & Sichtermann, 1978; Chait & Standing, 1981), and fast atom bombardment (FAB, Barber, Bordoli, Sedgwick & Tyler, 1981; Barber *et al.*, 1982). The similarities and differences of these methods have been summarized by McNeal (1982) and Macfarlane (1982).

The way in which heavy ions or atoms lose energy when colliding with and passing through an absorber has been studied for many years (Bohr, 1948; Lindhard & Scharff, 1961; Ormrod, MacDonald & Duckworth, 1965; Dearnaley, 1973). For ions with energies less than a few keV/u, the predominant mechanism for energy loss is by elastic collisions between the incoming particle and the atomic nuclei in the target. The maximum for this energy loss occurs at ~ 1 keV/u (Ormrod *et al.*, 1965). For ion energies greater than a few keV/u loss through electronic interactions predominates, passing through a maximum at ~ 1 MeV/u. In the former case a collisional cascade is formed in the target, whereas in the latter case electronic excitations are ultimately transferred to lattice vibrations. Thus, in both cases, a 'thermal spike' is formed in the substrate with a temperature of ~ 10^4K (Macfarlane, 1982), resulting in desorption from the surface of ions, some of which have not suffered decomposition. For molecular ions to be produced, the lifetime of the heated spot must be less than the time for vibrational excitation of any of the bonds. Macfarlane has suggested that this accounts for the similarity in the mass spectra produced by these methods and also by laser desorption.

In plasma desorption, Macfarlane & Torgerson (1976) have used the fission fragments from ^{252}Cf, which have energies of ~ 100 MeV, as the bombarding particles. In Håkansson *et al.* (1981), a variety of heavy ions of comparable energy are substituted for the fission fragments. The yield of ions has been studied and shown to increase with energy, with atomic number ($\propto Z^2$), and with charge state q ($\propto q^4$).

In the low energy region, a beam of alkali metal ions at, say, 20 keV (SIMS, Chait & Standing, 1981) or a 'fast' atomic beam (FAB, Barber *et al.*, 1981, 1982), with energies of ~ 2 keV, have also been used to bombard the sample. Again there is an increase in the yield of ions with energy and, dramatically, with the atomic number of the bombarding particle. These two methods involve the low energy region and are equivalent, inasmuch as the mechanism for energy loss in the sample and substrate is the same for both. A wide range of particle flux on the target has been used. At high fluxes the effect of radiation damage in the target may be minimized by the use of a liquid phase (for example, glycerol) containing the sample, so that fresh material is being bombarded at all times.

While the ^{252}Cf–PDMS source by its nature is associated with time-of-flight instruments, HIIDMS and SIMS sources may be adapted to either time-of-flight or deflection mass spectrometers. FAB sources have been used with deflection instruments.

All four of these modes of ionization are relatively gentle, although less so than field desorption, with the remarkable result that not only parent ions appear, but also those formed by the sequential loss of weakly bound components. Both positive and negative ions are formed. Macfarlane and his associates have observed cationized molecular ions at $M = 6980$ (McNeal, 1982) and an organic dimer at $M = 12\,637$ (McNeal & Macfarlane, 1981).

3.11 Philips ionization gauge ion source (PIG)

In 1937 Penning described a vacuum gauge which employed a cold discharge in a magnetic field (Penning, 1937). Using this principle, Backus (1949) developed a source of positive ions which has come to be known as the Philips ionization gauge (PIG) type source, and which has subsequently

Fig. 3.13. The Philips ionization gauge (PIG) type of source showing the anode, *A*, the cathodes, *C*, the drawing-out plate, *D*, the focusing plate, *F*, the grounded collimating slits, *B*, the beam-centring plates, *J* and *K*, and the poles, *N* and *S*, of the permanent magnet. (After Inghram and Hayden, 1954.)

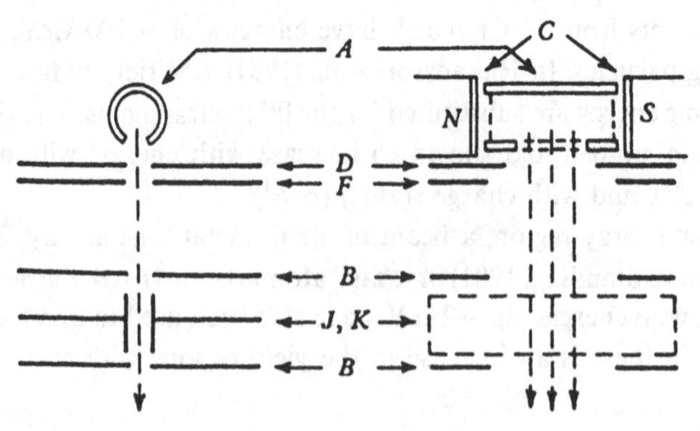

been modified and/or studied by Thomas (1947), Barnett, Stier & Evans (1953), Jones & Zucker (1954), Ehlers (1962), Kohno *et al.* (1968), McFarlin, Wilson, June & Chapman (1971) and Fuchs (1972). A form of this source is shown in two views in fig. 3.13.

Here electrons are pulled from the cathodes toward the cylindrical anode maintained at $\sim +2000$ V. Because of the magnetic field, the electrons move in tight spirals along the axis. As they approach the opposite cathode, their direction is reversed, with the result that they oscillate back and forth, enjoying an exceptional opportunity of producing positive ions by collision. The positive ions, which are usually drawn through a slot in the side of the anode, have been found by Backus to possess an energy spread of approximately 25eV under these conditions. With an alternative arrangement, in which the ions are extracted through a small opening in one of the cathodes, the energy spread is much greater.

This type of source has been used for isotope separators and as an ion source for particle accelerators, in particular for cyclotrons where the magnetic field required is provided by the cyclotron magnet. The ability of the source to provide multiply charged ions was noted by Jones & Zucker, whose high current source was developed for this purpose. This source has been changed only slightly in design since that time and has proved singularly well suited for its intended purpose. Although the cathode is 'cold' it is heated by the discharge at high power and emits electrons thermionically: this avoids the use of a heated filament. Bennett (1972) achieved higher instantaneous powers using a pulsed source and in this way increased the production of higher charge states, particularly for heavy ions. Thus, for xenon, he found the following charge states: 1 + (0.8%), 2 + (7.1%), 3 + (20.9%), 4 + (21.0%), 5 + (19.4%), 6 + (14.6%), 7 + (11.2%), 8 + (4.9%) – a shift toward higher charged states from that found in the duoplasmatron (§3.9). Non-gaseous material can be placed on the anode, to be sputtered into the discharge. In this way Trt'yakov, Pasyuk, Kul'kina & Kuznetsov (1970) obtained Ca^{9+} and Zn^{10+} ions.

Bennett (1964) found that each trapped electron in a nitrogen discharge produced eight ions. This high efficiency enables the discharge to be operated at low pressures, a fact which has led to the wide use of the source as a high vacuum *pump*. The normal pressure in the source is 1.3–1.3 × 10^{-2} Pa (10^{-2}–10^{-4} Torr) and standard ion currents are 10–50 mA. The source is much used in heavy ion accelerators and for ion implantation.

An extensive review of the Nielsen source, the duoplasmatron, the PIG source and other sources used in low energy accelerators has been given by Sidenius (1978).

3.12 Magnetron ion source

The classic magnetron geometry, involving radial electric and axial magnetic fields, has also been employed in positive-ion sources (Van Voorhis, Kuper & Harnwell, 1934; Cobic, Tosic & Perovic, 1963). Here, the electron emitter is a wire or ribbon mounted axially in a cylindrical anode and ions are extracted from a slit in the anode cylinder. With a heavy filament mounted close to the extraction slit, the plasma density is high, the plasma boundary is well defined and the source operation is not critically dependent upon the strength of the magnetic field. This feature is exploited in the Harwell ion source (Freeman, 1969) which operates at low fields and produces milliampere currents of most elements.

3.13 Photoionization source

As is well known, molecules can be ionized without dissociation by irradiation with ultraviolet light. The ionization potentials of most molecules lie in the range 7–16 eV, corresponding to photon wave lengths of 1250–775 Å, which lie mainly in the vacuum ultraviolet region. This method of ionization is gentle in comparison to the electron bombardment method with the result that, amongst others, molecular or 'parent' ions are obtained.

The first mass analyses of ions produced by photoionization were those of Terenin & Popov (1932) and Lossing & Tanaka (1956). The first use of a monochromator to provide the photoionizing radiation was by Hurzeler, Inghram & Morrison (1958). In this way, an energy spread in the photon beam of ~0.5 eV was achieved which made possible the accurate study of numerous ionization and appearance potentials (Weissler, Samson, Ogawa & Cook, 1959; Morrison, Hurzeler, Inghram, & Stanton, 1960; Frost, Mak & McDowell, 1962). Reviews of photoionization in mass spectrometry have been given by Marr (1967) and Reid (1971).

In order to reach the upper part of the energy range with an intense light source, the synchrotron radiation from an electron accelerator or storage ring has also been used in photoionization sources (Parr & Taylor, 1973; Guyon, 1976). In this case the source is a broadband one and, accordingly, is passed through a monochromator before impinging on a molecular beam.

The use of tunable lasers in photoionization sources is described briefly at the end of §3.14.

3.14 Laser ion source

The development and availability of lasers in the early 1960s provided a new method for evaporating and ionizing small areas of solid samples. The first use of a mass spectrometer to analyse the pulses of positive ions produced by the action of lasers was that of Honig & Woolston (1963) and Honig (1966). Not unexpectedly, the ions produced in this manner have a wide energy speed (~ 1000 eV–Linlor, 1963; average of 150 eV – Isenor, 1964); thus, a double focusing mass spectrometer is best suited to this type of source. As with the vacuum spark source, the relative ionization efficiencies of a laser source for various elements do not vary by more than a factor of 5–10. Also, as with the vacuum source, numerous multiply charged ions are produced.

In a typical laser source, a laser beam is passed through a telescope system and focused by means of an objective lens onto the surface of the material to be analysed, whose position is controlled by a micro-manipulator. Ions produced in the laser-induced plasma are accelerated by an extraction electrode and focused on the entrance slit of the mass spectrometer. Some developments and uses of laser ion sources are given by Vastola & Pirone (1966), Ban & Knox (1969), and Knox (1972).

Tunable lasers have found a special use in photoionization sources. Thus, Huber *et al.* (1977) have used such a laser with wavelengths near the threshold of molecular dissociation to enhance the sensitivity of the source for the precision study of molecular states.

3.15 Negative ions

The importance of negative ions in electrical discharges, in upper atmosphere phenomena, in stellar and solar atmospheres, in particle accelerators and in fusion reactions has stimulated much study of their formation and behaviour. As these subjects are complex and not much related to mass spectroscopy, we give first a brief description of the principal processes used for the production of negative ions, bearing in mind that most work has been aimed at elucidating the processes themselves, rather than at producing a negative ion beam *per se*. The various processes have been summarized by Massey (1976) as follows:

Radiative processes

(1) $XY + \mathrm{e}^- \rightarrow XY^- + h\nu$ – radiative capture of electrons,

(2) $XY + h\nu \rightarrow X^+ + Y^-$ – polar photodissociation.

Dissociative processes

(1) $XY + \mathrm{e}^- \rightarrow X + Y^-$ – dissociative attachment,

(2) $XY + \mathrm{e}^- \rightarrow X^+ + Y^- + e^-$ – polar dissociation.

Capture of bound electrons

(1) $X + Y \rightarrow X^- + Y^+$,

(2) $X^+ + Y \rightarrow X^- + Y^{2+}$.

The first study of *polar photodissociation* in which a monochromator and mass analyser were used was that of Morrison, Hurzeler, Inghram & Stanton (1960). Of the two *dissociative processes*, dissociative attachment occurs at lower energy than polar dissociation. Thus, for O_2, the former peaks sharply at an electron energy of 6.5 eV, whilst the second has a threshold of ~ 16 eV (Briglia & Rapp, 1965). The *capture of bound electrons* is studied in hydrogen, for example, by passing beams of protons through gaseous targets. The species H^+, H and H^- are all formed, and in proportions which depend upon the proton energy and the nature of the target gas.

We now describe two important fields of research in which negative ions are produced for their own sake. In the first, as has been mentioned, sources of negative ions are widely employed in particle accelerators. In the tandem Van de Graaff, for example, the particles are accelerated first as negative ions, become positively charged in passing through a 'stripper', and are then re-accelerated: thus, the same potential difference is utilized twice. In cyclotrons, on the other hand, the acceleration of negative ions, terminating in a 'stripper', simplifies extraction of the ions. Thus, ions accelerated as H^- emerge from the stripper as a well defined proton beam with a trajectory curving *out* of the magnetic field. For the former application, duoplasmatron sources have been used to deliver up to 100 μA of negative ions, whilst in the latter, PIG sources have yielded several mA of H^- ion current. The sputtering process is also used to produce significant numbers of negative ions, particularly when cesium ions are used as the bombarding agent (Middleton & Adams, 1974; Middleton, 1977).

More recently, an interest has developed in the production of *much* higher beams of negative hydrogen ions for possible application in fusion reactors, in high energy accelerators and in storage rings. In this work, the negative hydrogen ions are (a) extracted directly from plasmas, (b) are produced from positive hydrogen ions in a charge exchange chamber or (c) are formed by surface conversion of positive hydrogen ions. Three important developments based on (a) and (b) have been outlined by Prelec & Slyters (1973). The first is a hollow discharge duoplasmatron (Golubev, Nalvaiko, Tokarev & Tsepakhin, 1972), which exploits the discovery (Collins & Gobbett, 1965; Lawrence, Beauchamp & McKibben, 1965) that the highest density of negative hydrogen ions in the duoplasmatron is slightly off the axis. Beams of 11 mA of H^- have been obtained in this way from hydrogen and 60 mA of H^- from a hydrogen–cesium mixture (Kobayashi, Prelec & Slyters, 1976). In the second development, ~ 20 mA of negative ions have been extracted from a magnetron type of source

(Bel'chenko, Dimov & Dudnikov, 1972). In the third development, D^- ions are produced by charge exchange between D^+ and cesium vapour (Hamilton–Gordon & Osher, 1972). Process (c) – surface conversion – has been used by Ehlers & Leung (1980) in a multiline-cusp source which delivers more than 400 mA of H^-, for the production of high energy neutral beams.

4
Detection of positive ions

4.1 Fluorescent screens

The fluorescence of glass under the impact of positive rays led to their discovery by Goldstein (1886). Other materials were found to give a brighter fluorescence than glass, in particular, willemite, a natural silicate of zinc. Very uniform fluorescent screens were prepared by J. J. Thomson (see Thomson, 1913) by allowing a fine suspension of willemite in alcohol to deposit slowly on glass. Thomson found that the brightness of the fluorescence diminished with continued exposure to positive rays, and eventually disappeared, necessitating occasional renewal of the screen. Also, some materials deteriorated more quickly than others, for example willemite provided a longer, though less spectacular, source than zincblende, a naturally occurring zinc sulphide.

About 1910, the fluorescent screen was largely supplanted as a positive ion detector by the photographic plate. The recent revival of scintillator materials in connection with Daly detectors will be discussed in §4.6.

4.2 Photographic detection

The sensitivity of photographic plates to positive rays was discovered by Koenigsberger & Kutchewski (1910) and, independently, by J. J. Thomson. It was shown by Thomson that the darkening of the plate depends strongly upon the penetrating power of the particles which, for a given energy, decreases rapidly with increasing mass. As ion energy lost in the gelatin film itself does not contribute to the blackening, sensitive plates were sought among those with thin films and high silver content. Of these, Schumann plates, with a minimum of protective gelatin, were found to be the most sensitive, an observation subsequently verified by Aston (1925), Bainbridge (1931) and others. These plates were invaluable in the search for new isotopes, and were also useful in investigations in which the positions, rather than the intensities, of weak lines were of interest, as in atomic mass comparisons. However, their lack of uniformity made them unsuitable for photometry. They were seldom available commercially, but were occasionally prepared by dissolving the gelatin from ordinary plates in dilute sulphuric acid, a technique developed by Duclaux & Jeantet (1921).

Most of Aston's early work on the accurate photometric determination of isotopic abundances was done with the less sensitive, but uniform, 'Paget Half-Tones'. These were superseded in 1934 by 'Q' plates, which were developed by Messrs Ilford Ltd in collaboration with Aston, and which are still the plates most widely used in mass spectrometers employing photographic detection. These plates are low-gelatin plates and are available commercially in three speeds, Q1, Q2 and Q3, Q1 being the slowest and Q3 the fastest. The grain size and contrast vary with the speeds, Q1 being the finest and most contrasty. For many years Q plates were also used in atomic mass comparison work, where their sensitivity, which is only slightly inferior to Schumann plates, and fine grain proved of great value.

Other low-gelatin plates have also been developed, such as the ORWO UV2 and Kodak SWR (short wave ray). In another effort further to reduce energy loss in the gelatin, part of the gelatin layer has been removed after centrifuging the silver bromide to a lower level (Kodak–Pathé SC5, SC7 and Kodak 101–01), whilst gelatin free plates have also been prepared by direct vacuum evaporation of silver bromide onto a glass backing. In a quite different approach (Eastman Kodak III-O, UV) an ordinary emulsion has been coated with a thin film which fluoresces under the impact of positive rays, and which is then dissolved before development. When these plates are introduced into the vacuum system there is substantial evaporation of vapour from the emulsion, which limits the pressure for some time, an effect which is not important for low-gelatin plates. The characteristics of the various low-gelatin plates are discussed and compared by Honig (1972).

In the course of his isotopic abundance work, Aston made a thorough study of the blackening curves (opacity vs log exposure) of photographic plates for ions of different mass. Typical curves obtained by him for 38 keV ions are shown in fig. 4.1. These curves are to be compared in general form only; their positions relative to one another on the log scale are not significant. The highly penetrating hydrogen molecules do not give a linear plot until an appreciable density has been achieved, but thereafter the curve possesses a long linear section. Eventually saturation is reached and, beyond that (when the exposure is great enough), solarization. The curve for krypton is closely linear for a considerable range of density, that for xenon has a shorter linear range, and that for mercury ions, which are the least penetrating, has such a short linear range that intensity measurements are very difficult.

Fig. 4.1. Photographic blackening for positive ions of similar energy but different mass. (From Aston, 1942.)

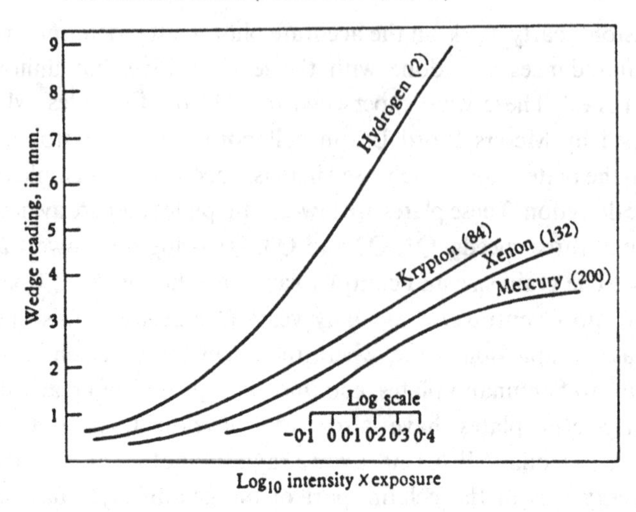

Fig. 4.2. Sensitivity of Eastman X-ray plates for ions of different energy and mass for a constant exposure of 5×10^{10} ions per mm^2. (From Inghram & Hayden, 1954, after Bainbridge, 1931.)

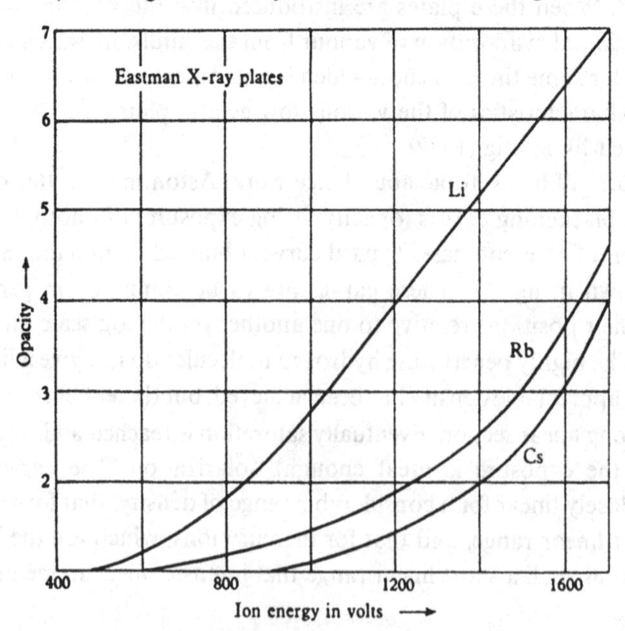

Bainbridge (1931) investigated the variation in sensitivity of Eastman X-ray plates with energy and mass of the impinging ions. The results of this work, shown in fig. 4.2, emphasize the advantage of using high energy ions and, also, demonstrate the pronounced mass discrimination exercised by the plate. The curves in fig. 4.2 were obtained with one-minute exposures at a current density of 1.32×10^{-8} A/cm^2. These curves are of interest in that they typify the behaviour of photographic plates with respect to these variables, and not for the information they present concerning Eastman X-ray plates, which are much inferior to Q and Schumann plates for both low (Bainbridge, 1931) and high (Aston, 1942) energy ions. The manner in which the sensitivity varies with ion energy and mass is also discussed in detail by Honig (1972), drawing upon recent experimental and theoretical studies of the energy loss of ions as they traverse condensed matter.

Higatsberger (1953) has drawn attention to the danger of the photographic emulsion becoming surface charged, thereby causing a distortion of the mass spectral lines. He has shown that the resistance between the ends of a strip of emulsion 90 mm in length and 30 mm in width is $1.2–4.8 \times 10^{13}$ ohms for Q1 plates and $1.0–2.5 \times 10^{12}$ ohms for Q2 plates. Ion currents of the order of 10^{-10} A may, therefore, give rise to voltage drops of ~ 1000 V. For currents less than 10^{-13} A there is no significant effect.

The useful datum has been given by Inghram & Hayden (1954) that, for 10 keV ions of mass 200 u, 10^{-10} C/mm^2 produce developable images on Eastman III-O plates. This minimum would be less for Q2 or similar plates.

Currently, the principal use of photographic plates is in the trace analysis of solids using vacuum spark source mass spectrometry. In this work, described in Chapter 12, ions over a wide mass range are recorded simultaneously on a photoplate, at the same time as the total ion current is being monitored. Thus, the photoplate records the integrated isotopic currents, whilst the monitor provides the relative exposures. The elemental concentrations are then deduced from the photographic density of the isotope lines.

The most important parameters influencing the photographic density–ion exposure relationship are the energy and mass of the impinging ion. The pioneer observations of Thomson, Aston and Bainbridge relating to these parameters have been greatly extended by others and have been rationalized in terms of the theory governing the energy loss of ions as they traverse condensed matter (Lindhard & Scharff, 1961). The photographic sensitivity (or blackening), as a function of ion energy or ion mass, has been expressed in the form $S = kE^x$ or $S = kM^x$ respectively. For Q2 plates, in studying the energy effect, it has been found that $x \cong 2$ for medium energies (3–10 keV), whilst for the mass effect (for $M > 10$) it has been found that x ranges between -0.5 and -0.8. The evidence for, and limitations to, these generalizations are discussed by Honig (1972).

4.3 Electrometer amplifiers

The first arrangements for the electrical detection of positive rays employed electrometers. These were later superseded by the so-called 'electrometer' valves, which grew out of the careful study by Metcalf & Thomson (1930) of the sources of grid current in thermionic valves. In the latter the positive-ion current, collected in a Faraday cup, develops across a large (grid leak) resistor a dc voltage which is electronically amplified. This requires the grid–cathode resistance of the first amplifier tube to be at least two orders of magnitude greater than that of the shunting grid leak, which is normally 10^9–10^{11} ohms. These valves are operated at very low anode voltages (~ 6 V) in order to prevent ionization of the residual gas, and contain a positively charged screen grid, between the cathode and control grid, which prevents passage to the latter of positive ions emitted by the filament. Under these conditions the grid resistance is about 10^{16} ohms. Various electrometer valves have been available commercially, in particular the diminutive 'acorn' types which could be mounted in vacuum adjacent to the Faraday collector.

Modern 'electrometer' amplifiers employ MOSFETs (metal oxide silicon field effect transistors) in which the high input impedance and low bias current are achieved by the oxide film that separates the gate electrode from the source and drain electrodes of the semiconductor. An integrated circuit operational amplifier (IC-OP-AMP) with MOSFET input yields the required high impedance input and current amplification.

The dc amplifier–electrometer tube, or MOSFET-OP-AMP, combination has been developed into a moderately stable detector of positive ions. Its overall stability is $\sim 30\,\mu$V over a half-hour period. This corresponds, with a 10^{11} ohm input resistance, to a current of 3×10^{-16} A. It is usually not practicable to increase the input resistance much beyond this value, since the circuit becomes too sluggish, especially if automatic recording is used.

4.4 Vibrating reed electrometer

Hull (1932) and Gunn (1932) first drew attention to the usefulness, in the measurement of small dc voltages, of a vibrating condenser acting as an electrostatic generator. This method was developed by Le Caine & Waghorne (1941), Palevsky, Swank & Grenchik (1947), Scherbatskoy, Gilmartin & Swift (1947), Thomas & Finch (1950) and others into a highly satisfactory detection technique for positive ions.

The usefulness of the vibrating reed electrometer is based on the fact that amplification is much more easily accomplished with ac than with dc. The input dc potential, arising from the passage of the ion current through a large resistor is, therefore, converted to ac by applying it through a series resistor to a capacitor whose capacitance is periodically varied with amplitude ΔC. An ac voltage is thus developed, of amplitude $\Delta V = V(\Delta C/C)$, which is proportional to the input dc voltage, and which may be fed into a conventional ac amplifier. The vibrating reed electrometer eliminates grid (bias) currents in parallel with the input current.

In practice, the output of the ac amplifier is generally rectified, using a phase sensitive device, and applied to the input as negative feedback of sufficient amount to cancel the input voltage. With this null technique, as used in all operational amplifiers (OP-AMPs), variations in the gain of the amplifier are of secondary importance. The amount of the negative feedback indicates the value of the input voltage.

Commercial vibrating reed electrometers, available both in the United Kingdom and the United States, possess input resistances exceeding 10^{15} ohms and input capacitances of approximately 40 pF. For rapid response applications (~ 0.1 s) these may possess a sensitivity of 1000 divisions/pA. This figure may be increased by at least an order of magnitude if a slower time response can be tolerated. As suggested above, such detectors are distinguished for their stability. The use of parametric amplifiers has also been tried (Russell & Ahearn, 1974).

4.5 Electron multiplier

The electron multiplier, which originated with Zworykin, Morton & Malter (1936), was developed by Bay (1938, 1941) and Allen (1939) as a detector of individual positive ions, electrons and photons. It was first seriously employed as a detector in mass spectroscopy by Cohen (1943). The remarks which follow are largely taken from a comprehensive description by Inghram & Hayden (1954).

In the original form of this detector, shown schematically in fig. 4.3, the positive ions impinge upon the first plate, or conversion dynode, giving rise to secondary electrons. These are accelerated and focused onto the second dynode, giving rise to a second, more numerous, generation, and so on, through ten to 14 stages, resulting in enormous gains. The dynodes are connected to successively higher positive potentials. Since, in the first instance, the positive-ion current is converted into an electron current, whereas, in succeeding instances, the electron current is simply multiplied, the conversion dynode serves a unique function.

Inghram, Hayden and Hess (Inghram & Hayden, 1954) demonstrated that the conversion efficiency of this first dynode is dependent upon a number of factors. (a) For a given species of positive ion, the secondary electron yield rises with increasing energy, as shown in fig. 4.4 for singly charged ions striking a silver–magnesium (1.7%) plate at a 45° angle. For low energies, the relationship is linear, and the range of linearity increases with mass of the ion; Barnett, Evans & Stier (1954) showed that the yield continues to increase at higher energies (5–250 keV). In most mass spectrometers the energy of the ions as they pass through the analyser is such that they must be further accelerated before striking the conversion dynode. (b) In general, for a given energy, as can also be seen from fig. 4.4, the efficiency decreases with increasing positive-ion mass. This observation

Fig. 4.3. The electron multiplier. (From Inghram & Hayden, 1954.)

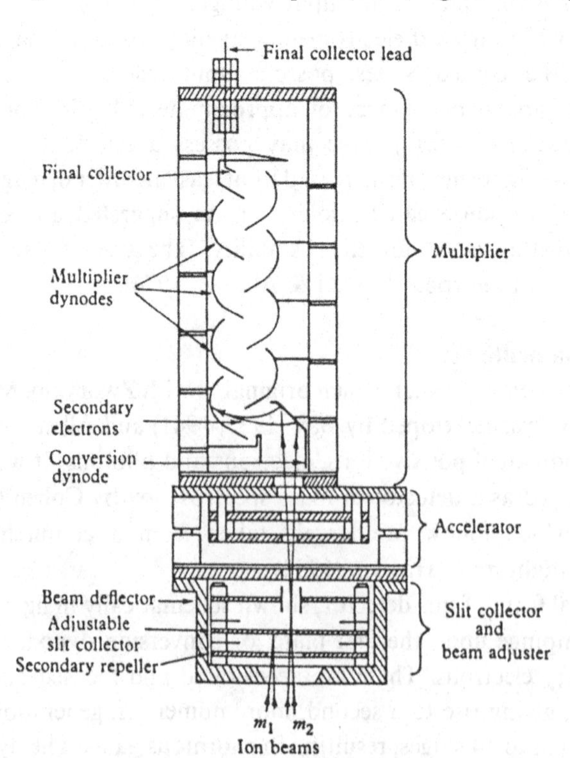

was further borne out by the work of Higatsberger, Demorest & Nier (1954) which indicated that for impinging ions of the rare gases of the same energy (in the range 2–6 keV) the secondary electron emission varies inversely as the square root of the mass. Barnett *et al.* (1954) found this trend to be reversed at higher energies; here, with few exceptions, the gain of the electron multiplier increases with the mass of the incident ion. (c) The number of secondary electrons decreases with increasing ionization potential of the ion. This may be regarded as a dependence upon the chemical nature of the incoming ion. (d) The secondary electron production increases with the angle of incidence of the impinging ion, as does, also, the extent of incident ion reflection. These two effects jointly lead to an optimum angle of about 70°. (e) The efficiency for negative impinging ions is greater than for positive ions, whereas, neutral and singly charged (Berry, 1948) and, also, doubly charged positive ions of the same energy produce substantially equal effects. (f) Molecular ions produce a greater number of secondary electrons than do atomic ions of the same mass.

The effectiveness of the remaining stages of the electron multiplier depends upon the geometry of the plates, the interstage voltage, the material and stage of activation of the plates, and the degree of magnetic

Fig. 4.4. Secondary electron yield at the conversion dynode of an electron multiplier as a function of energy and mass of the impinging positive ion. (From Inghram & Hayden, 1954.)

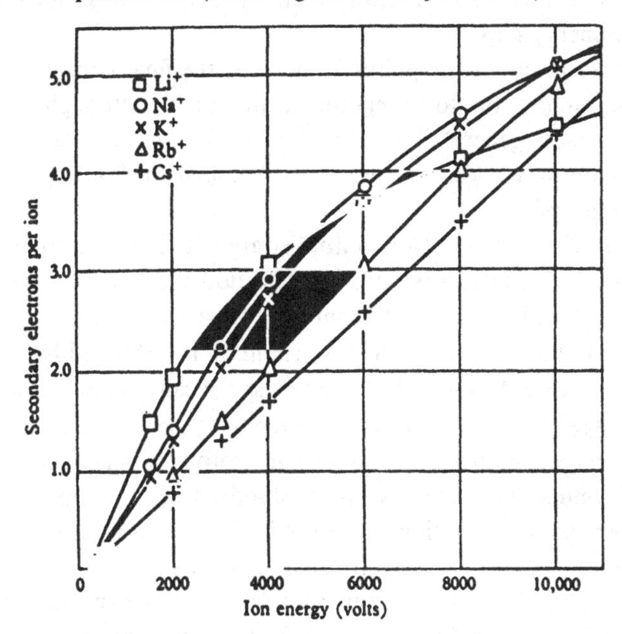

shielding. Four geometric arrangements have been used, namely, linear electrostatic focusing (Bay, 1938; Allen, 1939), circular focusing (Allen, 1950), the venetian blind (Allen, 1950) and magnetic focusing (Smith, 1951). In the Allen linear focusing design, which became the most common arrangement, the gain increases with interstage voltage up to 300–500 V, beyond which it changes only slowly. Two dynode materials are widely used, copper–beryllium (2%) and silver–magnesium (2–4%). These are activated (Allen, 1947) to obtain a thin beryllium (or magnesium) oxide surface layer which (a) is stable at atmospheric pressure, (b) possesses a multiplication factor significantly greater than unity (~ 4) and (c) possesses a high work function, thereby limiting the dark (or background) current to a negligible value. Adequate magnetic shielding is usually simple to achieve but, if not, Smith's magnetic focusing multiplier is designed to operate directly in a magnetic field of up to 0.10 T (1.0 kG). Indeed, in Smith's arrangement the magnetic field is utilized in combination with an applied electric field to create a crossed field electron multiplier. The electric field exists between two semiconductor plates or two glass plates coated with aluminum oxide. Thus, the secondary electrons travel in a series of cycloidal paths, multiplying in number each time they strike a plate. Another special purpose electron multiplier is the 'Johnston' or 'mesh' type. In this transmission electron multiplier a suppressor grid is operated at a negative potential with respect to the first dynode. (see, for example, Stickel *et al.*, 1980). This arrangement offers advantages for the detection of low energy ions.

One may choose to count the pulses arriving at the final collector of the electron multiplier, or to integrate them (§4.8). Although the latter method is in more general use, the former, if properly accomplished, substantially reduces mass effects and preserves, if desired, the fast response inherent in the multiplier.

The principal merits of the electron multiplier are its extreme sensitivity and its fast response. The former permits the detection, if desired, of single ions, whilst the latter allows a rapid scanning of the mass spectrum. It would appear, from the mass and other discriminations exhibited by the conversion dynode, that the multiplier must be specifically calibrated for all isotope work (§§7.4, 7.7). This limitation may frequently be avoided, as in the many recent studies of variations in isotopic abundance, by referring all determinations to a laboratory standard of the element in question. In other cases, calibration, if required, is a modest price to pay for sensitivity which makes certain investigations possible for the first time. It should be emphasized, however, that for most experiments the extreme sensitivity of the electron multiplier is not a significant advantage.

4.6 The continuous channel electron multiplier (CEM)

This compact version of an electron multiplier (Goodrich & Wiley, 1962; Wiley & Hendee, 1962; Evans, 1965; Schmidt & Hendee, 1966; Becker, Dietz & Gerhardt, 1972), frequently called CEM, comprises a hollow tube with a semiconducting inner surface, and contact leads at the ends. A potential difference of ~ 2 kV between the ends creates a uniform axial field due to current flow on the inner surface. Ions incident on one end (which is often funnel shaped to collect them) generate secondary electrons, and their transverse velocity causes further impacts with the inner surface, whilst the secondary electrons are carried along the tube by the longitudinal field. The amplified current is collected at the other end of the tube which, to save space, can be in the form of a spiral. Other advantages of the CEM are its simplicity, ruggedness and chemical inertness.

A further development has been the channel electron multiplier array (Somer & Graves, 1969; Brown *et al.*, 1970), abbreviated as CEMA. This consists of a large number of CEMs fused to form a regular array. Currently available CEMAs are ~ 100 mm in length with tubes of size ~ 50 μm, and their development makes feasible the simultaneous detection of the mass spectrum, thus combining the advantages of the photoplate and single ion electrical detection. The straight focal plane of a Mattauch–Herzog instrument is ideal for this purpose (Carrico, Johnson & Somer, 1973); the electrons can also be accelerated onto a scintillator screen and the ion images focused on a vidicon using fibre optics. Diode arrays and charge coupled devices can be employed in a similar fashion.

4.7 The Daly detector

The Daly detector (Daly, 1960), in its original form, employed a scintillator to detect secondary electrons emitted when ions strike a conversion electrode. This detector was an outgrowth of earlier attempts to use scintillators plus photomultipliers for the direct detection of positive ions. In this latter work (for example, Richards & Hays, 1950) the luminescent efficiency of the scintillator was observed to deteriorate under ion bombardment, as a result of atomic displacements and lattice damage. The use of secondary electrons, rather than ions directly, obviated this problem as the effect of electrons on the luminescent properties of scintillators is negligible. The manner in which Daly accomplished his purpose is shown in fig. 4.5.

The main feature of the detector is the conversion electrode (V_2 in the diagram), which is at a high negative potential (~ −40 kV) and which attracts the ions that emerge at ground potential from the exit slit of the mass spectrometer. The high energy positive ions impinge on the central portion of the conversion electrode (usually polished aluminum with a thin oxide film) and secondary electrons are emitted (~ 6 electrons per ion over a wide mass range). The secondary electrons are accelerated by the electric field produced by the conversion electrode potential and strike an organic plastic scintillator which is covered by a thin aluminum film at ground potential. The scintillations produced in the plastic phosphor are detected by an optically coupled photomultiplier located outside the vacuum system.

In addition to the advantage mentioned earlier – that the phosphor does not deteriorate – the Daly detector has the further advantage that its pulse

Fig. 4.5. The Daly detector. (After Daly, 1960.)

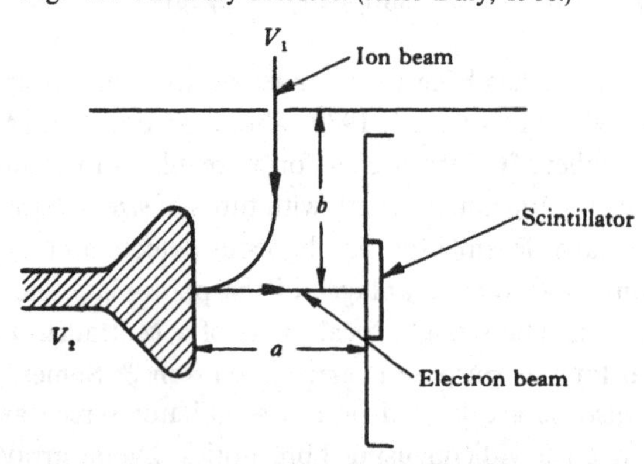

Fig. 4.6. Pulse height distribution curves for ^{205}Tl detected by *a*, a Daly detector and *b*, an electron multiplier. (After Daly, 1960.)

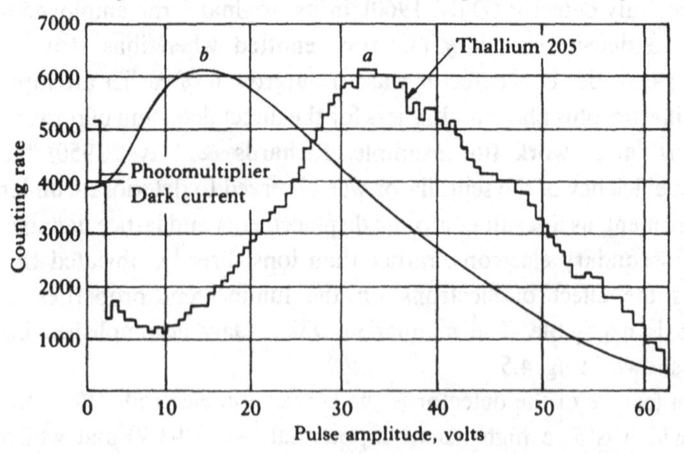

height distribution is more favourable for discriminating between ion counts and dark current than is the corresponding distribution for the electron multiplier. This point is illustrated in fig. 4.6. Whilst the maximum for the electron multiplier is above the dark current of the multiplier, thus making possible the detection of individual ions, an appreciable fraction of the ion pulses still lie at or below the indicated discriminator threshold setting. This leads to a loss in detection efficiency. The distribution maximum for the Daly detector occurs at a significantly higher value. Thus, for the same discriminator setting, much higher efficiencies in the counting mode or lower detection levels in the 'integrating' or 'current' mode of operation are obtained.

The fact that the ions impinging on the conversion electrode have such high energies (40 keV plus the source energy) increases the secondary electron emission over that found in electron multipliers. Thus, Daly found that the average number of secondary electrons per incident thallium ion was five, which is to be compared to the figure of 2.2 for the electron multiplier. Detailed studies of the secondary electron yields from the thin film oxide conversion electrodes have been made by Dietz & Sheffield (1973). In its normal pulse counting mode, the Daly detector has been used by Stoffels, Lagergren & Hof (1978) for ion currents in the range 10^{-20} A – 10^{-13} A and, with certain modifications, this range has been extended up to 10^{-10} A.

The basic principle of the detector has been used by Daly, McCormick, Powell & Hayes (1973) to enhance the metastable peaks in a mass spectrum. In this application, ions passing through the exit slit move towards the scintillator whose potential is variable but close to ground. Normal ions are stopped in the metal coating of the scintillator and produce no scintillations. The less energetic metastable ions, however, fail to reach the scintillator and ultimately impinge on the back of the exit slit where they produce secondary electrons which are accelerated to the scintillator and detected as in the original Daly detector. The geometry of this arrangement is linear and so differs from the original version.

In another utilization of scintillator detectors, Dietz & Hanrahan (1978) have combined an electron multiplier with a scintillation detector to count positive *or* negative ions. This arrangement is useful, for example, in observing the ions emitted in a microprobe (§12.2). In this tandem detector the ions are accelerated to the conversion electrode of a six-stage multiplier, whilst the electrons from the final stage of the multiplier are accelerated to the scintillator. This arrangement is also linear and, of course, the accelerating potentials are different for positive and negative ions.

4.8 Acquisition of data with electrical detection

Methods for recording the outputs of the various electrical detectors described in §§4.3, 4.4, 4.5, 4.6 and 4.7 are given in the references in which the detectors themselves are described. However, some general remarks may be made regarding the manner in which the signals are processed and recorded.

The desired information on a given peak may be its height (see Chapter 7), position (see Chapter 8) or, somewhat less frequently, its shape. As will be evident from the detailed discussion in §§7.4, 7.7 and 8.6, the primary requirement for the detector electronics is that the recorded signal be rigorously proportional to the ion current arriving at the detector, that is, the system must be linear. In addition the electronic apparatus must have a sufficiently high sensitivity and signal-to-noise ratio for the particular range of signals being used.

The ion current arriving at the detector is inherently digital in form. The recorded information, on the other hand, may be in either analog or digital form. Thus, in earlier work, electrometers (§§4.3, 4.4), which produce an analog signal corresponding to the ion current averaged over a short time interval, were used to drive a strip chart recorder whilst the mass spectrum was being scanned. In this system it is the gain of the electrometer that must be linear. If a digital voltmeter is added to monitor the analog output of the electrometer, the readings it produces may be recorded in a computer memory. In this case the information is converted from digital to analog and back to digital form.

The detectors which indicate the arrival of individual ions (§§4.5, 4.6 and 4.7) may be operated in a pulse-counting mode to preserve the digital character of the signal (Ihle & Neubert, 1971; Raznikov, Dodonov & Lanin, 1977). As discussed in §4.7, the output pulses have a pulse height distribution. The output of the detector is passed through a discriminator which produces a logic pulse (that is, a pulse of fixed height and length) for each input pulse which is greater than some predetermined level. Non-linearity of such an arrangement occurs if the pulse height distribution (fig. 4.6) shifts to the left with increasing count rate, an effect which occurs, for example, with continuous-strip electron multipliers. Moreover, in all detectors for individual ions, as the count rate increases, the dead-time counting losses become more significant (see, for example, Evans, 1955) and, finally, a maximum rate is reached beyond which the detector will not respond (Nguyen & Goby, 1978).

In modern instruments, the general availability and versatility of computers has led to their wide application in the recording of data from electrical detectors along with information on, say, the corresponding magnetic field. In those experiments in organic mass spectrometry (§11.3) in which a mixture is passed through a gas chromatograph (so that the components arrive at the source at different times) the mass spectrum may be scanned very rapidly (for example, at 50 kHz) and each scan recorded by the computer. The computer may then be used to calculate the masses as well as the time variation of the intensities (peak areas) of all of the components of the spectrum.

As implied in the above, computers are being used to control the operation of mass spectrometers, especially for routine analysis. They are also being used to develop large library compilations of the mass spectra of organic compounds (§11.2).

5

Deflection-type instruments

Modern deflection-type mass spectroscopes are either single (direction) focusing or double focusing, and will be discussed in this order. In the following we describe certain instruments which are widely used or which illustrate properties of special interest.

Direction focusing mass spectrometers

5.1 Sector versus semicircular instruments

For several years after magnetic field focusing theory had made feasible the use of sector fields, direction focusing instruments continued to employ, exclusively, the Dempster (1918) semicircular deflection. In 1940 Nier described the first instrument incorporating 60° deflection and, in 1942, Hipple made first use of the 90° sector field. Apart from a few exceptions (for example, 120° used by Bainbridge & Ford, 1950) relatively little use has been made of other angular deflections.

Most present-day instruments employ sector fields, the 60° arrangement being the most popular. This requires a smaller magnet than does the 180° analyser while, at the same time, the long object and image arms provide field free spaces for the source and detector structures. On the other hand, in sector instruments, the ions twice traverse the fringing field, where they experience deflections which are difficult to calculate. This necessitates provision in construction for empirical adjustment. With modern vacuum techniques the longer ion path in sector instruments is not a significant detriment. It will be recalled from §2.1 that the resolution is independent of the angle of deflection, provided $l'_m = l''_m$.

5.2 Modern sector-type instruments

Many of the current sector-type mass spectrometers are patterned after Nier's 1940 instrument and its lineal descendant (Nier, 1947). These instruments (fig. 5.1), with a 15 cm radius of curvature and 60° angular deflection, achieve a resolution of approximately 1 part in 100, and are used for gas and isotope analysis. Because these instruments are used to determine the relative current at a particular mass number, they are normally operated with the collector slit wider than the image width so that flat-topped peaks are obtained.

All analyses of the heaviest elements, uranium and the transuranics, require a resolution of at least 1 part in 250. This may be achieved, not only by reducing slit widths, but also by increasing the radius of curvature and by shaping the entrance and exit boundaries of the magnetic field so as to eliminate second order image aberrations (see §§2.9, 2.10). Such an instrument, described by Inghram, Hess & Hayden (1953), achieved a resolution of 1 part in 1000 and has served as a model for several others.

In the design of such an instrument, the relative size of the first order velocity dispersion should be compared with the size of the second order aberrations (see Chapter 2, especially §2.10) so as to assess the net benefit to

Fig. 5.1. Schematic diagram of a classic 60° magnetic sector instrument. (After Nier, 1947.)

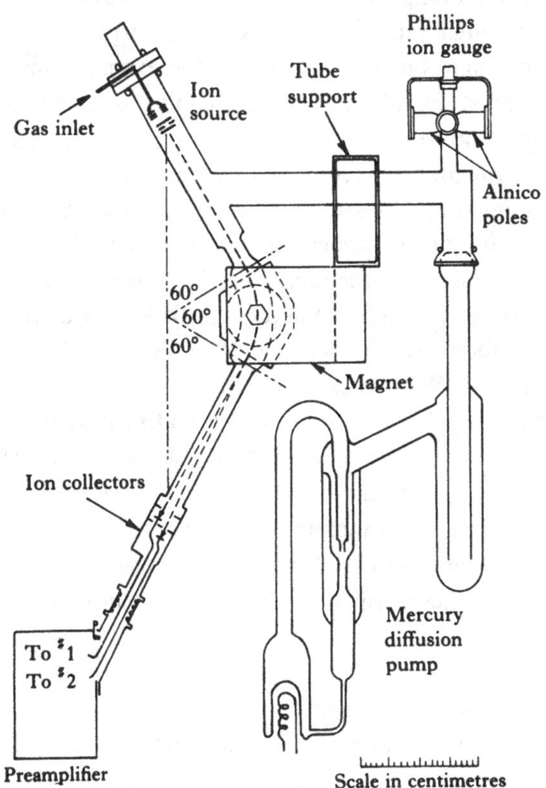

be derived from correcting, say, the α_r^2 aberration. The estimate of $\beta = \Delta V_a/2V_a$ should include the inherent energy spread of the source plus the ripple on the source voltage supply.* The requirement on the stability of the magnetic field may be calculated from equation (1.4) which gives

$$\Delta B/B = -\Delta r_m/r_m. \quad (5.1)$$

Although not exploited extensively, axial focusing, by means of inclined entrance and exit boundaries (§2.6) or by a non-uniform magnetic field (§2.8), may be used to increase the transmission of a sector-type instrument.

It is imperative in abundance work that background peaks arising from residual gases and hydrocarbon contaminants be reduced to a low level. This may be accomplished by standard modern vacuum techniques for the construction and operation of oil free systems. Usually, stainless steel is used for the vacuum chambers, with permanent joints made by argon arc or helium arc welding and demountable joints sealed by metal gaskets (for example, oxygen free high conductivity (OFHC) copper, indium, gold, aluminium). The components of the vacuum system are carefully cleaned following construction. Pumps which do not require oil as a working fluid are used both for rough pumping (for example, sorption pumps, gas aspirators) and for high vacuum operation (for example, ion pumps, turbomolecular pumps). Finally, the entire chamber is outgassed under vacuum at high temperatures (up to 400 °C). Although the details of these vacuum techniques are not properly included here, they are of the greatest importance in design and construction, and are described in many published articles and notes, principally in the *Journal of Scientific Instruments, Review of Scientific Instruments* and *Vacuum*, and in a number of standard references on vacuum techniques (Guthrie & Wakerling, 1947; Dushman, 1962; Roberts & Vanderslice, 1963; Roth, 1966; Redhead, Hobson & Kernelsen, 1968; Robinson, 1968).

The detector is commonly a dc amplifier or vibrating reed electrometer. When maximum sensitivity is required, a single ion detector, such as the electron multiplier, is employed (see §4.5). The requirement for a low background is here much more exacting, and must be met if such a detector is to be employed advantageously. Many of the experiments described in Chapter 9 have been made possible by the development of ion counting detectors.

5.3 Two-stage magnetic analyser

Even at very low pressures (Blears & Mettrick, 1947; Blears & Hill,

* V_a is the accelerating voltage.

1948; Ioanoviciu, 1973; Gall, Pliss & Shcherbakov, 1980) there is a scattering of the ion beam by the residual gas. There may also be scattering of mass components far from the central beam by solid parts of the instrument, for example the wall of the analyser tube. This results in peak 'tails', which are further enhanced by whatever energy spread characterizes the ion beam, and possibly, also, by electrostatic repulsion within the ion beam itself (Becker & Walcher, 1953). The extent to which an instrument is free of these defects is sometimes described by the abundance sensitivity, defined as

$$\text{abundance sensitivity} = \frac{\text{peak ion current (at mass } M)}{\text{ion current at } M+1}. \tag{5.2}$$

This abundance sensitivity is, in the best single focusing instruments, somewhat greater than 10^5 at mass number 100 (Ridley *et al.*, 1966). With such instruments the detection of faint isotopes, adjacent to abundant ones, is thus limited to those which are present to at least 1 part in 10^6 of the abundant species.

Inghram & Hess (1953) increased this abundance sensitivity to approximately 10^8 by utilizing a combination of two consecutive magnetic analysers. At the position of focus of the first analyser the ions passed through a discriminating slit, and were further accelerated before passing through the second analyser. In this way, ions which passed through the discriminating slit by virtue of scattering were resolved from the principal peak in the second analyser. Also, ions which passed through the discriminating slit with similar momenta but dissimilar masses received dissimilar momentum increments in the second acceleration, thereby ensuring their resolution. The analysers were arranged so that the deflections were in the opposite sense, that is, an 'S' configuration. The use of such an instrument in the search for rare or hypothetical isotopes is described in §7.2.

Subsequent instruments have exploited the advantages of tandem magnetic analysers (fig. 5.2), but without intermediate acceleration, in both 'S' (Wilson, 1963; Wilson, Munro, Hardy & Daly, 1961; Moreland, Rokop & Stevens, 1970) and 'C' (White & Collins, 1954) geometries. In the former the mass dispersions of the two analysers add and relatively high resolution may be achieved. Moreover, the 'S' geometry must be used when simultaneous recording of several masses at the collector is required. In the 'C' geometry the mass dispersion vanishes, so the intermediate slit must be smaller than the separation between adjacent mass numbers. Provided this is the case, the 'C' geometry produces more favourable peak shapes and

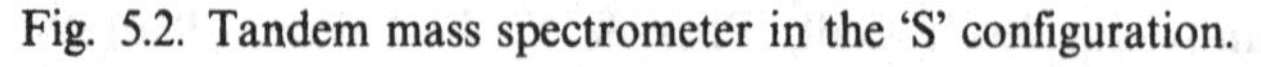

Fig. 5.2. Tandem mass spectrometer in the 'S' configuration.

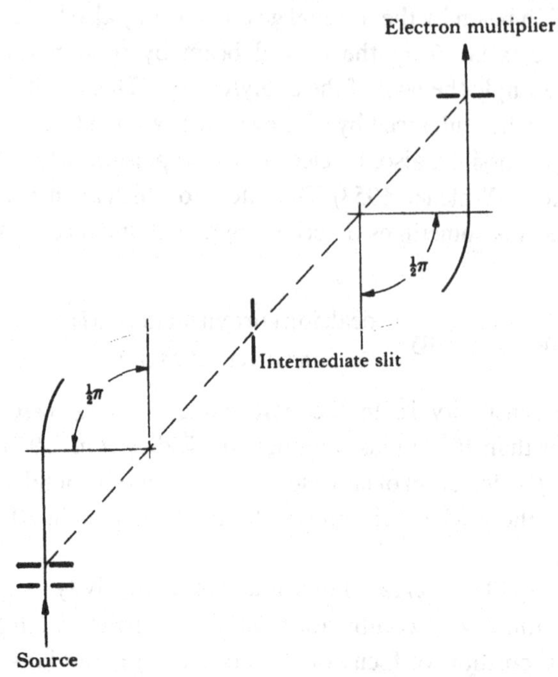

requires less stringent stability of the source potential, V_a, and of the magnetic fields.

For both geometries, mass scanning is more readily carried out by varying V_a. When magnetic scanning is used, the two fields may be made to track each other by means of nuclear magnetic resonance (NMR) probes (Sheffield & White, 1958) or by operating the coils in series and regulating the current.

Experimental values for the peak tails after one stage and after the two stages of a tandem instrument (Ridley *et al.*, 1966) are given in table 5.1. Here $r_m = 63.5$ cm and $\Phi_m = 90°$ for each stage and the 'S' configuration is used.

Table 5.1. *Experimental values for peak tails in vicinity of* ^{238}U[a]

$M =$	237	237.5	238	238.5	239
1 stage	2.5×10^{-6}	7.0×10^{-6}	1.0	2.9×10^{-6}	1.3×10^{-6}
2 stages	1.3×10^{-6}	4.1×10^{-6}	1.0	6.8×10^{-8}	3.6×10^{-8}

[a]Ridley *et al.*, 1966.

Gall, Pliss & Shcherbakov (1980) have calculated the behaviour of the peak tails for single and two-stage magnetic analysers, with the latter in both 'S' and 'C' configurations. The investigation is based on the scattering of the ion beam from the residual gas in the vacuum chamber. Agreement with the experimental values of table 5.1 is fairly good for the single analyser, but much poorer for the tandem, perhaps because scattering of other masses from the vacuum chamber walls was not included in the calculation. The scattering is asymmetric about the central peak for both cases shown in the table. This reflects the fact that, in *any* collision between particles in the ion beam ($M = 238$) and residual gas molecules (for example, $M = 32$), the kinetic energy of the ion will be decreased. Hence the tail on the low mass side of the peak must be larger than on the high mass side.

Other, more elaborate, instruments have also been used to attain high abundance sensitivities. White, Sheffield & Rourke (1958) have described an 'S' geometry to the end of which was added a 90° cylindrical electrostatic analyser having the same radius of curvature as the magnetic sectors. An abundance sensitivity of 10^7–10^8 for $M = 100$ was realized. The use of two double focusing instruments as the components of a tandem spectrometer makes possible a reduction in the size of the instrument by a factor of ~ 2. Nier (1963) and White & Forman (1967) have described arrangements in which two double focusing mass spectrometers were run back-to-back in the 'S' and 'C' configurations respectively. However it appears, in general, that the addition of more stages does not result in further dramatic changes in the abundance sensitivity.

The form of the peak tails for three particular double focusing mass spectrometers has also been calculated by Gall, Pliss & Shcherbakov (1980). As with the conventional two-stage tandem, there is an advantage in operating double focusing instruments in 'reverse geometry'. In this case, the ion beam traverses the magnetic analyser first, and only the desired mass enters the electrostatic sector to be deflected according to its energy. Thus, only the scattering of the ion of desired mass is significant.

Stevens (1963) has suggested that a retarding grid between the collection slit and the final detector can be operated at a potential only slightly below that of the source in order to discriminate against ions which have suffered collisions. Fenner *et al.* (1974) have described a 90° single focusing instrument with $r_m = 30$ cm which has been equipped with such a grid. With the principal peak located at $M = 235$, abundance sensitivities of 1.7×10^6 at $M = 234$ and 1.1×10^6 at $M = 236$ have been determined. Similarly, Moreland, Rokop & Stevens (1970) have used a retarding grid in conjunction with a two-stage 'S' geometry tandem instrument and have achieved an abundance sensitivity of 3.3×10^7 in the region about $M = 240$. If necessary, the energy resolution of the retarding grid may be made very high by the use of two grids operated at the same potential and spaced by an amount that is many times the dimensions of the hole size in the grids.

Recently, there has been interest in the direct determination, by mass spectroscopic methods, of the relative abundance of undecayed ^{14}C ($T_{1/2} =$ 5730 yr) for carbon dating of samples of historical interest. Inasmuch as ^{14}C is present in modern samples at a level of $\sim 1/10^{12}$, a corresponding abundance sensitivity is required, as well as the choice of an appropriate molecular ion. Anbar and his colleagues (Schnitzer *et al.*, 1974; Schnitzer, Aberth & Anbar, 1975; Schnitzer & Anbar, 1976) attempted to use $(^{14}C^{15}N)^-$ ions in an instrument comprising a Wien filter followed by a 90° magnetic sector. Ultimately they fell short of the necessary abundance sensitivity by about an order of magnitude because a background contamination, mostly ^{29}Si from the anode surface, was present.

This work is currently being pursued through an accelerator based technique (Bennett, 1979; Bennett *et al.*, 1979) which involves both mass and charge spectroscopy (MACS) and proves to be an extremely sensitive way of detecting particular isotopes (§7.11).

5.4 Isotope separators

Magnetic analysers for the separation of isotopes, of which the *calutron* is the most celebrated representative, constitute a distinctive mass spectrometric family. These instruments were developed initially for the large scale production of separated isotopes for the Manhattan Project. In such an instrument, a very large ion current, originating in an arc source, emerges from a long object slit, and is inevitably subject to scattering and space charge defocusing, the latter prompting the use of techniques for space charge neutralization. The necessary resolution is secured primarily by the use of large radii of curvature and massive magnets.

In the early large ('alpha') calutrons at the Oak Ridge National Laboratory (Keim, 1955; Normand, 1956), the ion beam emerged from an arc 36 cm in length, experienced a 35 kV acceleration, and was dispersed in a semicircular analyser 122 cm in radius. With favourable source elements the isotopes current exceeded 100 mA. The magnet and the vacuum housing were designed to accommodate, simultaneously, four such beams. The present Oak Ridge facility (Newman *et al.*, 1976) employs three types of separators. The first is a 61 cm radius ('beta') calutron, an improved version of the original design (Dawton & Smith, 1958). The second is a 225°, two-directional focusing instrument, having a field index (n in equation (2.35)) of 0.5 and a dispersion 50% greater than the 'beta' units. In both of these types of separator, the source and collector are located in the magnetic field. The third type has a 180° magnetic field of radius 61 cm and field index of 0.8. For this arrangement, the source and detector are external to the magnetic field and a dispersion of about seven times that of the 'beta' units is achieved. This last instrument is used for making enriched targets and for ion implantation studies, while the two former types are used for production of separated isotopes. The theory of non-uniform magnetic fields is outlined in §2.8.

The units at Harwell, having radii of 61 cm and 122 cm, were similar to the calutrons and roughly equivalent in productive capacity (Smith, 1956; Dawton, 1956). Only marginally inferior in ion current to the preceding machines was the Amsterdam separator (Kistemaker & Zilverschoon, 1953) which was the first to exploit a magnetic field shaped radially in accordance with the theory of Beiduk & Konopinski (1948; see §2.8). In this case, the field shape had the form given by equation (2.47), permitting the use of an ion beam of 15° half-angular spread in the radial direction, and proving axial focusing at the same image location.

In the years following 1945, a number of isotope separators were developed with ion currents in the range between the calutron and the conventional analytical mass spectrometer. These were developed in the period up to about 1960 for comparatively small scale separations, especially for the production of targets for nuclear physics experiments (see Chapter 9). The early instruments were located in Copenhagen (Koch & Bendt-Nielsen, 1944; Koch, 1953, 1956), in Stockholm (Bergström, Thulin, Svartholm & Siegbahn, 1949), in Uppsala (Andersson, 1954), in Goteborg (Almén, Bruce & Lundén, 1955, 1958) and in Paris (Bernas, 1953; Bernas, Kaluszyner & Duraux, 1954; Bernas 1956). In recognition of the geographical location of much of this work, these low intensity separators are frequently referred to as 'Scandinavian-type' isotope separators.

Typically, these instruments employ a 90° sector uniform magnetic field with relatively large radius of curvature, 1.5 m, but without magnetic axial focusing (Nielsen, 1970). A resolving power of ~ 1500 is readily achieved so that high enrichment factors can be realized. More recently, certain applications have prompted the pursuit of high transmission. Accordingly, instruments which accept a large angular spread and also provide axial magnetic focusing have been constructed. An example of the latter is the instrument at the Chalk River Laboratory (Schmeing *et al.*, 1981) in which a 1 m radius, 135° deflection inhomogeneous magnetic field with a concave boundary is used. Here, n is nominally $\frac{1}{2}$, but may be changed electrically by varying the current in shim coils located at the magnet gap, so that the position of the radial (and axial) focus is adjusted. Independently, the second derivative of the field may be altered in a similar fashion so that the α_r^2 image aberration may be minimized.

For the many technical details of these impressive machines, the reader is referred to Koch (1958) and to the *Proceedings of the International Conference on Electromagnetic Isotope Separators* (EMIS), at Harwell (Smith, 1956), Amsterdam (1957), Vienna (1960), Orsay (1962), Aarhus (1965), Asilomar (1967), Marburg (Wagner & Walcher, 1970), Skovde (Andersson & Holmen, 1973), Kirijat Anavim (Amiel & Engler, 1976) and Zinal (Ravn, Kugler & Sundell, 1980). The applications to which isotope separators have been put will be described in greater detail in Chapters 8, 9 and 12.

Double focusing instruments

Double focusing mass spectrographs were developed independently by Dempster, Bainbridge, and Mattauch & Herzog for the purpose of making precise atomic mass comparisons. These instruments, and others subsequently built, will be described below, primarily in terms of the general theory of Herzog, outlined in §§2.1–2.4. Some general remarks concerning double focusing instruments will then be made.

5.5 The Dempster double focusing mass spectrometer

This instrument (Dempster 1935; Barber *et al.*, 1964), shown schematically in fig. 5.3, consists of a 90° radial electrostatic analyser followed by a semicircular magnetic analyser. In Dempster's original instrument, the electrostatic analyser was used asymmetrically, but in the large instrument in Duckworth's laboratory, a symmetric geometry was used. In the latter case $l'_e = l''_e = 0.350\, r_e$ and the image of the electrostatic analyser is located at the effective boundary of the magnetic field, making $l'_m = 0$: the final image is formed at the exit boundary of the magnetic field ($l''_m = 0$). Velocity focusing occurs (equation (2.23)) for one radius of curvature in the magnetic field, namely, $r_e = r_m$. From equation (2.24) the resolution is $\Delta M/M = 2S_1/r_e$ for electrical detection; from equation (2.25) the dispersion $d = 0.010\, r_m$ per 1% mass difference, while the magnification $b''_m/b'_e = 1.0$. The slit S_2 guarantees that only those ions whose energies lie within the narrow energy range permitted by the Herzog theory enter the magnetic analyser.

This geometry has the merits of relatively small second order coefficients and overall simplicity. However, it suffers somewhat from the absence of axial focusing.

In Dempster's original instrument, $r_e = 8.5$ cm and $r_m = 9.8$ cm, corresponding to a resolving power of 3000 for a 0.0025 cm principal slit, and a dispersion of 0.098 cm/1% mass difference. The large instrument, originally at McMaster University and later at the University of Manitoba ('Manitoba I', Barber *et al.*, 1964) has $r_e = 273$ cm, and a working resolving power (at 10% of peak height) of 100 000. It has been used extensively for atomic mass determinations (Chapter 8) and has provided values having a precision of up to $3/10^9$ of the mass of the doublet. Other original-scale versions have been adapted for abundance measurements (Walker & Thode, 1953) and for chemical analysis (Gorman, Jones & Hipple, 1951).

5.6 The Bainbridge–Jordan instrument

This instrument, designed by Bainbridge in England in 1934, and completed at Harvard University in 1936 (Bainbridge & Jordan, 1936), is shown schematically in fig. 5.4. It consists of a linear combination of a $\pi/\sqrt{2}\,(127.3°)$ electrostatic analyser and a $\frac{1}{3}\pi(60°)$ sector magnetic field. Both fields are employed symmetrically, that is, $l'_e = l''_e = 0$ and $l'_m = l''_m = \sqrt{3}r_m$. The condition for velocity focusing is $r_e = r_m$. Thus, as in the Dempster type

Fig. 5.3. Dempster-type double focusing mass spectrometer. (After Barber *et al.*, 1964.)

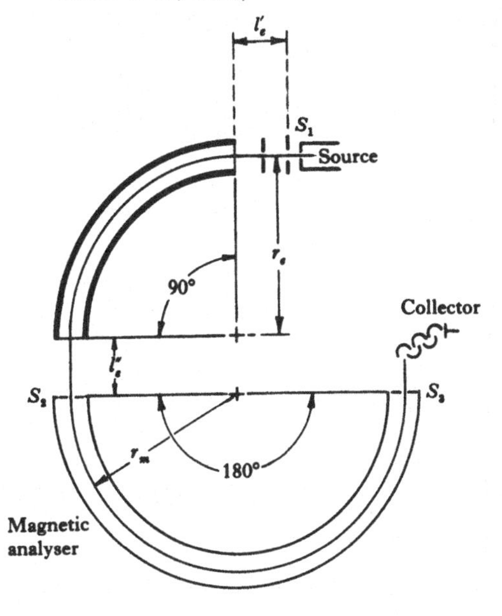

instrument, double focusing occurs only at one point. The resolution, dispersion and magnification are S_1/r_e, $0.02r_m/1\%$ mass difference and unity respectively.

In the first instrument, $r_e = r_m = 25$ cm. Thus, for a 0.0025 cm principal slit, $\Delta M/M = 1/10\,000$; the dispersion is 0.50 cm/1% mass difference. A similar instrument, built in Japan (Asada, Okuda, Ogata & Yoshimoto, 1940), was the subject of considerable improvement (Ogata & Matsuda, 1953*a*, *b*), which resulted in the attainment of resolutions ranging from 1/40 000 to 1/60 000 with photographic detection.

5.7 The Mattauch–Herzog instrument

This remarkable instrument (Mattauch & Herzog, 1934; Mattauch, 1936), the first of which was completed in Vienna in 1935, possesses double focusing for all masses. It will be seen from fig. 5.5 to consist of a

Fig. 5.4. The Bainbridge–Jordan double focusing mass spectrograph.

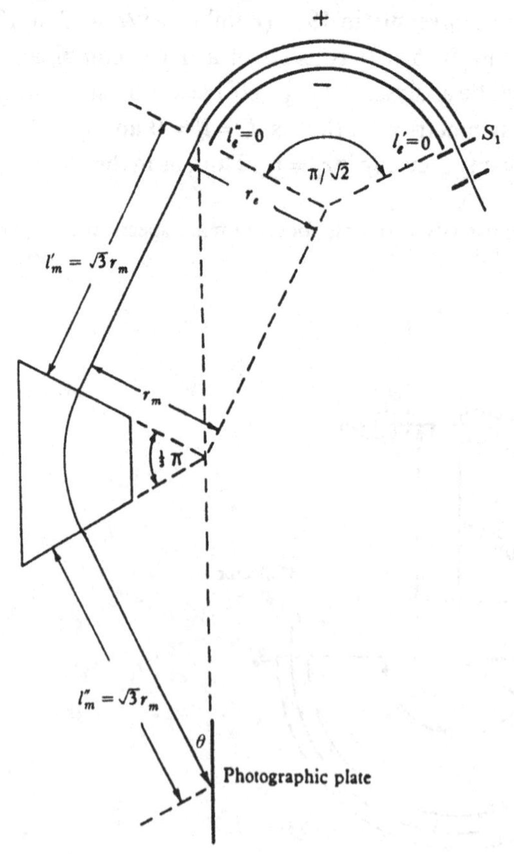

$\pi/4\sqrt{2}$ (31.82°) electrostatic analyser followed by a $\frac{1}{2}\pi$ (90°) magnetic analyser, in neither case employed symmetrically. The design of the instrument has the principal slit located at the principal focus of the electrostatic analyser, giving rise to a parallel emergent beam. Consequently, the final image is located along the principal focus of the magnetic analyser. As noted in §2.4, the velocity focusing condition, in these circumstances, may be satisfied through equation (2.27) independently of r_m. That is, the instrument is double focusing for all masses along the straight-line photographic plate indicated in fig. 5.5.

To simplify stray field problems, the photographic plate is placed at the exit boundary of the magnetic analyser, so that $\Phi_m = 90°$. From (2.27), the angular deflection in the electric field is $\pi/4\sqrt{2}$, in consequence of which $l'_e = r_e/\sqrt{2}$. The resolution, dispersion and magnification are $2S_1/r_e$, $0.007r_m/1\%$ mass difference, and r_m/r_e respectively.

Fig. 5.5. The Mattauch–Herzog double focusing mass spectrograph.

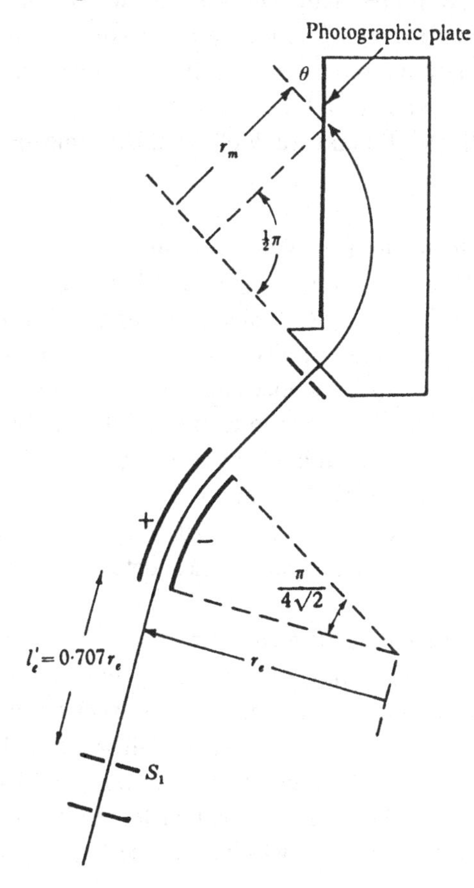

In the first instrument $r_e = 28$ cm. Thus, for $S_1 = 0.0025$ cm, $\Delta M/M =$ 1/5600. The dispersion and magnification vary with the position of the line under study; the maximum values, corresponding to $r_m = 24$ cm, are 0.17 cm/1% mass difference and 0.86, respectively. This double focusing arrangement has served as the basis for several other mass-comparison instruments. Both Mattauch (Mattauch & Bieri, 1954; Everling, 1957) and Ewald (1946, 1953) constructed instruments in which, by painstaking attention to detail, resolutions of ~1/100 000 were obtained (fig. 8.1). A later version, approximately eight times as large as the original, was constructed at Harvard University (Collins & Bainbridge, 1957; Bainbridge & Moreland, 1960) and a resolving power of up to 250 000 (at half-maximum with *electrical* detection) was attained. In connection with the achievement of high resolving power, it should be noted that, at the point on the photoplate where $r_e/r_m = 1.683$, second order angle focusing occurs (Hintenberger & König, 1959).

This geometry is particularly appropriate to experiments where velocity focusing is required and where the simultaneous collection of more than one mass is desired. Accordingly, it has seen very extensive commercial use in analytical work and has been employed in such diverse areas as spark source analysis of solids (Chapter 12) and study of the upper atmosphere (Chapter 14).

5.8 The Nier–Johnson double focusing mass spectrometer

Here (fig. 5.6), a 90° electrostatic analyser is followed by a 60° magnetic analyser (Nier & Roberts, 1951, Johnson & Nier, 1953). The electrostatic analyser is employed symmetrically and the magnetic analyser asymmetrically in order to obtain direction focusing to second order. This property of the instrument makes it possible to employ a relatively large divergence angle, thus securing good ion intensity without sacrificing resolution. The condition for velocity focusing is $r_m = 0.808\ r_e$. The resolution, dispersion and magnification are S_1/r_e, $0.0084\ r_m$ cm/1% mass difference and 0.68 respectively. In practice, an electrical detector is placed at the position of double focus.

The dimensions of the original instrument were $r_e = 18.87$ cm and $r_m = 15.24$ cm. Thus, for $S_1 = 0.0025$ cm, the resolution would be 1/7500 for photographic detection, whereas the dispersion is 0.13 cm/1% mass difference. In practice, with $S_1 = S_4 = 0.001\,25$ cm, the full width at half maximum (FWHM) of the peaks corresponded to a resolution of 1/14 000. The FWHM resolution is $\sim\frac{1}{2}$ the figure for complete resolution. A second, enlarged version of this instrument was built at the University of

Minnesota (Quisenberry, Scolman & Nier, 1956; Nier, 1957) with $r_e =$ 50.31 cm and $r_m = 40.64$ cm and has been one of the most productive instruments in the study of precise atomic mass differences (Chapter 8). With $S_1 = S_4 = 0.001$ cm, a resolution of 1/75 000 has been obtained, although the working figure generally lies in the range 1/30 000–1/60 000 (all FWHM values). Atomic mass differences have been determined with this instrument to a precision of up to $2/10^9$ of the mass at which the doublet occurs (Kayser, Halvorson & Johnson, 1976).

Despite the absence of axial focusing, this geometry has the favourable property that, in addition to second order angle focusing, the remaining second order aberrations, B_{12} and B_{22} (equation (2.66)), are relatively small. Indeed, even when the small effect of the fringing fields is taken into

Fig. 5.6. The Nier double focusing mass spectrometer. (After Quisenberry, Scolman & Nier, 1956.)

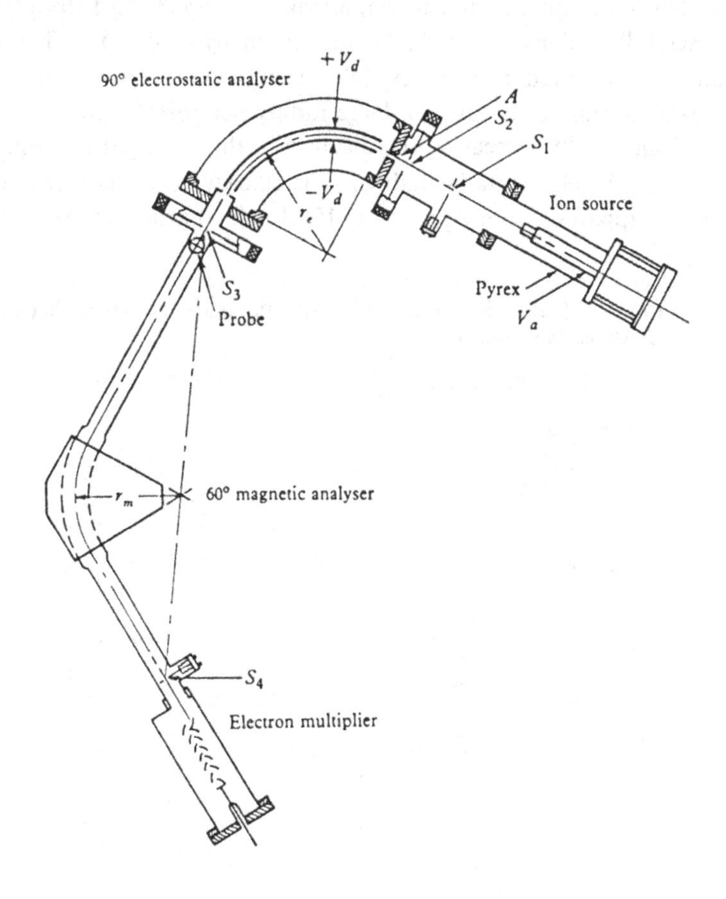

account (Matsuda, 1976), all of the second order coefficients in equation (2.66) are $\leqslant 0.32$.

Finally, the arrangement of the instrument is a convenient one in that provision may be made easily for focusing adjustment, usually through the movement of the magnet relative to the electrostatic analyser (Nier, 1957). Accordingly, the instrument has been extensively copied and is available as a commercial instrument.

A variation of this instrument, in which the magnetic sector has $\Phi_m = 90°$, has been described by Craig & Errock (1959). The other parameters are $\Phi_e = 90°$, $l'_e = 0.405\ r_e$, $l''_e = 0.299\ r_e$, $r_m = 0.800\ r_e$, $l'_m = 1.142\ r_e$, $l''_m = 0.561\ r_e$, $\varepsilon' = \varepsilon'' = 0$. The scale of the instrument is determined by the choice of r_e; in this case it was 19.1 cm (7.5″). A similar instrument with $r_e = 38.1$ cm (15″) has been described by Evans & Graham (1974).

5.9 The Ogata–Matsuda double focusing mass spectrograph

At Osaka University, Ogata & Matsuda (1955, 1957) constructed a very high resolution instrument employing a $\sqrt{2}\pi/3$ (84.85°) electrostatic analyser followed by a $\pi/3$ (60°) magnetic analyser (fig. 5.7). The high resolution is achieved in three ways: (a) by using a very narrow principal slit ($S_1 = 0.0023$ mm), (b) by using a large radius electrostatic analyser ($r_e = 109$ cm) and (c) by increasing the distance of the principal slit from the entrance to the electrostatic analyser. The efficacy of this last-mentioned change is apparent from equation (2.19). In this particular instance, by

Fig. 5.7. Ogata–Matsuda double focusing mass spectrograph. (After Ogata & Matsuda, 1957.)

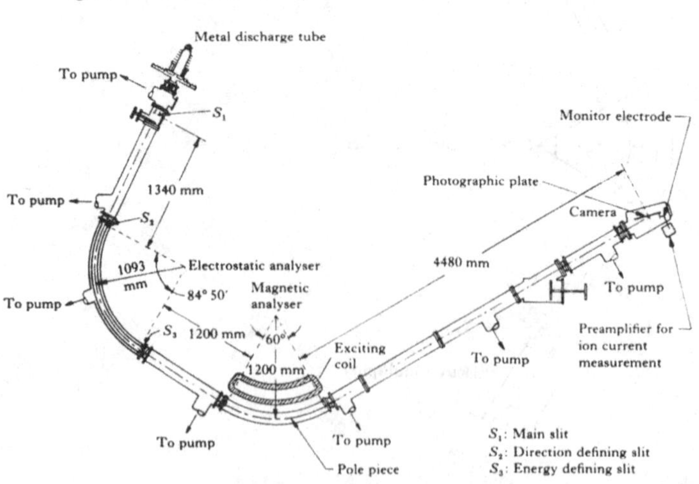

setting $l'_e = 1.225\ r_e$ and $l''_e = 0$, the resolution is improved by 50% over the value associated with the $l'_e = l''_e$ configuration.

Because of the large radius of the path in the magnetic analyser ($r_m = 120$ cm), and the small angle (17°) which the ion beam makes with the surface of the photographic plate, the dispersion of the instrument is also high. This was further increased by using the magnetic analyser asymmetrically, namely, $l'_m = r_m$ and $l''_m = 3.73\ r_m$. The calculated dispersion is 7.6 cm for 1% mass difference. The magnification of the instrument is 1.37 and the total path length is 10 m. A resolution of 1/900 000 was achieved with photographic detection (Ogata & Matsuda, 1957) prior to 1961 when the instrument was extensively damaged by flooding. It was subsequently moved and rebuilt, and operated with electrical detection at a resolving power of 500 000 (FWHM) (Katakuse & Ogata, 1972).

5.10 The Ewald astigmatic focusing mass spectrograph

A double focusing instrument consisting of a toroidal condenser ($\Phi_e = 29.47°$) followed by a magnetic analyser ($\Phi_m = 90°$), as shown in fig. 5.8, has been described by Ewald (1960) and Ewald, Konecny & Opower (1963). This makes use of the astigmatic focusing properties of the toroidal condenser mentioned in §2.2 to secure axial focusing together with Mattauch–Herzog-type radial focusing (§5.7), that is,

Fig. 5.8. Ewald astigmatic focusing mass spectrograph. (After Ewald, 1960.) The diagrams show the radial (upper) and axial (lower) focusing. S_1 is the principal slit, while the remaining slits limit the angular extent of the beam in the two directions. The electrostatic analyser is toroidal.

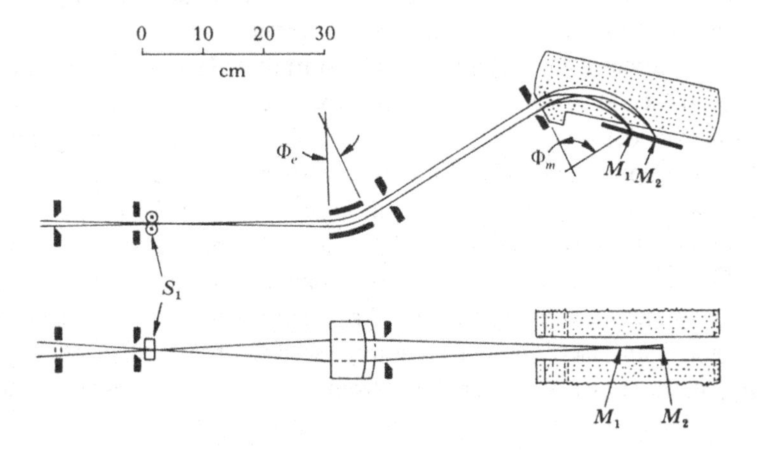

first order double focusing along a straight line with second order angle focusing at one point. For this combination of fields the resolution is given by

$$\frac{\Delta M}{M} = \frac{S\left[2 - \left(\frac{r_r}{r_z}\right)\right]\left[\frac{f_r}{l_r' - g_r}\right]}{r_r\left[1 + \left(\frac{f_r}{l_r' - g_r}\right)\right]}. \tag{5.3}$$

where the symbols have the same significance as in equations (2.12) and (2.13). This equation is identical with equation (2.24) except for the factor $2 - (r_r/r_z)$, which indicates that a double focusing instrument employing a toroidal, rather than the usual cylindrical electrostatic analyser, possesses a significantly higher resolving power. Thus, for example, with a spherical analyser the resolution is improved by a factor of two. Ewald and his colleagues have chosen $r_r/r_z = 1.26$, in which case the factor of improvement in the resolution is expected to be 2.70. This shape of toroid provides the axial focusing mentioned above.

This particular instrument was developed to analyse directly the abundances and charge states of primary fission fragments from heavy elements such as uranium. These particles have net charges of $\sim +20e$ and high kinetic energies so that deflecting voltages on the electrostatic sector of 300–400 kV are required. The desired high sensitivity leads to a modest resolving power of ~ 1000.

5.11 The Stevens double focusing mass spectrometer

A large ($r_e = 2.54$ m) versatile instrument has been constructed at the Argonne National Laboratory (Stevens *et al.*, 1960; Stevens *et al.*, 1963; Stevens & Moreland, 1967) and used for both high sensitivity isotopic analysis and for precise atomic mass determinations.

The geometry (fig. 5.9) is a modified Mattauch–Herzog version in which the object slit of the spherical electrostatic analyser is located at the principal focus so that the image is at infinity. The inclined entrance boundary of the magnetic field produces a strong convergence toward the median plane so that an axial focus occurs at about the 75° point in the magnetic field. The exit boundary produces divergence, but, with a sufficiently long collector slit, high transmission is achieved. Second order radial direction focusing is obtained by a curved entrance boundary to the magnetic analyser. The relatively large second order coefficients B_{12} and B_{22} (in equation (2.66)) are reduced from their nominal values of -6.69 and

− 10.3 by the addition of iron shims in the magnet (Stevens & Moreland, 1967).

As an analytical instrument, (Stephens *et al.*, 1963) a transmission of 95%, from the surface ionization source to the collector, has been reached with a resolving power of 10 000. Under these conditions, most impurity ions of the same mass number would be resolved, so that the limit in sensitivity is close to the background in the detector system. In this mode, the instrument is capable of detecting 4×10^{-15} g of Cm (10^7 atoms).

This spectrometer has also been operated with a resolving power of 5×10^5–7×10^5 (FWHM) for the precise determination of atomic mass differences (Stevens & Moreland, 1967, see Chapter 8).

5.12 The Barber–Duckworth double focusing mass spectrometer

At the University of Manitoba a high resolution mass spectrometer ('Manitoba II') has been constructed (Barber *et al.*, 1967; Barber *et al.*, 1971) according to the second order theory of Hintenberger & König (1959). This particular geometry, shown in fig. 5.10, is one of many suggested by them for spectrometers having $B_{11} = B_{12} = B_{22} = 0$ in equation (2.66). The desirable properties which prompted the choice were the following. An intermediate direction focus is formed by the electrostatic sector so that it may be used as an energy analyser. The total ion path is reasonably short, as are l'_e, l''_e, l'_m, and l''_m, so that problems from mechanical vibrations and stray magnetic fields are minimized. The magnetic field boundaries are straight.

As noted in §2.10, the calculations of Hintenberger & König did

Fig. 5.9. Stevens double focusing mass spectrometer. (After Stevens *et al.*, 1960.)

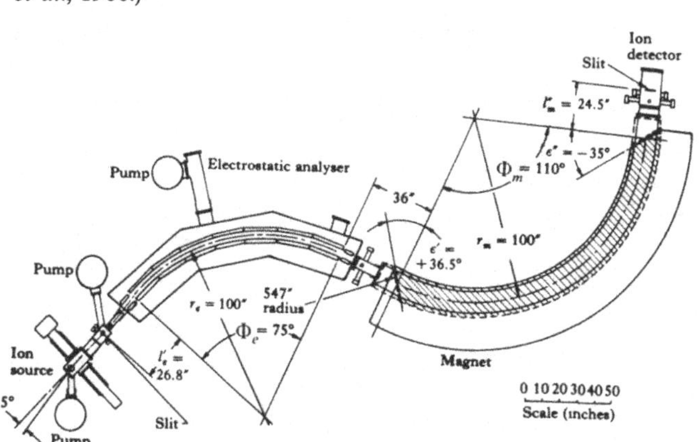

not take into account the effect of the fringe fields. However, subsequent calculations by Matsuda (1976) indicate that for this instrument the second order coefficients remain small (<0.31).

This mass spectrometer has been used extensively for precise atomic mass determinations and has been operated routinely with a resolving power of 3×10^5 (FWHM) or 1.5×10^5 at 5% of peak height. For favourable materials, higher resolving powers (9×10^5 (FWHM) or 4.5×10^5 at 5% of peak height) have been reached (Kozier *et al.*, 1980). The precision with which doublet spacings have been measured is up to $2/10^9$ of the mass at which the doublet occurs.

Although not a consideration in the original work, the geometry of 'Manitoba II' has the further desirable property that strong axial convergence of the ion beam occurs at the entrance boundary of the magnet. Despite divergence at the exit boundary, an overall net convergence occurs toward a point $1.93r_e$ from the final magnetic field boundary. Thus the use of a sufficiently long collector slit leads to high transmission.

As was the case with the Nier–Johnson spectrometer, the arrangement of this instrument is a convenient one. Routine minor focusing is normally carried out by a transverse displacement of the collector for velocity focusing and by adjustment of l'_e for direction focusing (Barber *et al.*, 1971).

Fig. 5.10. Barber–Duckworth double focusing mass spectrometer.

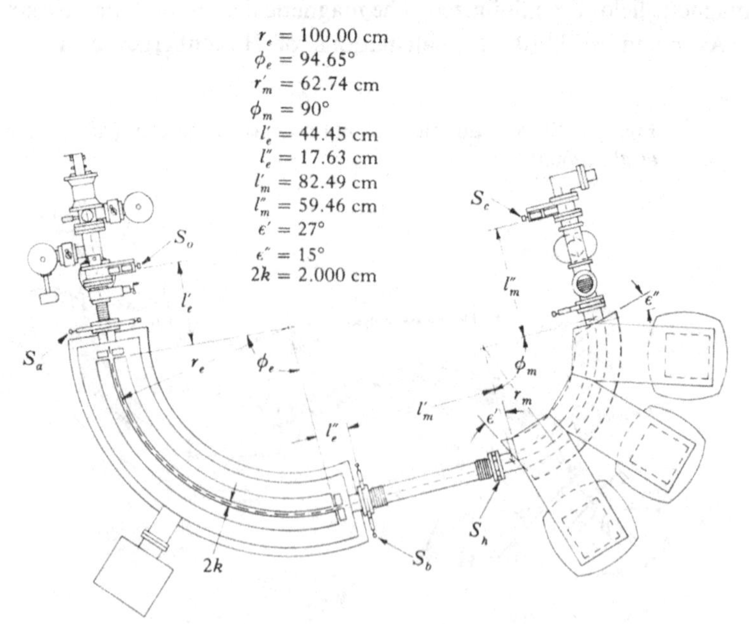

5.13 The Matsuda high dispersion mass spectrometer

A novel type of mass spectrometer, having a toroidal electrostatic field, followed by a magnetic field which varies as r^{-1} and finally by a uniform magnetic field, has been constructed at Osaka University ('Osaka II' Matsuda, Fukumoto, Kuroda & Nojiri, 1966; Matsuda, Fukumoto, Matsuo & Nojiri, 1966), as shown in fig. 5.11.

The object slit is located at the focal point of the toroidal condenser. Thus ions having a particular energy will emerge from it following parallel trajectories. The r^{-1} magnetic field has the property that, for ions of a given mass and energy, there is no direction focusing, and parallel rays remain parallel. Such rays then enter the uniform magnetic field and converge toward its focal point. The r^{-1} field produces a dispersion for both mass

Fig. 5.11. Matsuda high dispersion mass spectrometer. (After Matsuda, Fukumoto, Kuroda & Nojiri, 1966.)

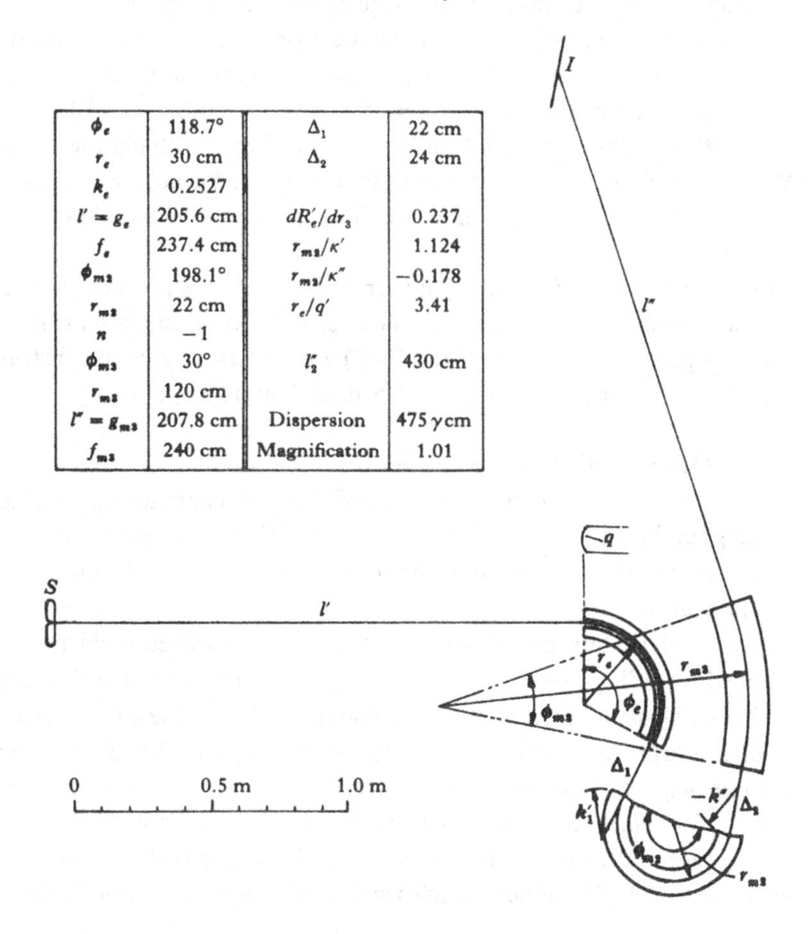

ϕ_e	118.7°	Δ_1	22 cm
r_e	30 cm	Δ_2	24 cm
k_e	0.2527		
$l' = g_e$	205.6 cm	dR'_e/dr_3	0.237
f_e	237.4 cm	r_{m2}/κ'	1.124
ϕ_{m2}	198.1°	r_{m2}/κ''	−0.178
r_{m2}	22 cm	r_e/q'	3.41
n	−1		
ϕ_{m3}	30°	l''_2	430 cm
r_{m3}	120 cm		
$l'' = g_{m3}$	207.8 cm	Dispersion	475 γcm
f_{m3}	240 cm	Magnification	1.01

and velocity. Velocity focusing is achieved by arranging the fields so that the velocity dispersion of the toroidal field is cancelled by the combined velocity dispersions of the two magnetic fields. Under those conditions there remains a net *mass* dispersion.

Axial focusing occurs in both the toroidal field and in the r^{-1} field, but not in the uniform magnetic field. The axial focus is located 4.30 m from the final magnetic field boundary.

In developing this instrument, it was originally assumed that the effective fields terminated abruptly and, on this basis, corrections for the second order aberrations were made. The boundaries of the r^{-1} field were curved, and the derivative of the radius of curvature of the equipotential line in the r–z plane of the toroidal field was chosen to have the appropriate value, so that B_{11}, B_{12} and B_{22} would be zero. Moreover, the entrance profile to the toroidal condenser was curved so that the α_z^2 aberration would also be zero.

Unfortunately, the effect of the fringe fields is quite significant in the instrument, as confirmed in later theoretical (see Chapter 2) and experimental work (Fukumoto & Matsuo, 1970) by the group at Osaka. Actual values for the coefficients were $B_{11}=48$, $B_{12}=37$ and $B_{22}=-73$.

Nevertheless, this instrument has been operated with both photographic (Matsuda, Fukumoto, Kuroda and Nojiri, 1966; Matsuda, Fukumoto, Matsuo & Nojiri, 1966) and electrical (Fukumoto, Matsuo & Matsuda, 1968) detection.

In the former case, a resolving power of 1.2×10^6 was achieved with a mass dispersion of 4.75 cm for 1% difference in mass. In the latter case, a resolving power of 2×10^5–10^6 (FWHM) was observed while the instrument was used for precise atomic mass determinations.

5.14 Other double focusing instruments

Clearly, the geometrical possibilities for constructing double focusing combinations are not exhausted by those in the preceding sections. Several other arrangements have been constructed, some with quite unusual features.

In particular, Matsuda (1976) has described two instruments which were designed according to the third order ion optics (§2.10), where the effects of the fringe fields have been taken into account. The first of these (Taya *et al.*, 1978) comprises a toroidal electrostatic sector ($\Phi_e = 85.2°$) followed by uniform magnetic field ($\Phi_m = 90°$, $\varepsilon' = 30°$, $\varepsilon'' = -9.6°$). An intermediate focus is formed, and overall stigmatic focusing makes possible a high transmission. In an instrument with $r_e = 21.1$ cm, a maximum resolving power of 83 000 (10% of peak height) and total transmission of 43% (from the principal slit to the collector) have been obtained.

The second instrument (Matsuda, 1976; Matsuda, Matsuo & Takahashi, 1977), shown in fig. 5.12, involves a cylindrical electrostatic sector and electric quadrupole lens coupled with a uniform magnetic field. Cylindrical electrodes are easier to machine than are toroidal ones, and so this instrument is more easily constructed, especially in a large version. An intermediate focus is formed at $l''_e = 35$ mm. A maximum resolving power of 240 000 (FWHM) has been achieved and the measured values of the second and third order coefficients are generally in good agreement with the calculated values.

An 'S' geometry arrangement involving two large, high resolution mass spectrometers, each with $r_e = 3.10$ m, has been constructed by Katakuse, Nakabushi & Ogata (1976) for the purpose of precise atomic mass determinations. In such a scheme the mass dispersions of the two instruments add. Although this instrument has been operated at a resolving power of 10^6 (FWHM), it has been susceptible to problems caused by stray magnetic fields from power lines and by building vibrations.

5.15 Some general comments concerning double focusing instruments

Double focusing instruments may be compared on several counts,

Fig. 5.12. Matsuda double focusing mass spectrometer (after Matsuda *et al.*, 1977) designed according to third order ion optics (TRIO). Axial focusing is shown in the lower section of the diagram. Dimensions in mm.

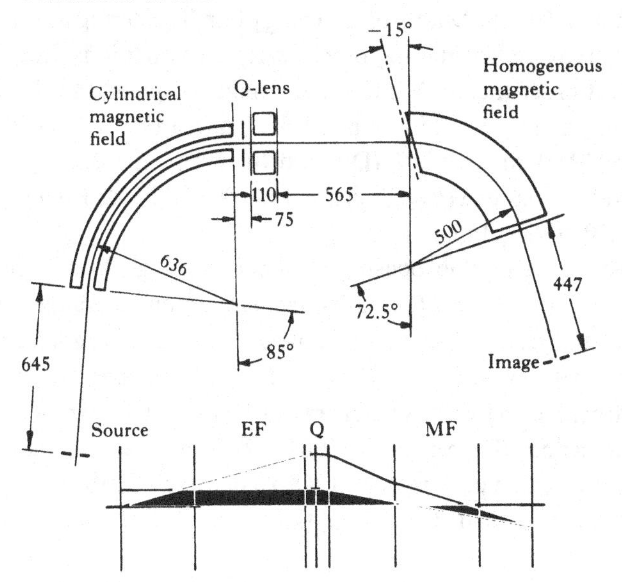

namely, resolution, dispersion, magnification, and other considerations (for example, Mattauch, 1953). Of these, the first-named is the most significant.

It will be recalled from §2.4 that the resolution of ordinary double focusing instruments is $\sim S_1/r_e$. This may be improved by a factor of two if one replaces the radial electrostatic analyser by a Wien velocity filter. On the other hand, the resolution deteriorates by a factor of two if the ions are caused to reverse their direction in mid-flight, as in the instruments which provide double focusing for all masses.

Resolution, dispersion and magnification are not mutually independent; in fact, for any instrument,

$$\text{resolution} \times \text{dispersion} = \frac{\text{slit width} \times \text{magnification}}{100}. \tag{5.4}$$

This result suggests the feasibility of a compact, high resolution, low dispersion, low magnification instrument. Unfortunately, this is not a practicable idea. In photographic detection the graininess of the plate requires a certain minimum dispersion if the theoretical resolution is to be fully exploited. In electrical detection a similar limit arises from the problem of making straight, parallel slit jaws. Thus the magnification is of the order of ~ 1. After the slit width, S_1, has been reduced to its workable limit, improved resolution is sought by increasing the dispersion, that is, by designing larger instruments.

In the absence of higher order focusing, the resolution of a given instrument can frequently be improved by reducing (a) the angular spread of the ion beam and (b) the 'width' of the energy bundle entering the magnetic analyser, thereby approximating more nearly the conditions described in Herzog's first order theory. Values of half-angular spread that have been employed in mass determination work range from $\alpha_r = 0.0003$ (see Bainbridge, 1953) to $\alpha_r = 0.02$ (Duckworth, 1950). Extreme values for fractional velocity spread ($\beta = \Delta v/v$) are $\beta = 0.0067$ (Nier & Roberts, 1951) and $\beta = 0.0003$ (Nier, 1956).

As shown in §2.4, double focusing at all points along a line implies that, for ions of a given energy, the paths are parallel on emerging from the electrostatic analyser, and an intermediate direction focus is not formed. In mass spectrometers where such an image is formed, the separate properties of the electrostatic analyser as an energy analyser may be exploited in the focusing procedure (Barber *et al.*, 1964, 1971) or in the application of Bleakney's theorem in peak matching (§8.3; Nier, 1957; Barber *et al.*, 1971).

In instruments intended for analytical purposes (Gorman, Jones &

Hipple, 1951; Shaw & Rall, 1947; Hannay, 1954), double focusing arrangements are employed, not primarily for the sake of high resolution, but to secure the velocity focusing property. In these instances, α_r and β will not greatly exceed the typical values mentioned above, but the slit width will be substantially increased.

5.16 Trochoidal-path instruments

The perfect double focusing properties of crossed electric and magnetic fields, first pointed out by Bleakney & Hipple (1938), and outlined in §2.5, have been utilized in several mass spectrometers. In these, the ions emerging from the source are constrained to follow trochoidal paths and are detected after travelling one or more cycles.

Experimentally, not all trochoidal paths are desirable; for example, a curtate orbit with a sharp cusp, or point of near-zero energy, may cause a space charge defocusing. Such paths are, therefore, baffled out as are, also, those followed by multiply charged ions of the species under study. This 'stopping down', however, is slight. It does not materially interfere with the high ion currents to be expected in an instrument of this type, in which there is no theoretical requirement to limit either the direction or energy spread.

The possibility of this focusing scheme was first demonstrated on a small scale (Bleakney & Hipple, 1938), following which, in the course of time, several larger instruments were constructed (Mariner & Bleakney, 1949; Monk, Graves & Horton, 1947; Hipple & Sommer, 1953).

In the Mariner & Bleakney apparatus, constructed at Princeton University about 1940, a slit-to-image distance of $b = 26.2$ cm was used, with source and collector slits of 0.05 and 3.17 mm respectively. The theoretical resolution of $\frac{1}{80}$ was experimentally realized. Also, the steepness of the sides of the peaks suggested that, with appropriately narrow slits, this could be improved to 1/1000 and, possibly, even to 1/25 000. This work was interrupted by the war and was not resumed.

The Hipple & Sommer instrument, located at the National Bureau of Standards in Washington, was very large, and made use of a trochoidal path of five cycles, with a distance of 2 m between the source and the final focus. It was converted, shortly before completion, to a time-of-flight instrument, and will be discussed as such in the following chapter.

At the other end of the scale, a very small instrument of this type with slit-to-object distance of 2.7 cm was developed by the Consolidated Electrodynamics Corporation (Robinson & Hall, 1956) for analytical work.

With a 10% energy spread and 6° half-angular divergence in the ion beam, a resolution of 1% was attained.

In recent years the potential advantages afforded by the focusing properties have been found to be outweighed by practical problems arising from space and surface charges and from the inconvenience of locating source and detector in the magnetic field (Voorhies *et al.*, 1959). Accordingly, this type of instrument now appears to be of historical interest only.

6

Time-of-flight and radio-frequency mass spectrometers

In this chapter we describe a rather heterogeneous group of mass spectrometers in which mass analysis depends on (a) the ionic velocity, and hence the time-of-flight, (b) the cyclotron frequency of the ion, or (c) the stability of the ion path within electric fields which vary at radio-frequency. Such instruments are sometimes described as 'dynamic' mass spectrometers, inasmuch as the time (frequency) properties of the motion of the ions in the fields are of fundamental significance. In this way they are distinct from the instruments described in Chapter 5, where mass analysis is accomplished by *static* fields.

Properly speaking, time-of-flight (TOF) instruments measure the time required for an ion to traverse a certain specified distance. This may be done either by a direct timing mechanism, employing pulsed ion sources and detectors, or by subjecting the ion to radio frequency (rf) fields. In either case the apparatus selects, from the ion beam, those ions having a certain velocity. If the velocity of the ion be characteristic of its mass, as is the case of singly charged ions which have fallen through the same potential, such a velocity filter may be used to effect a mass analysis. The first instance in which a velocity filter of this type was employed was the determination of the velocity of cathode rays by Wiechert (1899).

In certain other mass spectrometers the ions describe circular paths in homogeneous magnetic fields, with the cyclotron frequency

$$f = \frac{Be}{2\pi M} \tag{6.1}$$

which is independent of velocity, but linearly dependent upon mass. Here the mass is determined by measuring this cyclotron frequency, sometimes in terms of a high harmonic. These instruments will be described hereafter by the term 'cyclotron resonance'.

An entirely different effect on the paths of ions subjected to the combination of an rf electric quadrupole field superimposed on a dc electric quadrupole field was discovered by Paul & Steinwedel (1953). They showed that, in such a combination, ions of a certain mass execute stable oscillatory trajectories as they traverse the length of the field region. Ions having either lower or higher masses execute unstable trajectories of increasing transverse amplitude and are lost. Instruments employing this principle are called quadrupole field mass spectrometers.

Time-of-flight instruments

6.1 Smythe–Mattauch mass spectrometer

This (Smythe & Mattauch, 1932; Smythe, 1934), the first of the time-of-flight mass spectrometers, was built at the California Institute of Technology, following a proposal by Smythe (1926). In this scheme, alternating electric fields are applied at right-angles to an ion beam in such a way that only those particles having certain velocities emerge from the fields undisplaced and undeflected.

In the apparatus shown in fig. 6.1 the alternating voltage, of frequency f, is applied to the plates of the two condensers A and B, each of length $2a$. If the resulting deflecting fields stop abruptly at the physical boundaries of the plates, then those ions with velocity v, such that they pass through condenser A in an integral number of cycles n, receive as much acceleration in one direction as in the other and, on emerging, follow paths which are parallel to their original direction of motion, albeit displaced. In the second identical condenser, B, which the ions enter in the opposite phase to that in which they entered the first, the displacement is reversed, and the particles emerge from the combination both undeviated and undisplaced. Smythe showed that non-ideal fields may be employed, provided that their amplitudes satisfy the condition $f(x) = f(x - a)$.

For simplicity, Smythe & Mattauch made $f(x) = f(-x)$. This was done by forming both A and B from pairs of identical condensers placed in identical chambers. This simplification led to the transmission of an unanticipated number of velocities, corresponding to various ns. These 'ghosts', which limited the usefulness of the instrument, were later completely explained by Hintenberger & Mattauch (1937). In the original apparatus the distance D was variable. This provided a partially effective

Fig. 6.1. The Smythe–Mattauch mass spectrometer.

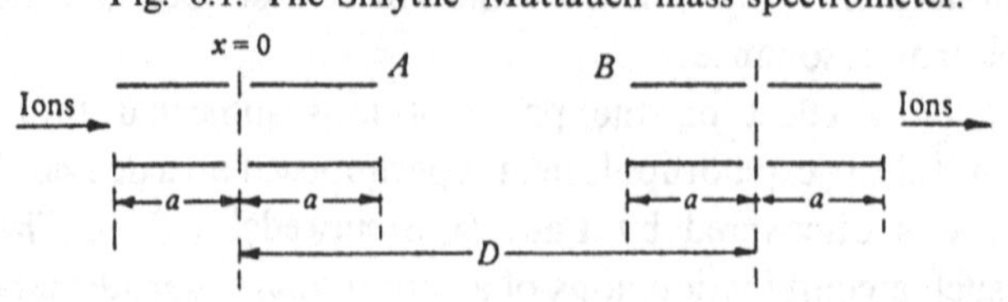

means of discriminating against ions of higher order, that is, $n > 1$.

The beam emerging from the filter was further analysed by a radial electrostatic field, and resolution was achieved which was adequate to permit a useful study to be made of the abundance of ^{18}O.

6.2 Pulsed-beam mass spectrometers

Remarkable advances were made during the Second World War in techniques for the generation of very short electrical impulses. These provided the basis for pulsed-beam mass spectrometers, developed independently by Stephens (1946) and Cameron & Eggers (1948).

Such instruments consist essentially of a comparatively long drift tube, with a source of ions at one end and a detector at the other. The source emits short bursts of ions, homogeneous in energy or momentum. These traverse the drift tube to reach the detector, which is sensitized for a brief instant to register their arrival. Since ions of different masses arrive at the detector at different times, the accurate measurement of the time between activating the source and sensitizing the detector gives information concerning the mass of the ions being detected, while the amplitude of the signal is a measure of the corresponding isotopic abundances.

If L be the length of the drift tube, the transit time for singly charged ions of constant energy, Ve, is

$$t = L(M/2Ve)^{1/2}, \tag{6.2}$$

and for those of constant momentum, p, is

$$t = LM/p. \tag{6.3}$$

If the collector be sensitized for a period Δt, at time t, the resolution becomes, for constant-energy ions,

$$\Delta M/M = 2\Delta t/t, \tag{6.4}$$

and, for constant-momentum ions,

$$\Delta M/M = \Delta t/t. \tag{6.5}$$

It should be noted that, in both instances, the resolution of this type of instrument is not mass dependent. Further, the arrangement employing a constant momentum source possesses, in theory, a factor of two advantage over the constant-energy type. However, it is not Δt which in practice limits the resolution, but rather ion thermal energies. In modern instruments the collector is kept continuously sensitive and the time of arrival of the ions is determined in other ways.

Early models of this type of instrument were attractive because they allowed a rapid panoramic display of the entire spectrum, despite the relatively poor resolving power, $\Delta M/M \sim \frac{1}{10}$, for the constant-energy variety and $\Delta M/M \sim \frac{1}{20}$ for the constant momentum type (Wolff & Stephens, 1953). In the latter case, the ions with constant momentum were obtained by turning off the accelerating voltage before the ions had passed through the entire accelerating field (Hays, Richards & Goudsmit, 1951).

Subsequent models possessing much higher resolution were constructed by Katzenstein & Friedland (1955) and Wiley & McLaren (1955). The Katzenstein & Friedland instrument incorporated two important modifications. In the first place, a grid in front of a continuously sensitive detector was pulsed rather than the detector itself. This eliminated the sluggish detector response caused by the capacitance of the collector assembly, and permitted an experimental resolution of $\frac{1}{75}$ to be achieved. Secondly, the ion source was designed to make use of the techniques, described in §10.1, of Fox, Hickman, Kjeldaas & Grove (1951) for obtaining beams of bombarding electrons homogeneous in energy, that is, the grid arrangements needed to pulse the electron beam into the ionization chamber and, at a slightly later time, to pulse the ion beam out of it. Provision was also made for modulating the electron beam as required by the Fox scheme. By employing an axial rather than a transverse electron beam, Katzenstein & Friedland obtained an exceptionally large ion current.

The Wiley & McLaren time-of-flight mass spectrometer still forms the basis of present day instruments having resolving powers of a few hundred. This is secured largely by use of an ion source containing two accelerating regions, as shown in fig. 6.2. While the ions are being formed, under pulsed

Fig. 6.2. Schematic representation of the Wiley and McLaren time-of-flight mass spectrometer. (From Wiley & McLaren, 1955.)

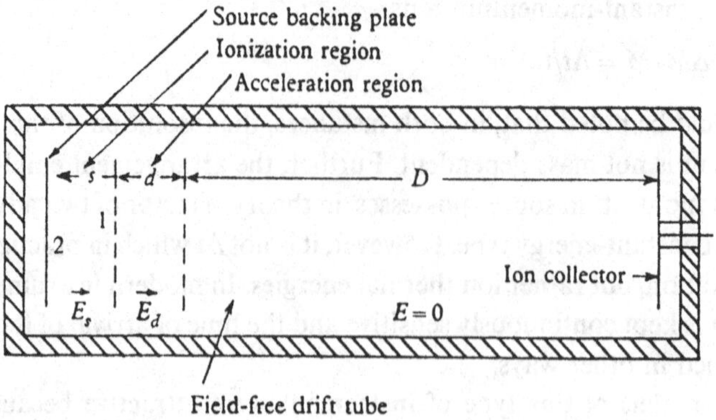

transverse electron bombardment, the potential of the source backing plate is the same as that of the first grid. At all times there is a field E_d in the second accelerating region, d, while the drift space, D, is field-free. Ions are pushed out of the source toward the collector when a positive pulse, giving rise to the field E_s, is applied to the source backing plate. The pulse lasts until all the ions have left the ionization region. This double field configuration introduces two new parameters, d and E_d/E_s, which are not available in the single field source.

Ions are created in different regions of the ionization space and with different initial velocities. Disregarding the velocity effect, those initially closer to the first grid acquire less energy than, and are eventually overtaken by, those initially closer to the source backing plate. This constitutes a space focusing or 'bunching' effect which, by adjustment of E_d/E_s, can be made to occur at the collector. The effect of initial velocities is reduced in this instrument in two ways: (a) by using larger than normal accelerating fields and (b) by introducing a time lag between ion formation and ion acceleration. An ion initially travelling away from the collector would normally arrive at the collector later than an ion initially travelling toward it. However, during the time lag they move to positions where, when the accelerating field is applied, the first acquires greater energy than the second and may overtake it (energy focusing). An electron multiplier having a plane conversion dynode is used for detection so that variations in the ion transit time that would be associated with a curved surface are eliminated.

Typically, a complete cycle lasts $\sim 50\,\mu s$, with the electron beam pulsed for $\sim 1\,\mu s$: thus, the duty cycle is $\sim 2\%$. The sensitivity may be enhanced by operating the instrument in a so-called 'continuum' or 'long-pulse ionization' mode in which the electron beam is left on for a much larger fraction of the entire cycle. The resulting ions are initially trapped in a potential well formed between the backing plate and the draw-out grid, and are then expelled by a voltage pulse applied to the grid. In an instrument having a flight path of 2 m, a resolving power of over 700 (at 10% valley) has been achieved (Rothstein, 1978).

6.3 ^{252}Cf plasma and ion-induced desorption mass spectrometer

We now describe time-of-flight instruments which have been constructed for the study of large organic molecules which are involatile and thermally unstable. These are similar to the mass spectrometers of the previous section, in that mass analysis depends on the flight time down a drift tube, but differ in the way the timing is done.

In the apparatus of Macfarlane & Torgerson (1976*a*, *b*), shown in fig. 6.3, the sample under study is placed on the right-hand side of a thin foil on the target holder. When a ^{252}Cf nucleus undergoes spontaneous fission, two heavy fragments are produced which travel in almost exactly opposite directions with energies of ~ 100 MeV each. If such a fragment travels toward the sample foil, it has sufficient energy to pass through the foil and cause the desorption of a molecular ion of the organic sample. The other fragment from the fission process is detected by the detector array to the left of the ^{252}Cf source and the pulse from this detector is used to 'start' a clock. The organic secondary ion is accelerated by ± 10 kV applied to the sample foil, and is detected at the end of the drift tube by the channel electron multiplier array which produces a 'stop' pulse. The time-of-flight is determined by the difference between the 'start' pulse and the 'stop' pulses which are applied to a time digitizer with a $\frac{1}{8}$ ns resolution and a 29 bit binary output to a memory unit.

Drift tubes of 1 m and 8 m have been used. In the case of the latter, a fine wire was aligned along the axis of the tube and maintained at ~ 10 V to improve the transmission (Oakey & Macfarlane, 1967).

A similar instrument has been described by Chait & Standing (1981). Here, however, short bursts of a primary beam of alkali–metal ions having energies of ~ 10 keV strike the sample foil as indicated in fig. 6.4(*a*). These are produced by the ion gun shown in fig. 6.4(*b*). The 'start' pulse is derived from the sweep applied to the deflection plates in the ion gun while the 'stop'

Fig. 6.3. The ^{252}Cf plasma desorption instrument of Macfarlane and Torgerson. (After Jungclas, Danigel & Schmidt, 1982.) Here MCP is micro channel plate, DA is delay amplifier, TSCA is timing single channel analyser, TAC is time to amplitude converter, and ADC is analogue to digital converter.

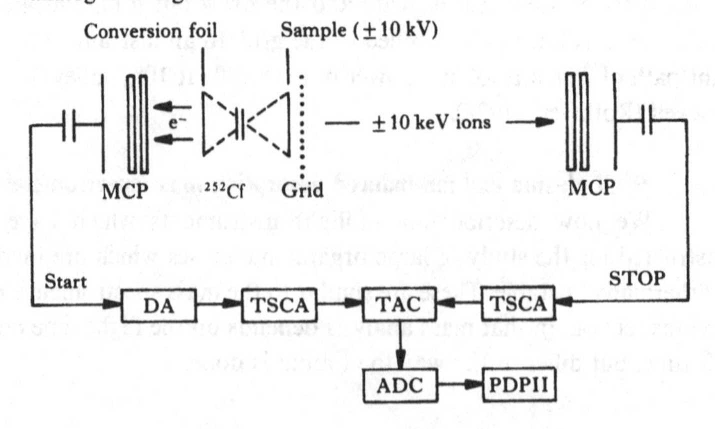

pulse comes from the final detector as before. The time-of-flight is determined in the same way.

The latter instrument has the advantage of the higher intensity in the primary beam, and flexibility of choice of primary ion type, pulse length and repetition rate. The similarities and differences in the formation of ions in these two techniques are discussed in §3.10.

Both of these instruments have an interesting property which differs from deflection instruments and which is of interest in the study of unstable ionic fragments. The velocity of an ion is determined by its mass and the kinetic energy which it acquires during its very short period of acceleration. Further, fragments of ions which dissociate during flight still appear at the mass position of the initial ion, since all of the fragments travel with essentially the initial velocity. A small broadening about this position will occur as a result of the kinetic energy acquired during fragmentation. The fragments may consist of neutrals or ions. The latter may be differentiated from the former by applying a voltage to deflecting plates placed at some intermediate point along the flight path.

Mamyrin, Karataeu, Smikk & Zagulin (1973) have described an instrument (the 'mass reflectron') which uses a grounded grid and plane reflecting electrode at the normal collector position. This reflects the beam back, almost along its original path. When the geometry is correctly chosen,

Fig. 6.4. (*a*) Time-of-flight apparatus of Chait & Standing (1981). (*b*) Details of the pulsed ion source.

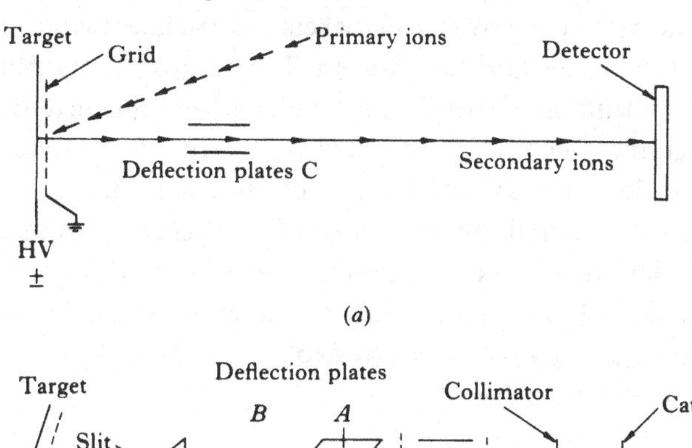

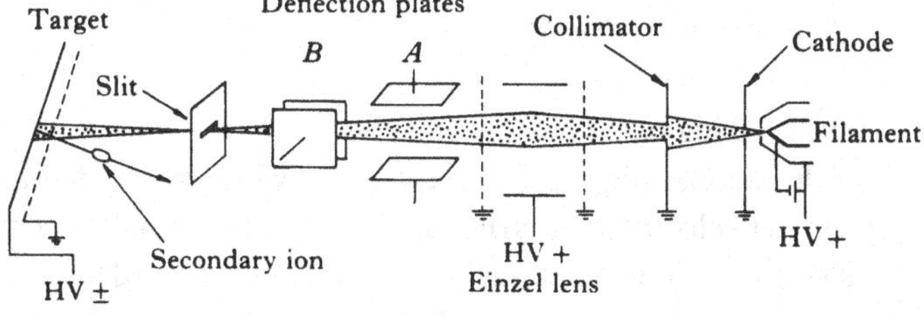

the arrangement is velocity focusing, with the result that remarkably high resolving powers ($\sim$ 3500) have been achieved. A similar instrument having linear geometry has also been described by Mamyrin & Smikk (1979).

6.4 Energy-gain mass spectrometers

In these instruments certain ions with a particular initial velocity are accelerated by periodically varying fields to the extent that they are able to overcome a dc potential barrier and reach a collector. Other ions experience insufficient acceleration in the rf field to surmount the final retarding voltage.

Bennett (1948, 1950, 1953) developed a mass spectrometer of this sort in which the ions, created by electron bombardment, fall through a modest fixed voltage before entering a three-stage resonance accelerator. Each stage consists of three grids and to the middle one in each case is applied the rf potential. As the acceleration experienced by an ion in each stage depends upon its velocity and the phase at which it enters, the three stages can be so placed relative to each other that ions with one particular initial velocity and phase experience a much greater cumulative acceleration than do any others. These, as mentioned above, can overcome the retarding voltage and reach the collector. A resolution of about 6% with a 5% transmission of the desired ions has been achieved.

It has been shown by Peterlin (1955) and Dekleva & Peterlin (1955) that the resolution of the Bennett spectrometer may be substantially improved by altering the shape of the applied rf voltage. Compact models of this instrument have seen extensive use in rockets and satellites for investigating the composition of the upper atmosphere (for example, Townsend, 1952).

Other instruments of the energy-gain type have been built in the form of a linear accelerator (Redhead, 1952; Boyd & Morris, 1955). Here the ion beam passes along the axis of a series of coaxial cylinders which are alternately connected to the two terminals of an rf generator. As above, the 'tuned' ion gains greatest energy from the rf field. With 20 stages Redhead has achieved a resolution of 1%. Analyses and evaluations of energy-gain mass spectrometers have been given by Redhead & Crowell (1953), Boyd & Morris (1955), and Kerr (1956).

6.5 Ion bunching

If the accelerating voltage increases with time, later ions will acquire a greater velocity than earlier ones, and will eventually overtake the latter, resulting in an ion 'bunching'. This eliminates the need for pulsing the ion source and leads to higher efficiencies than in the usual pulsed instruments.

First, the ions are subjected to a saw tooth accelerating voltage, following which they traverse a drift tube of proper length to allow the bunching to occur. Finally, they encounter a gating grid which is pulsed to allow the desired bunch to reach the collector. This scheme has been utilized by Wilson (1952) and Glenn (1952), the latter achieving a resolution of 1/250 with a transmission of about 20%.

Any further progress in mass spectrometry along this line will undoubtedly take advantage of the developments which have occurred in the design of accelerators in nuclear physics.

Cyclotron-resonance instruments

6.6 Helical-path mass spectrometer

This instrument (Hays, Richards & Goudsmit, 1951), proposed by Goudsmit (1948), was designed for the study of atomic masses. In it, ions from a pulsed source move in helical paths in a magnetic field of large volume. The pitch of the helix is sufficient to allow the ions to miss the source structure as they complete their first revolution. The detector is located directly beneath the source, and records the time of arrival of the ions as they complete their *n*th revolution. The cyclotron frequency associated with the ion in question is determined by measuring the time difference between, say, the second and eleventh revolutions.

It will be seen from equation (6.1) that the cyclotron frequency varies inversely as the mass. Thus, when based on frequency, the mass scale in this instrument is linear, and the precision, δM, is constant over the entire mass range. In practice, the elapsed time between the second and eleventh revolutions was about 10 μs per mass unit. Since this was measured to an accuracy of 0.01 μs, the precision of the mass determinations was about 10^{-3} u or 1 MeV. This instrument has been, therefore, of particular use with heavy atoms. It was calibrated by standard hydro- and fluorocarbons; for example, in the study of ^{208}Pb and ^{209}Bi the calibration masses were $C_3F_5(M = 131)$, $C_4F_7(M = 181)$ and $C_5F_9(M = 231)$.

The magnetic field was produced by a spherical air-core magnet, but, because of its extent, it was of low intensity. This dictated the use of low energy ions (25 eV for $M = 100$), which were strongly influenced by polarization of the walls of the vacuum chamber. It was not practicable to increase appreciably either the strength or the volume of the magnetic field, the former to reduce this polarization effect, or the latter to allow a larger number of spirals. Although the instrument was, for these reasons, not further exploited, it was used to make a timely and significant contribution to our knowledge of atomic masses.

6.7 Omegatron

The omegatron of Hipple, Sommer & Thomas (1949, 1950) is essentially a small cyclotron in which ions are accelerated if the frequency of the rf accelerating field coincides with the cyclotron frequency (equation (6.1)). This principle had been employed earlier by Alvarez & Cornog (1939), with an actual cyclotron, to demonstrate the existence in nature of ^{3}He. In recent times the cyclotron has again been used as a highly sensitive mass spectrometer for ^{14}C dating (Muller, 1977; see §7.11).

The maximum radius of curvature of the omegatron is 0.9 cm. The rf field is both uniform and weak, pervading the entire accelerating region, rather than possessing the dee-established shape characteristic of high energy accelerators. Ions are created in the central region by electron bombardment, and spiral out to the collector, making 3000–5000 revolutions en route. They are prevented by a dc field from escaping axially. The large number of revolutions results in a sharp mass discrimination, corresponding to a resolution of 1/10 000 for low masses. The mass spectrum is obtained by varying either the frequency or the magnetic field.

Although this instrument was used to make an important preliminary determination of the ^{2}D—^{1}H mass difference (Sommer, Thomas & Hipple, 1951), space charge difficulties have interfered with its use with heavier atoms. Its chief application has been in the determination of the proton moment in nuclear magneton units. This was done by measuring for protons the cyclotron frequency and the nuclear magnetic resonance frequency in the same magnetic field (Boyne & Franklin, 1961; Petley & Morris, 1968).

Bloch & Jeffries (1950) and Jeffries (1951) have used a small cyclotron to effect a resonant deceleration of ions. Here, 20 keV protons, initially following an orbit of 4.25 cm radius, reached the central collector after 500 revolutions. As with the omegatron, this equipment was used to determine accurately the proton moment. The width of the H^+ ion peaks that were obtained indicated a resolution of approximately 1/10 000 at half-maximum.

6.8 Trochoidal-path rf spectrometer

As mentioned in the preceding chapter (§5.13), the large, five-cycle, trochoidal mass spectrometer at the National Bureau of Standards (Hipple & Sommer, 1953) was converted to a cyclotron-resonance instrument. The ions, after traversing one cycle, passed between a pair of small closely spaced electrodes, to which the rf voltage was applied. If this rf voltage were approximately zero at the time of their arrival, the ion would be undeflected and would continue on a trochoidal path until they reached, four cycles later, a second similar pair of electrodes; otherwise, they would be deflected in the direction of the magnetic field. The radio-frequency f', which would permit the undeflected passage of the ions through the second pair of electrodes and, thence, to the collector, is simply related to the cyclotron or fundamental frequency, f, as follows

$$f' = \frac{n}{8} f, \tag{6.6}$$

where n, in this particular case, is the number of half-cycles of the radio-frequency. The cyclotron frequency, incidentally, is given by (6.1); it is not affected by the presence of the electric field.

In operation, the dc field was first adjusted to bring the desired ions to the collector, after which the rf field was turned on, adjusted, and its frequency measured. A similar procedure was followed for a second ion of neighbouring mass, thereby permitting the mass difference to be computed from (6.6). A resolution of 1/12 000 at half-maximum was obtained.

6.9 Mass synchrometer and rf mass spectrometer

This instrument, developed by L.G. Smith (1951*a*, *b*) at the Brookhaven National Laboratory, is shown schematically in fig. 6.5. It is an outgrowth of Smith's association with the helical-path device described in §6.6. In it the ions, while constrained by a uniform magnetic field to follow circular paths, are exposed in the 'pulser' to a local modulating field, S_3 being connected to the appropriate source of potential.

In the first model of the synchrometer this potential consisted of negative rectangular pulses of 1 μs duration, occurring at variable and measurable time intervals. With the first pulse, a group of ions in the neighbourhood of S_3 was decelerated sufficiently to miss the source housing and enjoy a free circulation in orbit 2. After these ions had made a number of such revolutions, a second pulse decelerated the group still further, with the result that they reached the collector a half-cycle later along orbit 3. With this arrangement, 250 eV ions of mass 28 were observed after performing 90 revolutions (72 m). The resolution at half-maximum was 1/24 000, but the intensity had reached a very low level.

In the second model (Smith, 1952, 1953), S_3 was connected to an rf oscillator. This effected an harmonic modulation of the orbit diameter. It was arranged that one cyclotron revolution took place in $n + \frac{1}{2}$ cycles of the rf. Thus, after one revolution, the ions reached the pulser in such a phase as to cause a demodulation of the diameter. In this way, most of the desired

Fig. 6.5. The mass synchrometer. (From Smith, 1953.)

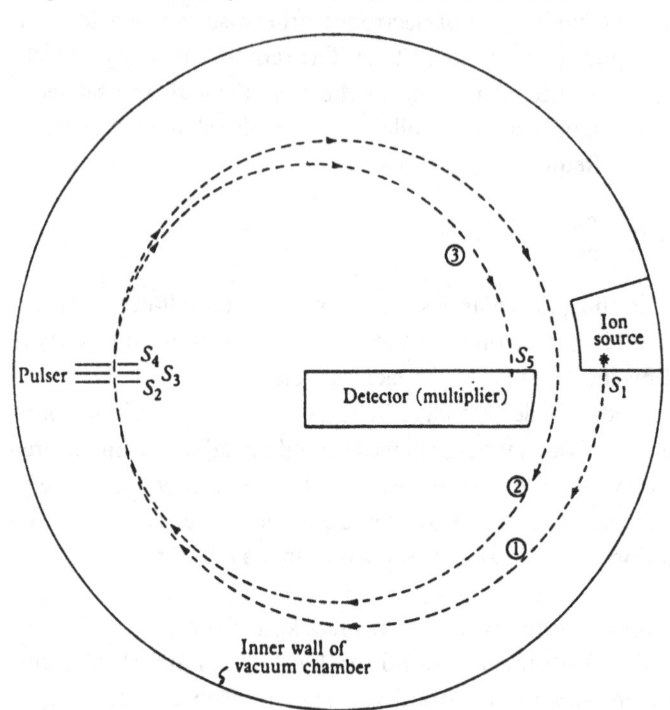

ions were reassembled for detection, and the long ion path of the original model was avoided. In practice, the demodulation at the end of one revolution was not as efficient as was hoped, and the actual resolution achieved was considerably below expectations. Also, the background current was large and there was serious interference from 'satellites' which had made three or more revolutions (Smith & Damm, 1956).

The instrument was further modified, and a detailed description of its operation in this final form was given by Smith & Damm (1956). Here the ions which were finally collected were decelerated by the modulator on each of three successive revolutions. In order to select these ions, two slits were located between S_1 and S_5 in fig. 6.5, say S_6 and S_7, with S_6 being the one nearer to S_1. After the first deceleration the ions passed through S_6, reaching the modulator again in such a phase as to be further decelerated by approximately the same amount. Their orbit then passed through S_7. After the third and final deceleration the orbit passed through S_5 to the detector, which was in this case a Faraday collector plus a dc amplifier, rather than the electron multiplier of earlier models. The choice of three revolutions was made to avoid satellites.

The cyclotron frequency was determined in terms of its *n*th harmonic, given by the frequency of the applied rf voltage corresponding to an ion peak. In a typical case ($M = 32$), n was 107 and the applied frequency was 14.9 MHz. For ion energies of 2500 eV modulating voltages of 180–320 V (rms) were employed. The diameter of the first orbit was 28 cm; this was decreased in three stages, as described, to a final value of 23 cm. The half-width resolution was adjustable electrically between the values 1/10 000 and 1/25 000 for all mass numbers below 250. In practice, the securing of satisfactory ion intensity was something of a problem.

This unusual instrument was exploited in the precise determination of atomic masses, especially amongst the light elements (Smith, 1958), by the technique described in §8.8. A similar instrument has been used by Mamyrin *et al.* (1975) in their precise determination of the magnetic moment of the proton.

The most recent version of this type of instrument was constructed at Princeton by Smith (1960, 1967) as shown in figs 6.6 and 6.7. In this latest rf mass spectrometer the ion beam passes through two 2.32 cm radius spherical electrostatic analysers, used asymmetrically and arranged in tandem to produce a direction focus with an energy dispersion at a slit (not shown in the diagram) located 2.54 cm outside the edge of the magnet gap. The beam is guided through the fringe field and focused on the entrance slit by the injector electrostatic deflection plates. The relatively large (5.33 cm) gap in the magnet provides sufficient room so that the ion beam can describe a helical path and, after one turn, miss the housings of the injector and ejector lenses. After two turns the beam passes through an identical arrangement of electrostatic analysers to the detector. The modulator is located, as before, diametrically opposite to the entrance slit, while a phase-

Fig. 6.6. Plan view of the Smith's rf mass spectrometer showing the complete ion orbit. (After Smith, 1967.)

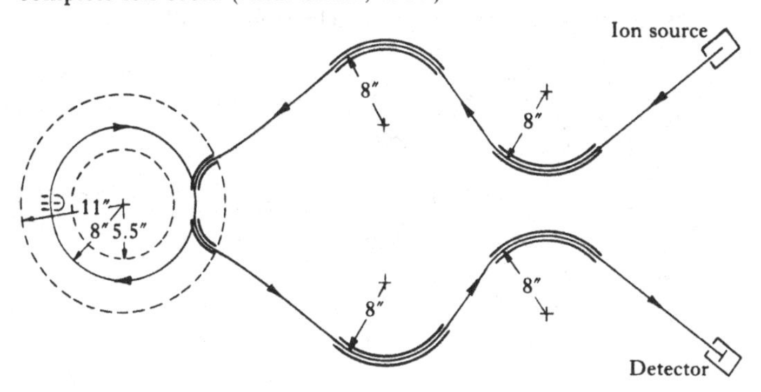

defining slit is located between the entrance and exit slits. The latter two slits are 1.27 cm above and below the median plan, respectively, while the nominal orbit diameter is 51.6 cm.

The greatly decreased size of the modulator relative to the orbit diameter, combined with the higher magnetic field and ion energy, permit the use of much higher frequencies and orders. For example, in a representative case for the ion CO^+, the accelerating voltage was 17.59 kV and the frequency was 249.53 MHz with the order $(n + \frac{1}{2})$ having the value 916.5 (Smith, 1971). Smith (1960) also showed that the ratio $\Delta D_m/D$, where ΔD_m is the amplitude of orbit diameter modulation and D is the orbit diameter, should be minimized in order that the intensity at a given resolution be a maximum. Further, when $\Delta D_m/D$ is reduced, the loss in resolution arising from inhomogeneities in the magnetic field is also reduced. Accordingly the new instrument employs orbits having the same diameters.

While the instrument was operating at Princeton, resolving powers in the range of $2–4 \times 10^5$ (FWHM) were achieved. Following Smith's untimely death in 1972 the instrument was moved to the Delft University of Technology where certain instrumental modifications have been made and resolving powers of up to 10^7 (FWHM) have been attained (Koets, 1976; Koets, Kramer, Nonhebel & Le Poole, 1980). This instrument has been used exclusively for atomic mass determinations (see §8.8) which are of unique value inasmuch as the experimental apparatus differs fundamentally from conventional deflection instruments.

6.10 Ion-cyclotron resonance mass spectrometry

In the conventional ion-cyclotron resonance (ICR) mass spectrometer (Baldeschwieler, Benz & Llewellyn, 1968), ions are formed in a cell and trapped by suitable electrostatic fields combined with a homogeneous

Fig. 6.7. Plan (*a*) and elevation (*b*) views of the section of Smith's rf instrument within the magnetic field. (After Smith, 1967.)

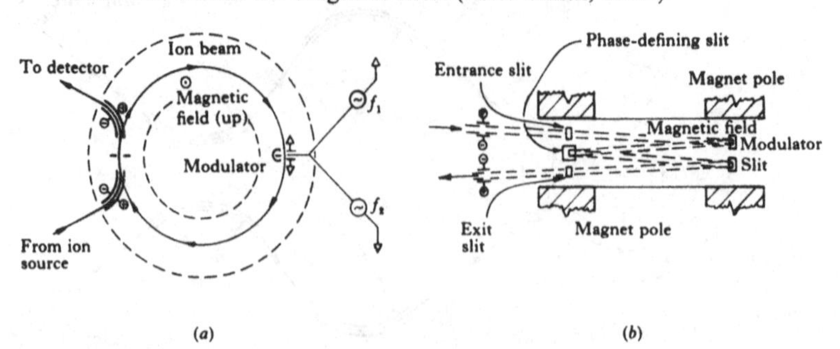

magnetic field. As indicated in fig. 6.8(*a*) the ion motion is excited by the electric field between the two plates which form part of the tank circuit in a marginal oscillator. The frequency of the oscillator is adjusted and the power absorption at the resonance frequency is observed. Typically a scan over the mass range from 15–200 u would require about 20 min.

An important variation of this technique is Fourier transform–ion-cyclotron resonance spectroscopy ('FT–ICR': Comisarow & Marshall, 1976, 1980; Comisarow, 1978*a*, *b*) for which the apparatus is shown schematically in fig. 6.8(*b*). As before, the ICR cell is an ion trap placed in a high, uniform magnetic field. Ion cyclotron motion is excited by turning on the oscillator and sweeping the frequency over a range which corresponds to the mass range of interest (equation (6.1)). The oscillator is turned off and ions, whose motions have been excited, will continue in their orbits until some loss process, such as collision with residual gas, occurs. A signal induced by the motion of the ions is detected at the lower plate and amplified by a broadband amplifier. This signal, which contains components having the cyclotron frequencies of the ions present, is mixed with the signal of fixed frequency from a local oscillator. The resultant signal passes through a low-pass filter and is converted to digital form by an analog-to-digital converter. This digital signal is stored in a computer memory. A quench pulse is applied to the cell and the cycle is repeated for a chosen number of times. The stored spectrum is then Fourier transformed so that the stored amplitude-modulated signal is converted to a frequency spectrum of the input signal.

As shown in fig. 6.9 for the particular instrument of Comisarow & Marshall (1980), the resolution of an FT–ICR instrument varies with mass. Significant improvements in performance have been realized by the application of superconducting magnets to this work, namely, high stability,

Fig. 6.8. (*a*) Conventional ICR spectrometer (the magnetic field is perpendicular to the diagram). (*b*) Fourier transform ICR spectrometer. (After Comisarow, 1978*b*.)

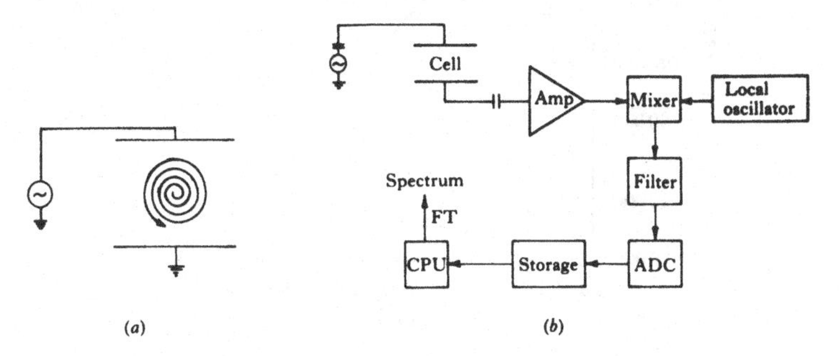

of the field in time coupled with high uniformity over the volume of the cell. Moreover the resolving power of the instrument at a particular mass number improves with higher fields. Accordingly, Allemann, Kellerhals & Wanczek (1983) have reported an instrument operated with a superconducting magnet (4.7 T) at a resolving power of $> 10^8$.

Both the conventional ICR and FT–ICR instruments have been used extensively in the study of ion–molecule reactions (see Chapter 10).

Quadrupole field instruments

6.11 Quadrupole mass spectrometer

The quadrupole analyser, first introduced by W. Paul and his group at Bonn University in the late 1950s, has become one of the most useful of mass spectrometers (Paul, Reinhard & von Zahn, 1958; Dawson & Whetten, 1969; Dawson, 1976*a*).

In this instrument (fig. 6.10), ions moving in the z direction enter a two-dimensional electric field in which the equipotential surfaces are rectangular hyperbolae of the form:

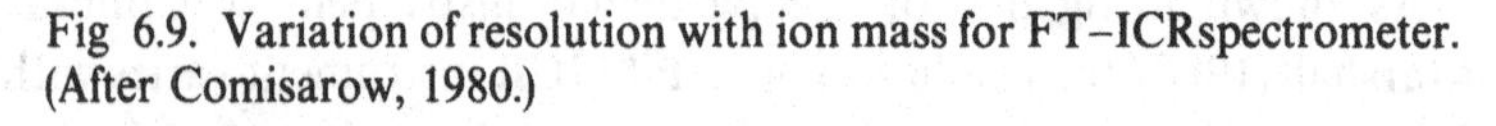

$$\Phi = \Phi_0(x^2 - y^2)/2r_0^2. \tag{6.7}$$

Such a field may be established by a set of four hyperbolic electrodes having potentials $\pm\frac{1}{2}\Phi_0$, where opposite electrodes have potentials of the same sign. In practice, a good approximation to the potential in the region near

Fig 6.9. Variation of resolution with ion mass for FT–ICRspectrometer. (After Comisarow, 1980.)

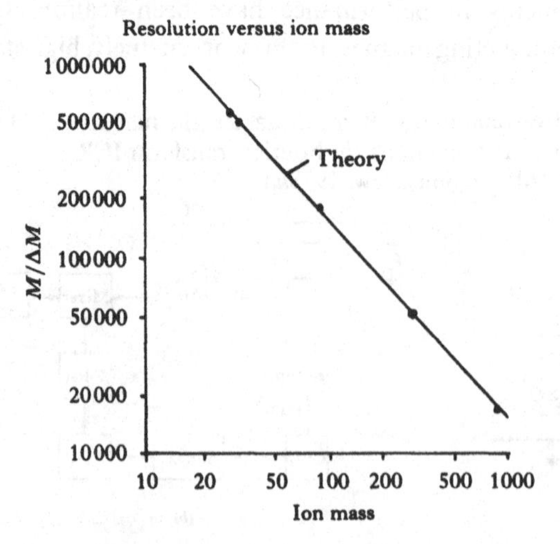

the axis may be achieved when rods with a circular cross section of radius $r = 1.16r_0$ ($2r_0$ = the distance between the electrodes) are used.

For a single quadrupole lens, where Φ_0 is constant, focusing takes place in one plane and defocusing in the perpendicular plane. However, when the dc voltage is combined with an ac voltage so that

$$\Phi_0 = U - V\cos\omega t, \tag{6.8}$$

where ω is the angular frequency of the rf component, an interesting situation arises. The equations of motion for a particle within such a field are

$$\left(\frac{d^2x}{dt^2}\right) + \left(\frac{e}{Mr_0^2}\right)(U - V\cos\omega t)x = 0, \tag{6.9}$$

$$\left(\frac{d^2y}{dt^2}\right) + \left(\frac{e}{Mr_0^2}\right)(U - V\cos\omega t)y = 0. \tag{6.10}$$

Now, with the definitions

$$\frac{4eU}{M\omega^2 r_0^2} = a_x = -a_y = a, \tag{6.11}$$

$$\frac{2eV}{M\omega^2 r_0^2} = q_x = -q_y = q, \tag{6.12}$$

and

$$\xi = \frac{1}{2}\omega t, \tag{6.13}$$

Fig. 6.10. Quadrupole mass spectrometer.

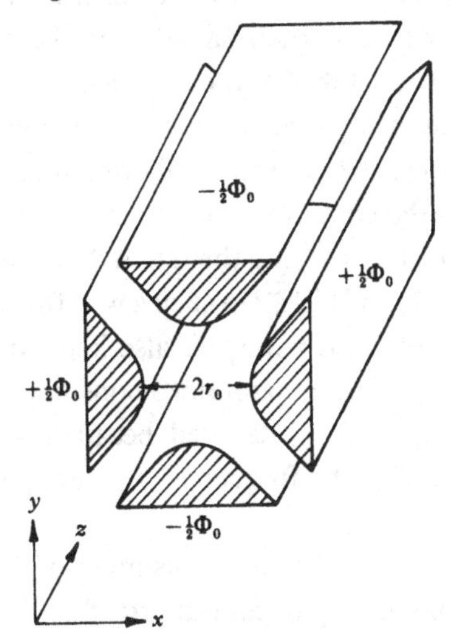

these equations take the same form:

$$\frac{d^2u}{d\xi^2} + (a - 2q\cos 2\xi)u = 0, \tag{6.14}$$

where u represents either x or y. This is known as the Mathieu equation and describes the ion trajectories. The essential feature of the equation is that the solution for u can be expressed as a series involving terms of the form $e^{\mu\xi}$ which may be either (a) oscillatory (when μ is imaginary) with finite amplitudes in x and y or (b) exponential (when μ is real) with amplitudes that increase rapidly with time.

Since μ is a function of a and q, the conditions for stable trajectories may be identified (fig. 6.11). The ratio $a/q = 2U/V$ is independent of the specific charge, e/M, and hence the operating points for all masses lie on a line that passes through the origin with slope $2U/V$. The points at which this line intersects the stability region determine the pass band of the mass spectrometer. Ions of lower mass are unstable in the x direction while those of higher mass are unstable in the y direction. In either case the ions are lost to the rods or the walls of the instrument.

A qualitative picture of the behaviour of the instrument is as follows. Lower masses tend to follow the rf component more than the higher masses. In the xz plane the dc component alone produces focusing. This may be seen for the special case where $V = 0$, from equation (6.9) which becomes

$$\frac{d^2x}{dt^2} + ax = 0 \tag{6.15}$$

and which describes a simple harmonic oscillation. When a sufficiently large rf voltage is applied, this feature is maintained, but the motion increases in amplitude and becomes unstable in the x direction. In the y direction, on the other hand, the dc field is defocusing, but a sufficiently large rf voltage opposes this and makes the motion oscillatory, intersecting the axis with a longer period. Thus the combination acts as a 'mass filter'.

The mass spectrum may be scanned by varying the rf and dc voltages (V and U, respectively, in equations (6.11) and (6.12)) in such a way that U/V is constant. In this case the frequency of the rf voltage is also kept constant. The alternate method, in which U and V are maintained constant and the angular frequency, ω, is varied, has been little used because practical difficulties are encountered in sweeping the frequency over an extended range.

The resolution may be varied electrically by an adjustment of the U/V ratio; as it is increased, the operating line in fig. 6.11 approaches the tip of the stable region located at $a_y = 0.23699$ and $q_y = 0.70600$. For an instrument with unlimited cross section and length, unlimited resolution appears possible in principle. However, in practice the finite physical dimensions of a real analyser impose limits on instrument aperture and attainable resolving power (~ 8000). The relationship between the finite dimensions and both the resolving power and the transmission have been discussed by Dawson (1976*a*). The resolving power depends on the fact that one must distinguish between the trajectories of masses M and $M + \Delta M$, that is, by the 'wavelength' of the trajectories. An empirical relationship expressed in terms of the number of cycles, n, is (Paul, Reinhard & von Zahn, 1958)

$$\frac{M}{\Delta M} \approx \frac{n^2}{12.2}. \tag{6.16}$$

The number of cycles depends on the analyser length L, the axial energy of the ions E_z, and on the applied frequency f, namely,

$$n = fL(M/2E_z)^{1/2}. \tag{6.17}$$

Special characteristics of the quadrupole instrument are the following: (a) small size and weight (since there is no magnetic field), (b) rapid scanning of the mass spectrum (see §11.3), (c) linear operation to relatively high pressure (10^{-4} Torr), (d) low source energy (< 10 eV) and (e) electrical variation of the resolving power (by varying the U/V ratio).

Measurements on the actual performance of quadrupole systems with regard to transmission, resolution and peak shapes, and comparison of

Fig. 6.11. Mathieu diagram.

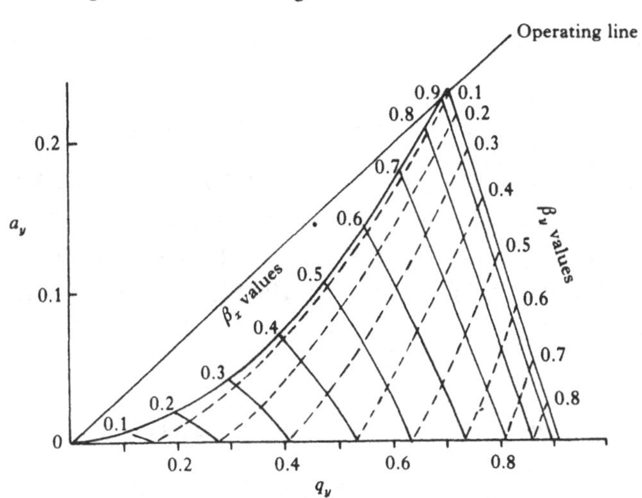

these properties with theoretical predictions, are given by Dawson (1976*b*). The principal factors which cause deterioration from ideal performance are (a) misalignment of the rods (adjustable arrangements are frequently used), (b) departure from hyperbolic fields (hyperbolic surfaces and improved approximations with wires may be used (Matsuda & Matsuo, 1977)) and (c) the build-up of surface charges which alter the voltages on the rod surfaces.

Ion sources for quadrupole instruments are somewhat modified in the light of the low ion energies and the acceptance of a wide energy band (Austin, Holme & Leck, 1976). Usually cylindrical cages formed by wires or fine mesh are used. The filament is located outside the cage. Ions are formed near the axis and extracted in the axial direction through a lens system which produces a relatively large beam cross section in the quadrupole filter.

Ions are extracted from the quadrupole field through a circular aperture and detected by a Faraday cup or electron multiplier. When the scanning voltages are swept at frequencies of ~ 0.5 kHz, the ac voltage across the output resistor may be fed into a lock-in amplifier and monitored, for example, by an oscilloscope.

The quadrupole rf voltage is provided by an rf oscillator and amplifier, the gain of the latter being modulated to produce the periodic mass scan. The dc voltage is derived by rectifying the rf voltage and the U/V ratio is adjusted by means of a potential divider.

The basic equations ((6.7) through (6.14)) permit the calculation of the major design parameters. The apex of the stability curve occurs at $a/q = 0.1679$. From the definitions of a, and q, the maximum values of the voltages are

$$V = 14.438\, M f^2 r_0^2 \tag{6.18}$$

and

$$U = 2.424\, M f^2 r_0^2 \tag{6.19}$$

for a quadrupole of radius r_0(cm), frequency f(MHz) and ion mass M(u).

The conditions for 100% transmission may be derived from the off-axial maximum displacement for ions corresponding to all phases of the rf field. These are related to the size of the entry aperture, D_m, and the expected resolution, such that

$$D_m = r_0(\Delta M/M)^{1/2}. \tag{6.20}$$

The maximum axial energy of the incident ions is calculated from (6.16) and (6.17) to be

$$E_z = 4.25 \times 10^{-2} f^2 L^2 M/R, \tag{6.21}$$

where E_z is in eV, f in MHz, L in cm, M in u, and R is the resolving power. The radial energy for which 100% transmission occurs can be computed (Dawson, 1976*a*) to be

$$E_r < 5.24 \times 10^{-3} r_0^2 f^2 M/R, \tag{6.22}$$

so that the maximum angle of acceptance, ψ_m, for the ion beam is given by

$$\tan\psi_m = 0.351(r_0/L). \tag{6.23}$$

Austin, Holme and Leck (1976) have shown that, if the constructional error in manufacturing the quadrupole electrodes is defined as

$$\text{error} = (2\varepsilon + \varepsilon')/D_m, \tag{6.24}$$

where ε is the manufacturing tolerance and ε' is any additional error deliberately introduced, then the resolution of the instrument can be shown to depend on these errors so that

$$R_{max} \propto (\text{error})^{-1.3}. \tag{6.25}$$

Variations of the basic quadrupole described here have included operation with a square wave (where no dc component is required: Richards, Huey & Hilter, 1973) and the use of a short set of quadrupoles which precede the main field, to which only the rf voltages are applied ('delayed dc ramp': Brubaker, 1968).

Quadrupole instruments have enjoyed widespread applications in vacuum technology as partial pressure analysers, in instruments which combine gas chromatographs with mass spectrometers ('GC–MS', see Chapter 11), in secondary ion mass spectroscopy ('SIMS', see Chapter 12), in upper atmosphere and space research (see Chapter 14), in plasma diagnostics (Halsted, 1974) and in other combinations (§11.10).

6.12 Monopole mass spectrometer

A simplified version of the quadrupole analyser, introduced by von Zahn (1963), consists of one quadrant of the quadrupole field. A 'V'-shaped electrode with two flat surfaces at right-angles to each other is located on the position of the surfaces of zero potential in the quadrupole field, and a cylindrical electrode is positioned to give a hyperbolic field in analogy to the quadrupole case. The negative dc potential, Φ_0, is applied to the rod; thus the y direction in this instrument is defined by the plane through the centre of the rod and the intersection of the plane electrodes (fig. 6.12).

Again the ion trajectories in the field are described by the Mathieu equation

$$\frac{d^2y}{dt^2} - \frac{\omega^2}{4}(a - 2q\cos\omega t)y = 0, \tag{6.26}$$

where

$$a = 8eU/M\omega^2 r_0^2 \tag{6.27}$$

and

$$q = 4eV/M\omega^2 r_0^2. \tag{6.28}$$

The monopole differs substantially from the quadrupole in its mode of operation, primarily because the 'V' electrode physically restricts the available region of space. The requirement that $y \geqslant |x|$ limits trajectories to small x deflections. The number of rf cycles which the ion may spend in the field is limited to less than the number of cycles in the characteristic half-beat length for the y direction. Further, ions may enter with initial conditions such that they quickly strike the 'V' electrode.

Equations (6.26) may be simplified by the approximation that $y = \bar{y}$ and the fact that a is always much smaller than $2q$, so that

$$\frac{d^2y}{dt^2} = -\left(\frac{1}{2}\right)\bar{y}q\omega^2 \cos \omega t, \tag{6.29}$$

where $\bar{y}$ is the average value of y, so that

$$y = \bar{y}\left\{1 + \left(\frac{q}{2}\right)\cos \omega t\right\}. \tag{6.30}$$

This describes small oscillations about the median distance, $\bar{y}$, from the z axis. From this, the average values of the forces exerted by the rf and dc components can be calculated. These indicate the 'strong focusing' by the rf

Fig. 6.12. Monopole mass spectrometer.

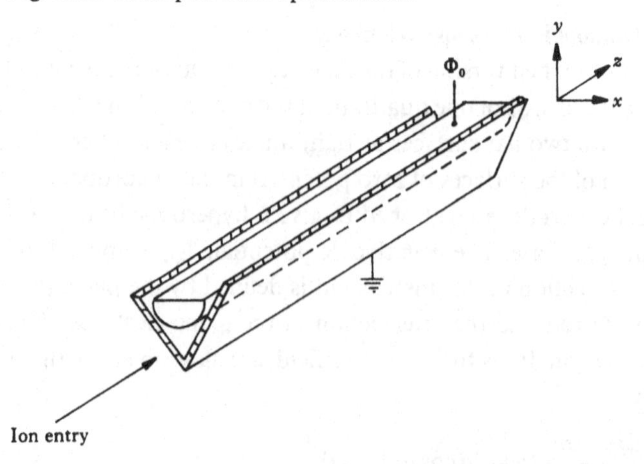

field and defocusing by the dc field and, at the transition between stable and unstable paths, produce compensating deflections (Herzog, 1976).

If we replace the actual forces by the average forces, and define

$$\beta^2 = (\tfrac{1}{2}q^2) - a, \tag{6.31}$$

equation (6.29) becomes

$$\frac{d^2y}{dt^2} = -\bar{y}\left(\frac{\beta\omega}{2}\right)^2. \tag{6.32}$$

This represents oscillatory motion if β^2 is positive, but if β^2 is negative it represents an unstable path. If the dc voltage corresponding to $\beta = 0$ is U_0, then

$$U_0 = U^*(U/V)^2 M, \tag{6.33}$$

where

$$U^* = \omega^2 r_0^2 m_0/e \quad \text{(volts)} \tag{6.34}$$

and $m_0 = 1.66 \times 10^{-27}$ kg so that, with U/V constant, a linear mass scale is obtained.

As with the quadrupole instrument, the resolution is determined by the number of cycles, n, the ion takes to traverse the field region and by the ratio (U/V). An experimental value is

$$M/\Delta M = 20(U/V)^2 n^2. \tag{6.35}$$

The length of the pole and the value of the ion accelerating potential determine n. Again there is a maximum permissible potential,

$$V_{z,max} = 0.5(U/V)^2 U^*(L/\pi r_0)^2, \tag{6.36}$$

which is usually ~ 100 V.

Then the number of cycles, n, is given by

$$n = fL(m_0/2e)^{1/2}(M/V_z)^{1/2}, \tag{6.37}$$

and the resolution (at half-maximum) is given by

$$R = (U_0/V_z)2.5(L/\pi r_0)^2, \tag{6.38}$$

an expression which reflects the increase in resolution with decreasing accelerating voltage and with increasing length. It should be noted that both of these factors also reduce the sensitivity of the instrument.

Typical values for the frequency (~ 1 MHz), field radius (~ 5 mm) and rod length (~ 20 cm) are similar to those for the quadrupole. The apertures are usually circular and located a suitable distance from the apex of the V-shaped electrode. The detector may be a Faraday cup or some variety of electron multiplier. Ion sources of the Nier electron bombardment type are used. Some work (Dawson, 1976*a*) indicates that alignment of the electrodes is less critical than is the case for the quadrupole.

Mass scanning has been performed by sweeping the rf voltage, keeping the ratio (U/V) constant, as with the quadrupole, and also by scanning the frequency. The latter method has the advantage of covering an extended mass range (von Zahn, 1963).

The principal advantages of the monopole mass spectrometer are the use of lower rf voltages and less critical dependence on the dc/rf ratios (most instruments are operated well below the critical limit of $U/V = 0.168$). The instrument has the disadvantage that it sometimes shows 'ghost' peaks near the main peaks.

6.13 Three-dimensional quadrupole trap

Another quadrupole field arrangement, having somewhat different properties, has proved to be useful for certain applications (Fischer, 1959). As shown in fig. 6.13., a three-dimensional field is produced by a ring in the shape of hyperboloid (single sheet) and two end caps, also in the shape of hyperboloids (two sheets). The field produced has rotational symmetry about the vertical axis. The field radii in the horizontal (r_0) and vertical (z_0) directions are not equal but, as required by Laplace's equation, have the relationship

$$r_0^2 = 2z_0^2. \tag{6.39}$$

A potential difference of the form given by equation (6.8) is applied between the end caps (which are connected together) and the ring electrode. For such an arrangement there is a stability diagram similar to that for the

Fig. 6.13. Quadrupole ion storage trap (QUISTOR, from Dawson, 1976*a*). The diagram is shown in cross section and has rotational symmetry about the *z* axis.

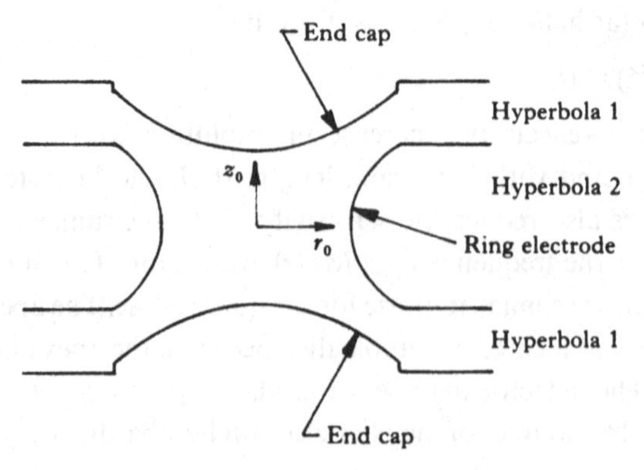

other quadrupole field instruments. Ions having a mass corresponding to the appropriate point on the (a, q) diagram are trapped in the field and circulate until removed though a collision. The properties of this arrangement, to which the name 'QUISTOR' (for quadrupole ion storage trap) has been given, have been described in detail by Dawson (1976*a*) and by Todd, Lawson & Bonner (1976).

The trap operates at a very low pressure and serves as both source and mass analyser. Electrons are injected horizontally through a hole in the ring electrode or along the z-axis through a hole in the end cap and cause ionization. The resulting ions, which have the appropriate mass, are trapped in the field and may be detected in either of two ways. One technique involves sensing the absorption of power at the ion resonant frequency in the rf circuit. In the alternate method a switched mode of operation is used in which the trapped ions are periodically pushed out of the field through a hole in the end caps by a potential applied to them. The ions are then detected, usually by an electron multiplier, in the region external to the quadrupole field.

Accurate construction of the surfaces in the QUISTOR require the use of programmable lathes, although electrodes that provide approximations to the hyperbolic field may be fabricated with wire mesh.

Ion–molecules reactions (see Chapter 10. II) have been studied, where the product ions are measured as a function of the trapping time (milliseconds). In one application the QUISTOR has been used as the source of a conventional mass spectrometer (Mather, Lawson, Todd & Baker, 1978). The reagent and reactant gases may be introduced as a mixture in which ion–molecule reactions take place. In this way the conditions for low pressure chemical ionization are effectively achieved. In this case mass spectra may also be obtained as a function of the storage time (that is, from a few μs to a few ms). Thus, there is a continuous transition from the conditions pertaining to electron impact to those for chemical ionization, as more and more ion–molecule reactions occur.

7

Determination of isotopic abundances

The investigation of the isotopic composition of naturally occurring elements has comprised, firstly, the identification of isotopes and, secondly, the determination of their relative abundances. The former has an aspect of finality in that an isotope may not be discovered more than once. On the other hand, the matter of relative abundance is one of continuing interest. For example, on-going efforts are being made to improve the accuracy with which ratios of isotopic abundances can be determined. The study of small variations in these ratios can assist in the elucidation of various processes in chemistry, geology, biology, hydrology, reactor physics, etc.

7.1 Identification of naturally occurring nuclides

The identification of naturally occurring nuclides may be regarded as almost complete if the paucity of new isotope discoveries be taken as a criterion, the last species reported being ^{180}Ta (White, Collins & Rourke, 1955).

From the point of view of nuclear theory, it is important to know which nuclear species are stable. In fact several of the premises of the modern theory of nuclear forces were inferred from regularities existing in the table of isotopes. Between hydrogen ($Z = 1$) and bismuth ($Z = 83$) there are 283 naturally occurring nuclides, subdivided as follows: 165 even (protons)–even (neutrons), 56 even–odd, 53 odd–even, and 9 odd–odd, several of the last group being weakly radioactive, albeit naturally occurring (§9.1). These statistics reveal the pairing tendencies of nucleons, and establish that there is no significant difference in stability between odd-proton and odd-neutron configurations.

The above statistics do not include the naturally occurring radionuclides ^{238}U, ^{235}U, ^{232}Th and their radioactive daughters, certain of which have been studied by mass spectrometry. Also not included are the nuclides ^{14}C, ^{10}B, ^{3}H, etc., produced by cosmic rays. While these latter are normally studied by their radioactivity, attempts to measure the abundance of ^{14}C and ^{10}B using the cyclotron as a mass spectrometer have been successful (§7.11), whilst ^{3}H was discovered using the cyclotron (§6.7).

7.2 Upper limits of isotopic abundance

An important aspect of the search for naturally occurring isotopes is the setting of upper limits for the abundance of hypothetical ones. This work was greatly furthered by the development of ion counting (§§4.8, 7.4) and, even more, by the advent of the two-stage ('tandem') magnetic analyser. With this latter type of instrument (§5.3), in which background is reduced to a very low level, White, Collins & Rourke (1956) examined a very large number of elements for hitherto undetected isotopes, and established, in many instances, limits of abundance of a few parts in 10^8. To give an unfavourable case, for example, although ^{180}Hf is 35% abundant, an upper limit of 3×10^{-9} has been set for ^{181}Hf.

7.3 Isotopic abundances by photometry

Most of the isotope abundance work done prior to 1932 employed photographic detection. Sufficient reference has already been made to the photographic effect of positive rays (§4.2) to suggest the extent of photometry and calibration which this work necessarily involved. In this activity Aston was by far the leading figure, and, undoubtedly, the tests of patience which he underwent contributed to the opinion, which he expressed on one occasion, that mass spectroscopy would die with him.

Since the positive rays affect only the surface of the photographic plate, Aston (1942) employed a very strong and rapid developer. Further, he adopted as routine a series of simple operations which could be repeated time after time without serious variation. For an element with many isotopes, such as tin or mercury, the blackening curve was obtained by making a number of exposures of different durations on the same plate. The abundances of these isotopes, once determined, were then useful as a comparison scale for studying other elements of approximately the same mass; for example, mercury provided the scale for osmium and tungsten. During the period 1930–35 Aston applied these methods of photometry to 58 elements. In 14 cases the chemical atomic weights deduced therefrom showed serious discrepancies with accepted chemical values. In these cases, that of tellurium excepted, Aston's abundances were, at least qualitatively, confirmed by subsequent revisions of the chemical values.

The technique of determining isotopic abundances by means of calibration lines simultaneously deposited on the photographic plate was developed to a high level of accuracy by Mattauch & Ewald (1943). Here, for example, a spark between brass electrodes provided both copper and zinc lines, the intensities of the former, ^{63}Cu and ^{65}Cu, lying between the more abundant ^{64}Zn and ^{66}Zn, and less abundant ^{67}Zn, ^{68}Zn and ^{70}Zn.

Since the isotopic constitution of zinc was accurately known from electrical determinations by Nier, the zinc lines provided a calibration for each plate, in the specific region of interest, by means of which the copper abundances could be derived (Ewald, 1944; Duckworth & Hogg, 1947). This method, which depends upon the existence of electrical abundance determinations for nearby elements, enabled Mattauch and his colleagues to give reliable isotope abundance figures for a number of heavier elements (Mattauch & Ewald, 1943, 1944; Ewald, 1944; Mattauch & Scheld, 1948). The stated precision of these determinations was generally in the range 0.3–1.0‰. The distribution of errors arising in the photometric determination of isotopic abundances has been comprehensively discussed by Mattauch (1951).

Whilst the photographic plate is no longer used in isotopic abundance determinations, its properties of integrating ion currents and of recording all ion currents simultaneously are being regularly utilized in quantitative studies of trace element concentrations using the vacuum spark source (§3.6).

7.4 Isotopic abundances by electrical detection

As mentioned above, modern isotopic abundance determinations invariably are based on some form of electrical detection (see §§4.3–4.7). In addition, they are made under the following stringent conditions:

(a) the accelerating voltage and magnetic field are constant to ~ 1 part in 10^5,

(b) the vacuum is 10^{-8} Torr or better, and

(c) the abundance sensitivity for adjacent masses (§5.3) is $\sim 10^6$.

Thus, the reproducibility of determination is almost entirely dependent upon the steadiness of the ion beam and the detector. Very high reproducibility of results can, in fact, be achieved. Thus, for the counting of N ions of an isotope, the reproducibility (expressed as the *statistical error*) approaches the theoretical limit of $1/\sqrt{N}$. For 10^8 ions (a current of 6×10^{-11} A integrated for 1 s), this corresponds to a statistical error of 10^{-4} (0.01%). For reasons to be discussed later, the *absolute error* may be considerably larger. Many experiments, however, are concerned only with variations in isotope abundance (between the sample and a standard), and are able to capitalize on the lower statistical error.

The sources of error in abundance measurements have been discussed by Bainbridge & Nier (1950), Bainbridge (1953), and de Bièvre (1978). Much of what follows stems from these discussions. The errors fall essentially into two categories: those associated with the method of introducing the sample into the mass spectrometer and those arising from discriminating effects within the source, analyser, and detector regions. The errors arising from introduction of the sample will be discussed separately for (a) the surface ionization source and (b) the electron bombardment source – the two sources most frequently employed for precision determinations of isotopic abundances. The other errors – which are less specific – will be discussed thereafter.

7.5 Errors caused by sample introduction

Thermal ionization sources

The main source of error arises from isotopic fractionation during evaporation. In this phenomenon, known as Rayleigh fractionation, the fractionation varies inversely as the square of the mass. Thus, for isotopes of mass m_1 and m_2 (where $m_2 > m_1$), the lighter isotope is preferentially evaporated by the factor $\sqrt{(m_2/m_1)}$ and the ratio of the lighter to heavier isotope decreases with time as the sample becomes depleted in the lighter component. This sequence of events was observed by Eberhardt, Delwiche & Geiss (1964) for Li, K and Rb. It has also been studied theoretically by Kanno (1971) with results that agree qualitatively with experiments (Moore, Heald & Filliben, 1978), but it is not yet possible to correct accurately for the effect in order to obtain true isotopic ratios. Standard synthetic isotope mixtures are now available for many elements of interest (for example, SRM 987 of National Bureau of Standards for Sr) against which the experimentally-observed ratios may be normalized (see de Bièvre, 1978). Alternatively, where three or more isotopes are available, for example ^{86}Sr, ^{87}Sr, ^{88}Sr and the ratio ($^{87}Sr/^{86}Sr$) (which is variable for different samples) is required, the experimental value can be corrected for discrimination due to evaporation (and certain instrumental effects also) by normalizing each ($^{87}Sr/^{86}Sr$) value with respect to the ratio ($^{86}Sr/^{88}Sr$) $= 0.1194$ (invariant).

The Rayleigh evaporation described above presupposes evaporation from a liquid with continuous mixing of the sample. In many cases (Inghram, 1948), such as with rare earth oxides, it is likely that the solid itself sublimates, with the result that fractionation effects are small.

Other experimental parameters that influence the observed abundance ratios are sample size, purity and acidity, ion source pressure and the length of time for the measurement (de Bièvre, 1978).

Electron bombardment source

Gases to be ionized are led from a reservoir to the ionization chamber via a constriction or 'leak'. If the pressure in the reservoir be low and the dimensions of the constriction be small (Honig, 1945), so that the mean free path of the molecules is substantially (say ten times) greater than the diameter of the constriction, the flow of gas through the leak is molecular or effusive (the molecules all having thermal energies), so that the velocity of flow varies inversely as the square root of the mass of the molecule involved. Under these conditions, the gas reaching the ionization chamber is not characteristic of the sample in the reservoir. In the ionization chamber itself, however, the pressure is such that conditions of molecular flow invariably exist. Thus, the rate of admission to, and the rate of withdrawal from, the ion source are both proportional to $1/\sqrt{M}$ and, by a fortuitous cancellation of discriminations, the composition of the gas in the chamber is the same as that in the reservoir. It should be noted that the composition of the gas in the reservoir changes with time, owing to its impoverishment in the lighter mass components. This change may be kept to negligible proportions by the use of a large reservoir. The type of leak suitable for molecular flow consists of one or more small leaks, either purposely made in a thin diaphragm, or existing in porous materials such as sintered glass discs.

If, on the other hand, the mean free path of the molecules in the reservoir be short compared with the dimensions of the leak, the flow is viscous and governed by Poiseuille's law. In this case, the rate of flow through the leak is independent of the mass of the molecule, and the ion peaks observed in the mass spectrometer must be corrected by the proper $1/\sqrt{M}$ factor to compensate for the molecular outflow from the source. This arrangement has the general advantage that the isotopic constitution of the reservoir remains constant and the particular advantage, important in certain analytical applications, that the reservoir pressure is relatively high. In practice, however, the flow will be viscous at the reservoir end, molecular at the source end, and more complex in the intermediate region. This results in a concentration of molecules at the high pressure side and hence back-diffusion. Halsted & Nier (1950) showed that this type of discrimination can be prevented by a capillary on the high pressure side of the leak and the use of suitably high pressures in the reservoir. In general, however, as for thermal ionization sources, the isotopic composition of a sample is compared to that of a standard, so that no errors are introduced if the fractionation remains constant.

7.6 Discrimination caused by sweeping the mass spectrum past the collector

A singly charged ion of mass M falling through the potential V and then entering a uniform magnetic field B follows a circular path in the magnetic field of radius given by

$$r^2 = 2MV/eB^2. \tag{7.1}$$

Thus, the mass spectrum may be swept past the collector (constant r) (a) by varying the accelerating potential (V) but keeping the magnetic field constant (voltage scanning), or (b) by varying the magnetic field (B) but keeping the accelerating potential constant (magnetic scanning).

Voltage scanning has the advantage of high speed. Thus, in a given period the data can be collected and averaged over several cycles. However, for a given ion group it is found that increasing the accelerating potential and bringing the ion group again to the collector by increasing the magnetic field leads to an increase in peak height. This is known as the 'voltage effect', that is, an increased transmission of ions caused by (a) the energy spread amongst the ions becoming relatively less important, resulting in an improvement in resolution, (b) improved extraction of ions from the source, resulting from changes in ion optics of the source and (c) reduction in lateral spreading of the beam. For ions formed by the dissociation of molecules, the first effect can be very large. Chait & Hull (1975) have proposed overcoming the voltage effect by separating the ion source into two discrete regions – one in which the ions are produced and the other in which they are accelerated.

Magnetic scanning avoids the voltage effect but its speed is slow, being determined by the response time of the analyser magnet. In addition, in mass spectrometers with an auxiliary magnet in the source to collimate the ionizing electron beam (electron bombardment source), the ions reaching the collector will not have traced identical paths, a desirable condition if discrimination is to be avoided (Bainbridge, 1947). The conditions for identical paths have been given by Swann (1931) and Bleakney (1936) (§8.3).

When the voltage effect is kept small by employing high accelerating potentials (2000–4000 V) and by avoiding ions which are the products of dissociation, the two methods of scanning yield results which agree to within 1% (Nier, 1950*a*, *b*).

7.7 Other sources of error in abundance determination

Background currents limit the precision of isotope abundance determinations. The problem is rendered more acute when extremely small quantities of sample are being analysed (for example, nanogram quantities). The 'analytical blank' in this case is high and variable and includes, for example, contributions from reagents (used to purify and isolate the sample), the environment (for example, lead from automobile exhausts) and from the source (for example, sample filaments). Proper laboratory procedures can control the external 'blank' contributions, and sufficient baking of the source (and sometimes analyser) can eliminate the instrumental contributions, as well as 'memory' effects that stem from previous analyses. The scattering of ions of neighbouring peaks from either the walls of the analyser or from molecules of the residual gas, as indicated by the abundance sensitivity (§5.3), must be reduced to sufficiently low levels.

Errors in abundance ratio measurement also arise in the ion detection process. Thus the Faraday collector, under ion bombardment, emits a significant number of secondary electrons. The extent of this emission is strongly dependent on the mass and energy of the incoming ions (§§4.3, 4.4). This leads not only to an enhancement of the observed ion currents but to a change in their relative values with changing conditions. This effect may be eliminated by a negative suppressor grid between the exit slit of the analyser and the collector. This grid repels the secondary electrons back to the collector. In a 180° analyser there is no problem as the secondary electrons return to the collector under the influence of the ambient magnetic field.

The input resistors of electrometer amplifiers may be non-linear, displaying a positive or negative voltage coefficient. If this phenomenon be present, the resistances may be calibrated *in situ* by the method of Bainbridge & Ford (1953). In this method accurately known induced currents are fed into the detection system, using the collector-to-suppressor capacitance. Further, care must be taken to insure the accuracy of amplifier gain changes (de Bièvre, 1978).

When an electron multiplier is used, the output current obtained by integration of the ion pulses is mass dependent, varying approximately as the square root of the ion mass ($\sqrt{M}$). It also shows irregular changes, especially on exposure to vacuum or to ion bombardment, as well as non-linearity effects. The mass discrimination can, in principle, be avoided by counting the ion pulses from the multiplier, small variations in multiplier gain being also avoided, so long as the pulses are greater than the threshold of the electronic discriminator. This procedure works even better for the Daly detector as the spectrum of pulse height amplitudes is practically all above the noise level of the discriminator. Two aspects of ion counting merit special comment: (a) the statistical error may limit the smallness of the sample that can be used for quantitative measurement, for example for an accuracy $\sim 10^{-4}$ one has to count $\sim 10^{8}$ ions; assuming a mass spectrometric sensitivity $\sim 10^{4}$ atoms/ion, one must have a sample of $\sim 10^{12}$ atoms (for example, ~ 0.2 nanograms of Sr^{+}); (b) the dead-time τ of the electronic counting system relates the actual (n_a) and measured (n_m)

counting rates by the expression

$$n_a = n_m/(1 - n_m\tau). \tag{7.2}$$

Thus, in the above example, if one measures 10^4 ions/s, the dead-time for 0.01% accuracy is $\sim 10^{-8}$ s (Nguyen & Goby, 1978).

7.8 Counteracting the effect of fluctuating ion beams

We describe briefly two systems that are employed for the precise determination of isotopic abundance ratios. These are, respectively, the double collector system and the beam switching system. The former is employed mainly with gaseous samples and the electron bombardment source, whilst the latter is usually employed with solid samples from the surface ionization source. Both methods, of course, can be employed with either type of source, the choice depending upon the sample under study.

The double collector provides simultaneous detection of two isotopic beams. It was first utilized by Straus (1941) to ascertain the isotopic constitution of nickel, employing a vacuum spark (§3.6) – a type of source which is notoriously unsteady. By collecting two isotopic beams simultaneously, Straus obtained a ratio of the two beams which was relatively insensitive to large fluctuations in the individual beams. Modern systems are employed even with stable ion beams in the accurate study of small variations in isotopic ratios. Typical of one of the analog methods employed for double collection is the one shown in fig. 7.1 (Nier, Ney & Inghram, 1947; Inghram & Heyden, 1954). Here, ions of mass numbers M_1 and M_2 are incident on a collector plate and a collector cup, respectively, each provided with its suppressor electrode, whilst a coarse diaphragm

Fig. 7.1. A dual-collection arrangement showing C_1 and C_2, the final and slit collectors, respectively, the guard electrodes, G, the electron suppressors, S, a portion of the spectrometer tube, T, the grid resistors, R_1 and R_2, the amplifiers, A_1 and A_2, the null indicator, N, and the balancing low resistance, R. (After Inghram & Hayden, 1954.)

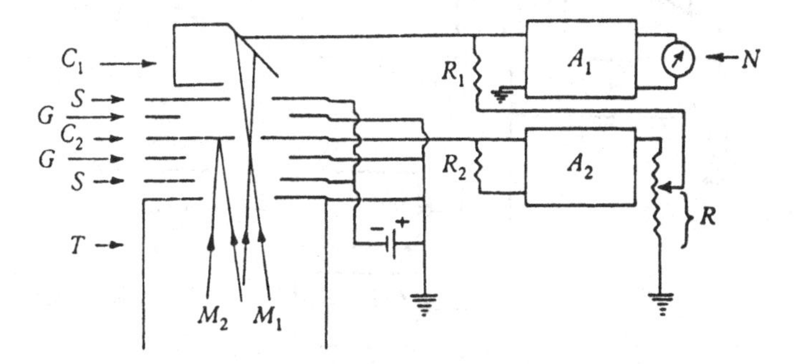

transmits the ions of the mass numbers desired. The ratio of the ion currents is determined as follows: the voltage developed by the larger ion current across its grid resistor R_2 is transformed by the feedback amplifier to a low resistance R, where a known portion of it may be used to balance the voltage developed by the smaller current across its own grid resistor R_1.

The introduction of electronic digital techniques has also provided considerable improvement in the accuracy with which isotopic ratios can be determined. The key component here is a voltage-to-frequency converter in which the output voltage due to the ion current is converted into a frequency which can, of course, be determined to a much higher accuracy (<1 part in 10^6). The principle of the digital measurement of ion beam ratios is indicated in fig. 7.2 (Nielsen, 1968). The voltage-to-frequency converters provide the output frequency f_A and f_B, which are directly proportional to the output voltages. As seen, f_A is fed to the normal input of the counter, while f_B is fed in place of the 'clock' (external time-base input) so that the counter records the ratio (f_A/f_B), which can be taped. The main advantage is that the signal voltages are integrated over the whole counting time (in contrast to measurements by a normal digital voltmeter).

A similar system, which relies on the capacitive integration of the ion beam for each Faraday cup of a dual ion beam system, has been described

Fig. 7.2. Principle of digital measurement of ion beam ratios. (After Nielsen, 1968.)

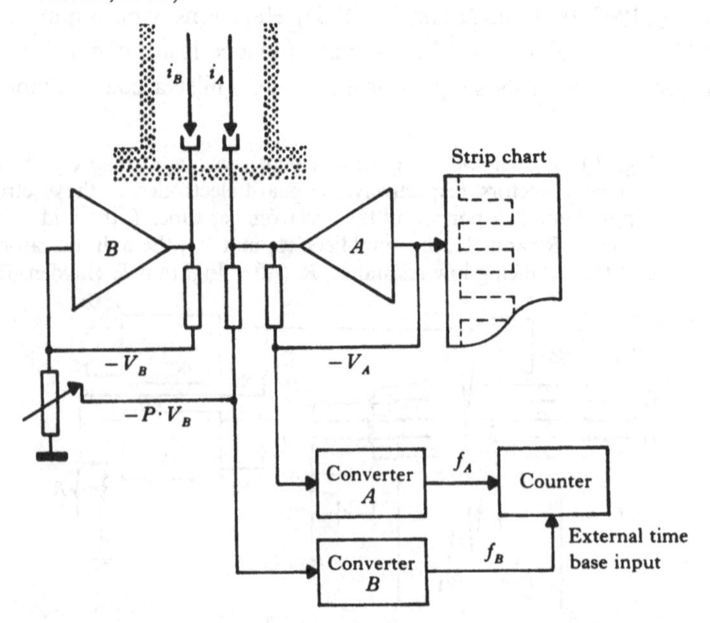

by Jackson & Young (1972). The capacitors are connected from input to output of the electrometers which have field effect transistor (FET) input stages. The capacitors charge linearly and the ratio of the electrometer output voltages is monitored by a digital voltmeter.

When one is concerned only with a change in the isotope abundance ratio between a sample and a standard, both of which are in vapour form and of identical chemical composition, a further improvement is effected by a *dual inlet system*, comprising two gas lines and two leaks connected to the source through a magnetically operated valve, by which each sample is led alternately to the source chamber or the waste vacuum (for example, McKinney *et al.*, 1950).

In the dual beam arrangement of Nielsen (1968), described above, the two collectors are Faraday cups, and the dual inlet valve is switched every 20 s, the counting time being 10 s (to allow for decay of memory effect). In the system of Jackson & Young (1972) the sample change time was 10–60 s, while the measurement time was 20–100 s.

The beam switching method is usually employed when the relative abundances of three or more isotopes are to be determined. As the name implies, the magnet current (or accelerating voltage) is changed, in discrete steps, so as to bring the desired ions, in turn, to the exit slit. The instrument is adjusted for flat-topped peaks, so that the peak heights of the selected isotopes are recorded digitally (along with the background values if required). For changing the magnetic field, one uses a Hall probe, whose output determines (via a feedback circuit) the required magnetic field settings. The currents or counting rates are integrated over precise time intervals (1–10 s, usually 1 s for each peak) provided by electronic counters. The readings of a large number of cycles of repetitive scanning are normally handled by a digital voltmeter interfaced with a computer.

When corrections are applied for (a) beam variation with time, (b) discrimination effects (for example, via normalization with respect to ratio of stable isotopes, as $^{86}Sr/^{88}Sr$ for calculation of $^{87}Sr/^{88}Sr$) and (c) interference with isobaric peaks (for example, ^{87}Rb and ^{87}Sr), present-day instruments routinely attain accuracies of 10^{-5} in the isotopic ratios.

7.9 Isotope dilution technique

The isotope dilution method was developed (see Inghram, 1954) for ascertaining quantitatively the presence of low-level elements in a sample.

The method involves spiking the unknown sample with a known number of atoms of one or more of the isotopes of the element under study. When the spiked sample is analysed in the mass spectrometer, and the effect of the spike taken into account, the number of atoms of each of the isotopes in the original sample may then be calculated. Thus, in the simplest case in which the spike is enriched in one isotope only, and the ratios of this isotope to a normal (invariant) one are S in the spike and N in the unspiked sample, respectively, a mixture of x atoms of spike and y atoms of sample yields the isotopic ratio

$$R = \frac{xS + yN}{x + y}. \tag{7.3}$$

This y (the unknown) can be ascertained from measurements of R, as values of x, S and N are known.

A great virtue of the method is its high sensitivity: $\sim 10^{-12}$ g of alkali elements have been observed, and even less for rare gases (Reynolds, 1956). Also, interference effects are small, inasmuch as the isotopic spike can usually be chosen to avoid them, or conditions in the ion source can be suitably adjusted (for example, for the disappearance of ^{87}Rb during measurement of ^{87}Sr, which is surface ionized at higher temperatures). The limitation to the use of the method is usually the background or contamination from chemicals or the atmosphere.

The main applications of the isotope dilution method are to nuclear physics and to geology. Thus, Thode and his colleagues (for example, Petruska, Thode & Tomlinson, 1955) used a spike of ^{128}Xe (produced by neutron absorption in ^{127}I) with a mixture of fission-product xenon to determine the absolute yields of the fission-product xenon isotopes. Several investigators used the method to determined half-lives and/or branching ratios of a variety of long lived radioactive species, for example, Reynolds (1950) determined the relative abundances of ^{40}Ar and ^{40}Ca resulting from the branching decay of ^{40}K. Other examples in nuclear and geophysics are given in Chapters 9 and 13, whilst a review of a much wider spectrum of applications of this powerful technique is given by de Bièvre (1978).

7.10 Absolute determinations of isotopic abundances

In 1950 Nier (1950*a*, *b*) reported redeterminations of the isotopic constitutions of a number of elements with an absolute precision of $\sim$0.1–0.2%. The mass spectrometers used in these studies were calibrated for mass discriminating effects by means of carefully prepared synthetic isotope mixtures of ^{36}Ar and ^{40}Ar. The results of this work were of great interest not only because of their high accuracy, but also because they provided irrefutable evidence concerning the absolute accuracy of conventional abundance determinations.

This new work agreed, in general, with older work within the 1% accuracy claimed for the latter, but indicated that the type of leak in common use had not been permitting molecular flow, as had optimistically been assumed. This led to the study of leaks by Halsted & Nier (1950) to which reference has been made in §7.5.

The elements studied by Nier were carbon, nitrogen, oxygen, neon, argon, potassium, krypton, rubidium, xenon and mercury. These elements were then available as substandards for calibrating other mass spectrometers.

Subsequently, synthetic isotopic mixtures were made for many other elements. Notable for accuracy of preparation were the $^{235}U/^{238}U$ mixtures prepared at the National Bureau of Standards (Garner, Machlan & Shields, 1971), which permitted the ^{235}U content of enriched samples to be determined to an accuracy of 0.1%. A census of elements whose isotopic compositions have recently been studied in this way is given by de Bièvre (1978), and includes hydrogen, lithium, boron, neon, magnesium, silicon, chlorine, potassium, vanadium, chromium, copper, bromine, rubidium, silver, cadmium, rhenium and lead.

Finally, it will be obvious that absolute isotopic abundances plus accurate isotopic masses (Chapter 8) provide the data needed for the calculation of accurate 'atomic weights'. Accordingly, the official body which recommends on atomic weights – the *Commission on Atomic Weights and Isotopic Abundances* of the International Union of Pure and Applied Chemistry – reviews biennially the available absolute isotopic abundance data with this purpose in mind (for example, *Commission on Atomic Weights*, 1984*a*, *b*).

7.11 Mass and charge spectroscopy

As suggested in §5.3, abundance sensitivity in conventional instruments is limited by several factors. In particular, these are: (a) unresolved molecular or isobaric ions, (b) unresolved ions having multiple charges, (c) small-angle elastic scattering of nearby intense ion currents, (d) metastable decomposition and large-angle scattering from walls or residual gas into the detector region, (e) low-probability charge change events by which undesired ions arrive at the detector. Because of these factors the abundance sensitivity in conventional instruments is limited, as described in §5.3.

Recently a specialized technique has been developed to determine the abundance of isotopes where the required abundance sensitivity is $\sim 10^{15}$. Several relatively long-lived isotopes are known to exist in nature at such low levels, the most familiar of which is ^{14}C whose decay provides the basis for carbon dating of organic materials. Other isotopes which have been observed are ^{10}Be, ^{26}Al, ^{36}Cl and ^{129}I. While the usual way in which the abundances of such isotopes have been studied is by their decay in a very low background environment, it was appreciated early that a method by which the *undecayed* atoms could be detected would have an inherent potential for improved sensitivity. A new mass spectrometer method (fig. 7.3), which incorporates an accelerator as one element in the system (Muller, 1977; Nelson, Korteling & Stott, 1977; Bennett *et al.*, 1977; Litherland *et al.*, 1981; Purser *et al.*, 1981) yields the required abundance sensitivity and is sufficiently related to the classical work to be described briefly here.

While both cyclotron and tandem Van de Graaff accelerators have been used in systems for this work, it appears that there are certain advantages to the Van de Graaff system illustrated schematically in fig. 7.3. Negative ions of the sample material are formed in the sputter ion source. The use of negative ions allows some isobaric separation to be made: for example $^{14}C^-$ is stable whereas $^{14}N^-$ is unstable. In the first magnetic analyser, elastic scattering of negative ions at large angles is orders of magnitude smaller than that for positive ions, inasmuch as an electron is almost always detached in a collision. In the acceleration and dissociation stage the ions pass through a gas stripper and emerge as *atomic* ions with a charge state of 3+. The electrostatic deflector determines the energy of

Fig. 7.3. Schematic diagram of system for mass and charge spectroscopy based on a tandem Van de Graaff. (After Purser *et al.*, 1981.)

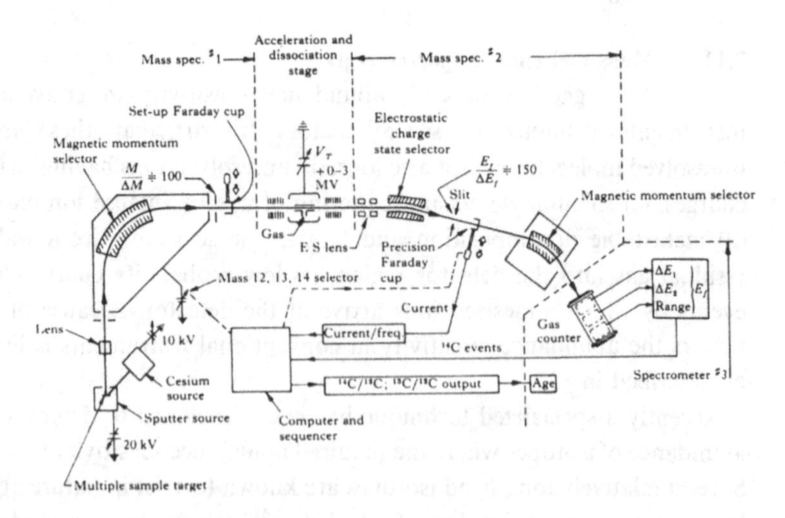

the ion that is transmitted through the system, and thus determines the charge state. In the final magnetic analyser the ions have energies in the MeV range where the differential elastic scattering cross sections are orders of magnitude lower than in the keV range. Moreover the operation of the stages in the MeV range gives rise to a substantial compaction in phase space and allows high transmission (several tens of percent). Finally the particles are detected by an array that records E, dE/dx and the range, so that a given particle is identified according to its atomic number (for $Z < 20$).

The detector background count rates for the entire system are typically much less than one count per hour. For the study of ^{14}C, this system detects a few percent of all the ^{14}C atoms present in the consumed sample at a concentration of a few parts in 10^{15}.

Ages of organic material determined by this method (sometimes called 'Accelerator Mass Spectrometry') are accurate to about one percent, and are particularly useful in archeology.

8

Determination of atomic masses

8.1 Atomic masses and nuclear stability

The mass of an atom is less than the combined masses of its constituent protons, electrons and neutrons. It is this difference between the deduced and observed masses, known as the 'binding energy', which accounts for the stability of the atom. Since most of the binding energy is associated with the nuclear constituents (protons and neutrons), or 'nucleons', it is customary to regard this quantity as a measure of the stability of the nucleus alone. A widely used concept, the 'binding energy per nucleon', is defined as

$$\text{be/nucleon} = \frac{[Z^1H + (A - Z)n] - {}^A_ZM}{A}, \tag{8.1}$$

where ${}^A_ZM$ represents the mass of an atom of mass number A and atomic number Z, and ${}^1H$ and n are the masses of the hydrogen atom and neutron, respectively. Since the binding energies of the orbital electrons, here practically neglected, are not only small, but increase with Z in a gradual manner, the be/nucleon gives a accurate picture of the variations and trends in nuclear stability. Furthermore, a knowledge of atomic masses is the most useful single type of information in predicting the details of nuclear transformations and disintegrations.

8.2 The doublet method of mass comparison

The mass spectroscopic comparison of atomic masses is usually accomplished by studying 'doublets'. A doublet is a pair of mass spectral lines produced by two species of ions whose specific charges are almost, but not quite, equal. Thus, for example, a doubly charged ion, which appears on a mass spectrum at a point corresponding to one-half its mass, may form a doublet with a lighter, singly charged ion. If the mass of one of the doublet members be known, the mass of the other may be computed from a knowledge of the doublet spacing and the dispersion of the instrument.

Prior to 1951, mass spectrographs were used exclusively in such studies, that is, the mass spectral peaks were recorded as lines on a photographic plate, as in fig. 8.1. In this case, the lines of interest were located with a travelling microscope, including those used to calculate the dispersion in the mass region under study. In principle, it was possible to compare the masses of atoms, even if their mass spectral lines were widely separated. However, in practice, it was difficult to achieve a uniform dispersion over a large distance, whereas it was much easier to do so for the region represented by the doublet spacing.

Mass determinations based on doublets recorded on a photographic plate played an important role in the history of atomic mass determinations for four decades (Duckworth, 1958). However, in the light of developments discussed below (§§8.3, 8.4), photographic detection has now been superseded almost completely. The last reported use of a mass spectrograph was that of Demirkhanov, Dorokhov & Dzkuya (1972).

In 1951, Nier and his colleagues (Nier, 1953) at the University of Minnesota introduced electrical detection for the determination of atomic mass differences. This was done in the 'Nier–Johnson' mass spectrometer described in §5.8. As electrical detection requires steady ion currents and low ambient electrical noise, an electron bombardment source was employed (§3.3). The precision achieved in this work led others to develop systems for electrical detection and sources appropriate to its use. And there was a corresponding shift to the use of singly charged ions almost exclusively. Typical of the doublets used in modern atomic mass compari-

Fig. 8.1. (*a*) Doublets at mass numbers 26 and 27 between hydrocarbon fragments and doubly charged metallic ions. The source of ions was a high frequency spark in hydrocarbon vapour between stainless steel electrodes. (Duckworth & Johnson, 1950.) (*b*) Multiplet at mass number 16 obtained with resolution of approximately 1/100000. (Courtesy of Professor J. Mattauch.)

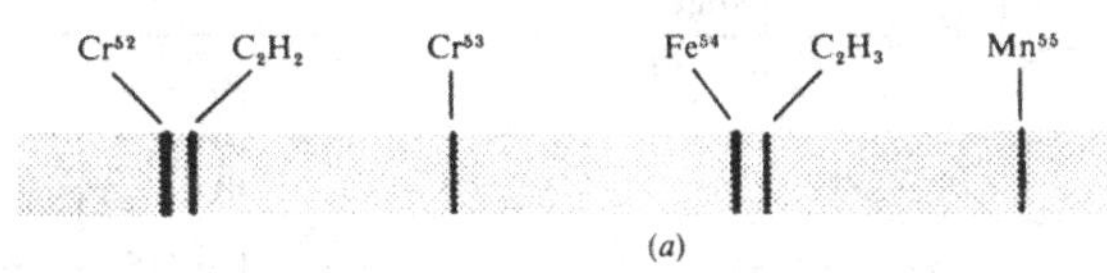

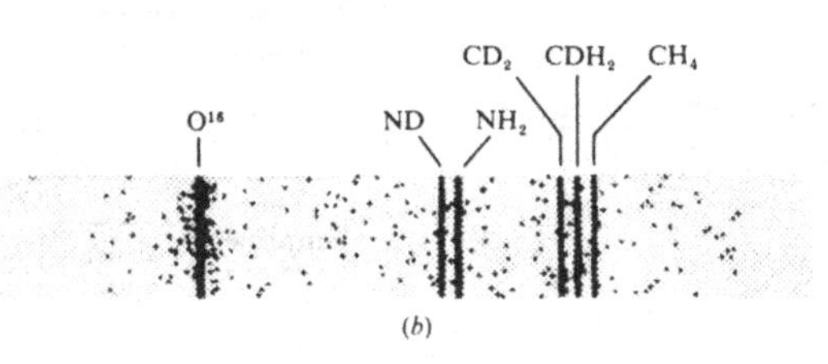

sions are those of the following family:

$$^{A}M^{+} - {}^{A}X^{+} = \Delta M_1, \tag{8.2}$$

where M is a hydrocarbon molecule and

$$({}^{A+2}X^{35}\mathrm{Cl})^{+} - ({}^{A}Y^{37}\mathrm{Cl})^{+} = \Delta M_2, \tag{8.3}$$

where X and Y may or may not be the same element. Doublet spacings are now routinely determined using some variation of 'peak matching'.

8.3 Peak matching in deflection instruments

The use of electrical detection (§5.8) makes possible the achievement of a dispersion law that is a function of a readily accessible voltage. This powerful technique depends on a general theorem by Swann (1931) and Bleakney (1936) which holds for any combination of electric and magnetic fields.

Suppose that an ion of mass M, initially at rest, traverses a particular path through the mass spectrometer. Then an ion of mass M', also initially at rest at the same point, will traverse exactly the same path through the instrument provided that all magnetic fields remain constant and all electric fields, E_i, are changed in magnitude to E_i', such that

$$ME_i = M'E_i'. \tag{8.4}$$

Accordingly, if a potential V_j is applied initially to the jth electrode, it

Fig. 8.2. Schematic diagram of peak matching apparatus (Kozier *et al.*, 1980).

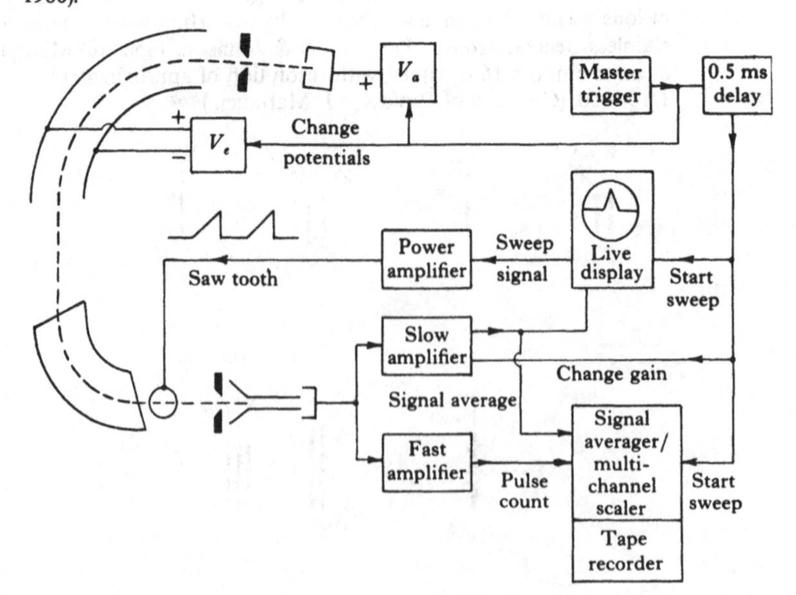

must be changed to V'_j so that

$$M V_j = M' V'_j. \tag{8.5}$$

This theorem forms the basis of the 'peak matching' technique, which was originally used by Smith & Damm (1953*a*, *b*) with their mass synchrometer (§6.9) and subsequently modified (Giese & Collins, 1954; Nier, 1957) for deflection instruments. A typical arrangement for peak matching is shown in schematic form in fig. 8.2. The master trigger produces regular pulses which start the oscilloscope sweep circuit. The saw tooth voltage from the oscilloscope controls the output current from the power amplifier. This current produces a small saw tooth magnetic field which modulates the ion beam across the collector slit. Ion current passing through the collector slit is detected by an electron multiplier whose output is amplified and displayed on the oscilloscope from which the saw tooth originated. In this way a peak corresponding to a given ion of mass M is generated.

If, in accordance with the Swann–Bleakney theorem, the relevant potential, V, is switched by the amount given by equation (8.5), then the peak corresponding to M' appears exactly superimposed on the reference peak, M. Before doing this, however, the height of the M' peak may be maximized by changing V_a by an amount

$$\Delta V_a = V_a - V'_a, \tag{8.6}$$

which can be calculated from an approximate knowledge of

$$\Delta M = M' - M. \tag{8.7}$$

(Since the mass spectrometer is double focusing, the peak *position* is insensitive to this change in the accelerating potential V_a.) Now the potential applied to the electrostatic analyser is changed by the amount

$$\Delta V = V - V', \tag{8.8}$$

in order to bring M' to the same position on the screen as M. Equation (8.5) may be rewritten, with the help of the definitions in equations (8.7) and (8.8), in the convenient form

$$\Delta M/M' = \Delta V/V. \tag{8.9}$$

Thus, at the matched condition, the ratio of $\Delta V/V$ is measured and the mass difference of the doublet is calculated.

Several techniques have been developed for the precision determination of the matched condition. We shall now discuss the more useful of these.

8.4 On-line methods of peak matching

In early peak matching work, the exceptional power of the eye to recognize lack of coincidence was exploited, originally with the mass synchrometer by Smith & Damm (1953*a*, *b*, 1956) and subsequently with deflection instruments by Nier and his coworkers (Giese & Collins, 1954; Nier 1957) and by others (Barber *et al.*, 1964; Moreland & Bainbridge, 1964; Matsuda, Fukumoto & Matsuo, 1967; Ogata *et al.*, 1967). On alternate sweeps of the display oscilloscope (fig. 8.2) the trace was displaced vertically and the gain of the slow amplifier was switched so that the peak heights of the two members of the doublet appeared the same (fig. 8.3). The operator adjusted the magnitude of ΔV until, in his opinion, the two members of the doublet were aligned on the oscilloscope screen. Permutations of the display and switching conditions (for example, direction of the scan of the spectrum) were made and the unweighted average of the values for ΔV was taken to be the value for a given 'run'. In this way one attempted to eliminate systematic bias on the part of the operator (Barber *et al.*, 1964) for the value of that run. A set of perhaps 20 runs was taken to establish the final value for the doublet spacing.

It should be added parenthetically that one of the important advantages of such a 'live' oscilloscope display is that it can be used in the focusing of the instrument (Barber *et al.*, 1964). The display is an instantaneous profile of the ion beam at the collector slit. In the double focusing instrument, for example, *velocity focusing* can be tested by changing, on alternate sweeps, the accelerating voltage by, say, 10–20 V and requiring that a peak remain

Fig. 8.3. Matched doublet with visual matching technique.

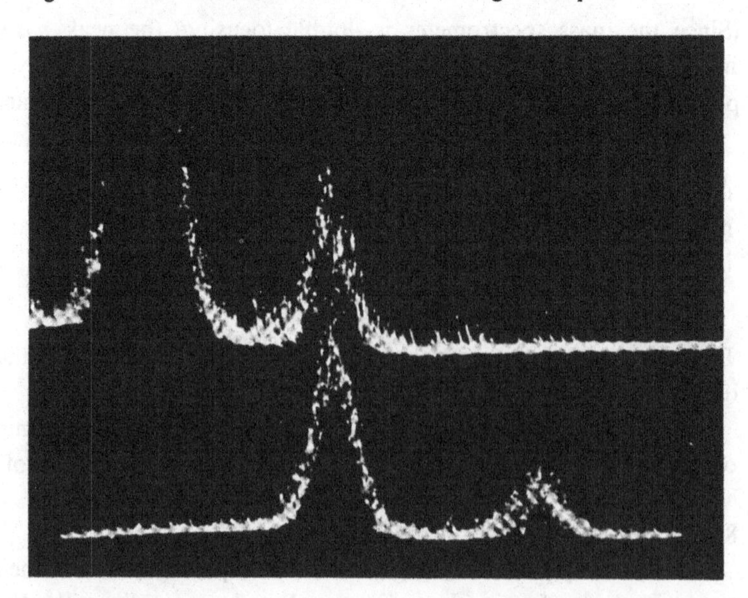

matched to itself (fig. 8.4), whilst the degree of *direction focusing* is indicated by the resolution (fig. 8.5), provided the spread in ion energy is not unusually large. In fig. 8.4, the instrument displays velocity focusing but not direction focusing, whilst in fig. 8.5 the reverse is true. The geometry of the instrument may be adjusted so that the two foci coincide (fig. 8.6). For example, a change in l'_e changes the position of the direction focus

Fig. 8.4. Velocity focusing without direction focusing. The ion energy for the upper trace is 10 eV greater than that for the lower one.

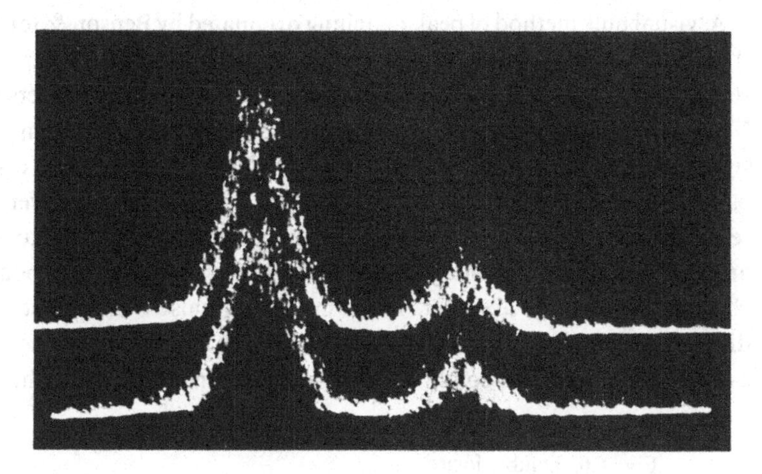

Fig. 8.5. Direction focusing without velocity focusing.

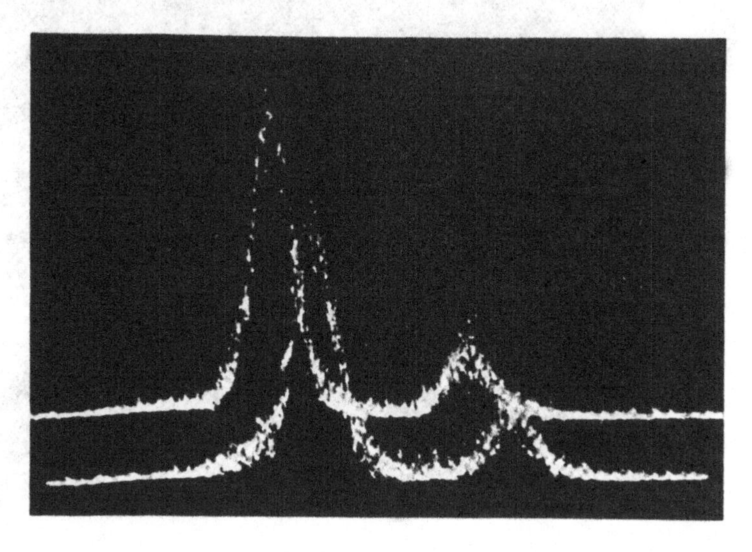

while leaving the velocity focus unaffected (see Barber *et al.*, 1964; Barber *et al.*, 1971 for the application of this to particular instruments). Similarly, the alignment of the principal slit, which involves a lengthy procedure when photographic detection is used (Ewald, 1953), becomes a relatively simple adjustment with electrical detection and a 'live' display.

In more recent work, the peak matching technique has been improved by signal averaging methods in which the spectral information is stored in a digital memory. These methods benefit from the fact that, as n scans of the spectrum are added to the memory, the signal-to-noise ratio is improved by the factor of $\sqrt{n}$.

A 'visual null' method of peak matching originated by Benson & Johnson (1966) was the first such technique in which signal averaging was exploited. It is still in use both at the University of Minnesota and the University of Manitoba (Sharma *et al.*, 1977; Kozier *et al.*, 1980), but with much improved signal averagers. In the current version of this method, the display oscilloscope trace (fig. 8.2) is divided into, say, 2048 segments. Over each segment the signal voltage is integrated and a number from -64 to $+64$, and proportional to the integrated voltage, is stored in the corresponding channel of the signal averager. The same timing pulse is used to start the display oscilloscope sweep and the sweep through the memory of the signal averager. The matched condition may be determined by adding the peak

Fig. 8.6. Double focusing.

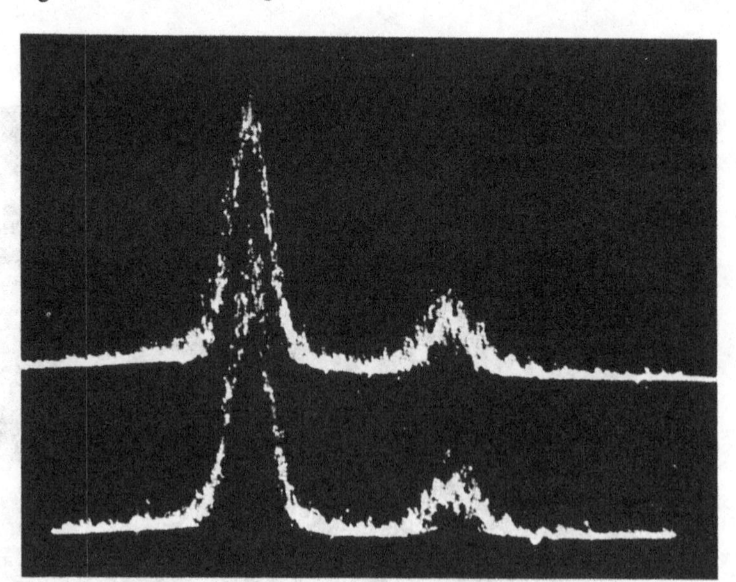

corresponding to M to the memory on odd-numbered sweeps, and then subtracting the peak corresponding to M' from the memory on even-numbered sweeps. If M and M' are normalized at the matched condition by switching the gain of the slow amplifier, then a symmetrical noise signal results (fig. 8.7). If M is displaced from M', then an 'S'-shaped error signal is accumulated (fig. 8.8). The appropriate change in ΔV required to achieve

Fig. 8.7. Visual null method, at the matched condition.

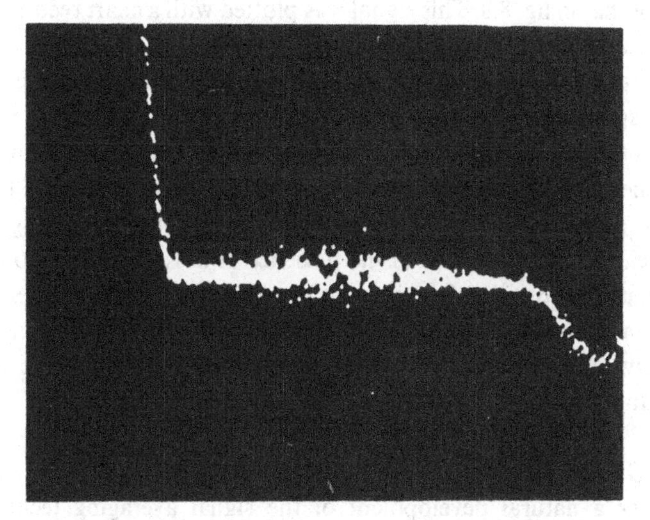

Fig. 8.8. Visual null method, mismatched by 10% of the peak width.

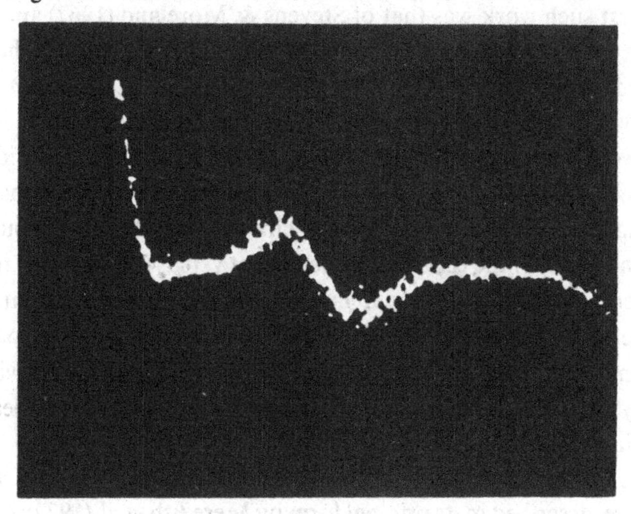

the matched condition is indicated by the amplitude and phase of the error signal. Again permutations of the switching and display conditions are used to eliminate, as far as possible, the effect of systematic biases.

A similar, on-line error signal method was used by Stevens & Moreland (1967). For a particular ΔV, the peaks corresponding to M and M' were accumulated, each in one of the two halves of a 400 channel multiscalar. A digital spectrum stripper was then used to normalize the peaks and take the differences between them and thereby to achieve an S-shaped error signal like that in fig. 8.8. This signal was plotted with a chart recorder and the amplitude of the S-shaped error signal was measured. A linear plot of the amplitude of the error signal as a function of ΔV was used to derive the ΔV corresponding to the matched condition.

Yet another method, used by Bainbridge and his colleagues (Bainbridge & Dewdney, 1967; Kerr & Bainbridge, 1970) involved the use of a lock-in amplifier rather than a signal averager. This device, which was designed to detect weak repetitive signals in the presence of noise, was used to locate and indicate the position of the centre of the ion peak relative to its position for zero modulating magnetic field. The ion peaks M and M' were thus alternately adjusted to arrive at the detector at the zero value of the modulating field.

8.5 Off-line methods of peak matching

As a natural development of the signal averaging techniques, computer analysis of the information stored in the digital memory has been exploited.

The first such work was that of Stevens & Moreland (1967) and was an alternate method to the 'spectrum stripping' described above. The procedure used the same raw data, with peaks corresponding to M and M' stored in the two halves of a 400 channel multiscalar when voltages V, $V+\Delta V$, respectively, were applied to the electrostatic analyser. The centroids of the two peaks were calculated, plus the difference in their positions, corresponding to the particular value of ΔV. This was repeated for various values of ΔV and a plot of the centroid displacement versus ΔV was used to derive the ΔV corresponding to the matched position. Second and third moments of the peaks were also calculated, so that peak shapes could be compared. Peak limits were established at five channels outside the points where the intensity was 10% of the peak height. The width of a peak was ~ 100 channels.

At the University of Manitoba extensive use has been made of a similar procedure, described in its original form by Meredith *et al.* (1972), and in an improved version by Sharma *et al.* (1977). In more recent work, the signal averager (fig. 8.2) is operated as a multiscalar, divided into four quarters of 1024 channels each. During one scan of the live display the averager sweeps through one quarter of the memory with a dwell time of 11 μs per channel, so that one scan takes 11.26 ms. Pulses from the electron multiplier pass through the fast amplifier and are entered directly as individual counts into the particular channel that happens to be open.

The reference peak, M_H, is scanned and stored in the first quarter of the memory (fig. 8.9). Then in the second, third, and fourth quarters, the 'unknown' peak, M_L, is scanned by changing the electrostatic analyser potential, V, by the amounts $\Delta V - \delta V_1$, ΔV, and $\Delta V + \delta V_2$ so that the three locations of M_L bracket the matched condition. The $\delta V_1, \delta V_2$ are chosen to correspond to $\sim \frac{1}{10}$ of a peak width, so that the relationship between peak position and voltage is linear to very high precision. Again, a linear least squares fit of the centroid positions to the voltages is made and the voltage difference corresponding to coincident centroids is calculated.

In the methods where the centroid is taken to define the peak position,

Fig. 8.9. Contents of the memory, computer matching method. M_H is the reference peak and M_L, the lighter member of the doublet, is being matched to M_H.

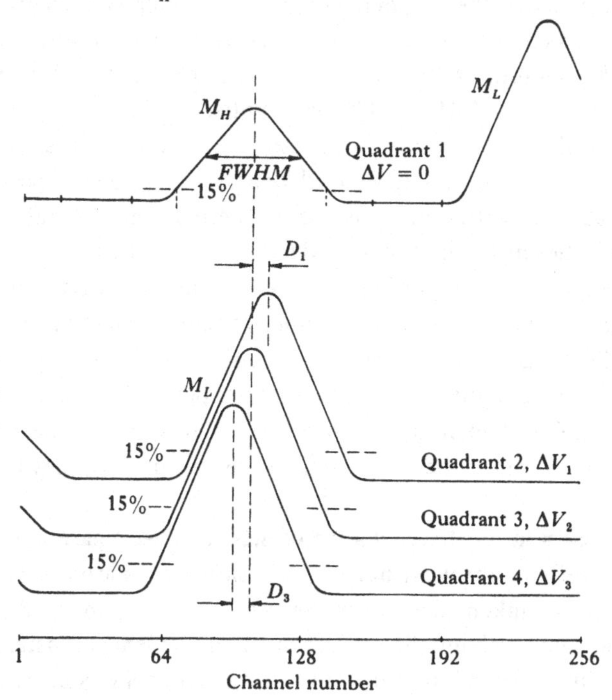

the information about the *shape* of the peak (for example, higher moments) is not normally used. Alternative procedures, which do use the information about the shapes of the peaks have been developed, but have not been used widely to date. Two such methods may be described by referring to the data illustrated in fig. 8.9. In each method, the peaks corresponding to M' are multiplied by a constant so as to match the intensity of peak M.

In the first method ('least differences', Barber *et al.*, 1976; Southon, 1973), the differences between, say, the peaks in the first and second quarters are taken to obtain the numerical equivalent of the error signal in the 'visual null' method (fig. 8.8) described earlier. The peak in the second quarter is then artificially displaced and the procedure is repeated for each of a set of these displacements. Then the displacement for which the sum of the absolute values of the differences is a minimum is calculated. This gives the displacement of the peak in the second quarter relative to the reference peak in the first quarter; it corresponds to the voltage $\Delta V - \delta V_1$. The calculation is repeated using the peaks in the first and third, and then first and fourth quarters. The displacements calculated for each of the second, third and fourth quarters are related in a linear fit to the appropriate voltages and the value for the matched condition is obtained.

In the second method, the peaks are matched in a similar fashion, but with the criterion that the sum of the *squares* of the differences between the normalized peaks be a minimum (Kayser, Britten & Johnson, 1972; Kayser, Halverson & Johnson, 1976; Kozier, 1979). The group at the University of Minnesota (Kayser *et al.*, 1972, 1976) has pointed out that, mathematically, unresolved peaks do not differ from resolved ones. A generalized peak matching procedure has been developed for a multicomponent spectrum in which Laplace transform theory is used to derive peak separations and relative intensities by a least squares calculation. In particular, the method has been demonstrated for partially resolved and unresolved doublets. It is required experimentally that the spectrum be swept across the collector slit by a highly linear saw tooth voltage which is added to the usual square wave voltage and applied to the electrostatic analyser plates. Thus, each channel of the multiscalar is identified with a particular value of the analyser voltage; the mass scale is then derived from the Swann–Bleakney theorem.

8.6 Precision in the determination of mass doublet separations

The steady improvement in the precision of atomic mass determinations which has taken place over the years is shown in fig. 8.10. The improvements up to about 1955 reflect the improvements in instrumental resolution which resulted from better focusing and greater size. Since that time, the improvements are largely related to the introduction of electrical detection and to the development of the peak matching methods described above.

The mass width, W, of a mass spectral peak is dependent on the resolution, $R = W/M$ of the mass spectroscope. Additionally, the position of the peak may be determined to some fraction, F, of its width. Thus, the uncertainty in an experimentally determined mass is:

$$\delta M = FW = FRM. \tag{8.10}$$

In the classical work with mass spectroscopes employing photographic detection, F was typically $\frac{1}{50}$ on the basis of a number of measurements. The grain size of the photographic plate ($\sim 10^{-4}$ cm) set the lower limit to the actual line width, W, with the result that the mass width could not be reduced indefinitely by improving the resolution alone. This limit, always given theoretical recognition, became a practical one in the work of

Fig. 8.10. History of the precision of atomic masses. Each principal investigator is indicated, with the resolution of the corresponding apparatus in brackets. See also Chapters 1 and 5 for references. (After Williams & Duckworth, 1972.)

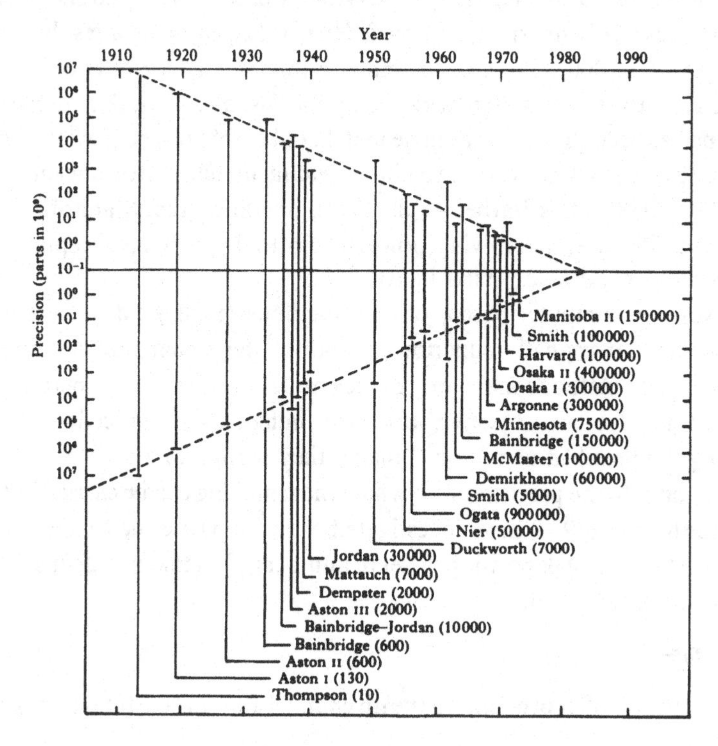

Mattauch and his colleagues (Everling, 1957) who secured, with careful adjustment, a resolution (R) of 10^{-5} with a mass spectrograph possessing a dispersion of only $\sim 0.2\,\text{cm}/1\%$ mass difference (see fig. 8.1(*b*)). This corresponds to an actual line width of $2\,\mu$m. Ewald and his colleagues (Ewald, 1953) also approximately reached this limit, which represents a precision in mass determination of 1 part in 5×10^6.

As noted previously (§2.1), the introduction of electrical detection leads to a reduction in the resolving power by a factor of two when the collector slit has the same width as the image. However, this loss in resolution is much more than offset by the improvement in the precision with which a line can be located.

In the original 'visual' version of the peak matching technique, an individual match could be made to within 1% of a peak width. As described earlier (§8.4), eight individual matches might be made in one 'run', with an average of, say, 20 runs yielding the final value for a given doublet spacing. The mean of these values represented an improvement in the precision by a factor of $(8 \times 20)^{1/2}$. Thus, for the final value reported for the mass difference, F was typically $\sim 1/1000$. On this basis, Smith (Smith & Damm, 1953*b*, 1956), with a resolution of $\sim 1/10\,000$–$1/25\,000$, was able to determine masses to an uncertainty of less than 1 part in 10^7, while Nier and his colleagues (Quisenberry, Scolman & Nier, 1956), enjoying a resolution of $1/30\,000$–$1/60\,000$, were able to improve this by a factor of two.

More recently, the precision achieved by the 'visual null' peak matching technique has been shown, for a single match, to be 0.6% of a peak width. In this case 18 runs, each comprising eight individual matches, are sufficient to given $F \sim 1/2000$. Thus Barber *et al.* (1971), operating the 'Manitoba II' instrument (§5.12) at a resolving power of up to 150 000, have reported a precision as good as 2.5 parts in 10^9.

The various computer-based, off-line methods (§8.5) yield a level of precision only marginally superior to that of the visual null method. However, these methods make more efficient use of instrument operating time and involve an objective, operator independent, procedure for obtaining the matched condition. Finally, they permit an estimate of the statistical limit to the precision with which the matching can be carried out.

Meredith *et al.* (1972) have investigated the case where the centroid is taken to define the peak position. A set of numbers, f_i, is taken to define the peak position, such that:

$$n_i = f_i N, \tag{8.11}$$

where n_i is the number of counts in the ith channel and N is the total number of counts in the peak. Then the standard deviation of the centroid position is:

$$\sigma = \frac{1}{\sqrt{N}}\left(\sum f_i X_i^2 - \bar{X}^2\right)^{1/2}. \tag{8.12}$$

In particular, for a high resolution mass spectrometer, the peak shape may be approximated by an isosceles triangle of width, W. Then the fraction, F $(=\sigma/W)$ of a peak width within which the centroid may be located can be shown to be:

$$F = (24N)^{-1/2} \approx 0.20N^{-1/2}. \tag{8.13}$$

Here the numerical constant depends on the peak shape, but does not change dramatically from this value. For example, the constant changes from 0.20 to 0.29 when rectangular peaks, such as those used in studies of isotopic abundances, are considered.

In practice, the error obtained for computer runs has been found to be a factor of about two larger than the purely statistical limit. This probably reflects the presence of small random variations in peak position which are of the same order of magnitude and which arise from residual instabilities in the electromagnetic fields or from physical vibration of the apparatus.

8.7 Sources of error in doublet determinations

The history of doublet determinations is studded with instances in which values for a given mass difference, obtained in different laboratories, were incompatible. However, in recent years the degree of consistency between different laboratories has become extremely good. Since the associated precision is high, it appears likely that the significant sources of systematic error have been recognized.

Some of the sources of systematic bias were peculiar to the classical instruments and have been described in detail elsewhere (Aston, 1942; Duckworth, 1958). In particular, the following problems were important: the non-linear response of photographic plates gave rise, in the case of asymmetric lines, to an apparent shift in line position with increasing exposure; operator judgement of a line position on a photographic plate was shown to be somewhat subjective; knowledge of the dispersion law along the plate was required to high accuracy; and the spark and discharge sources, which were used at that time, produced ions with such a wide range of energy that the electrostatic analyser plates were bombarded. This last effect could lead to the formation of insulating films which became charged and gave rise to large distorting fields (Stewart, 1934).

In modern instruments an effect somewhat reminiscent of this last problem has been encountered. It has been found that the spacings of wide 'doublets' such as

$$^{A}X^{37}\mathrm{Cl} - {}^{A}X^{35}\mathrm{Cl} = \Delta M, \tag{8.14}$$

when determined by the peak matching procedures described in §§8.3 and 8.4, may be systematically in error by up to 200 ppm, although the statistical precision associated with ΔM is, say, $\sim$ 6 ppm (Southon *et al.*, 1977). Such a systematic effect has been reported for instruments at the Argonne National Laboratory (40 ± 1 ppm, Stevens & Moreland, 1970), Harvard University (20 ± 2 ppm, Kerr & Bainbridge, 1970), Osaka University (36 ± 5 ppm, Nakabushi *et al.*, 1970), the University of Minnesota (20 ± 3 ppm, Kayser *et al.*, 1972), and the University of Manitoba ($\sim$ 40 ppm, Southon *et al.*, 1977). It was also found that, although the magnitude of the systematic error differed from day to day, it remained remarkably constant within a given day (Southon *et al.*, 1977; Stevens & Moreland, 1967).

It appears that this systematic error originates from the presence of small surface potentials which appear on the plates of the electrostatic analyser. Thus, the potential differences which are measured and applied to the electrostatic analyser plates are not precisely those that establish the electric field in the gap of the analyser. Since the mass difference is calculated from the *measured* voltages, a systematic error occurs.

The following mechanism has been suggested (Southon *et al.*, 1977). The primary ion beam, which does not itself strike the plates, collides with the residual gas in the analyser. The charged fragments from such collisions are accelerated to the analyser plates where they give rise to observable currents. Petit-Clerc & Carette (1968, 1970) have investigated potentials developed on metal surfaces in vacuum systems under electron and ion bombardment. They have observed potentials of up to 0.5 V developed by 8–43 eV electrons at current densities of 2×10^{-9} A/mm^2 and somewhat smaller residual potentials which persisted for hours. They conclude that charges are trapped in polymer films from organic contaminants. The collision cross sections leading to the dissociation of either the residual gas molecules or the molecular ions in the primary beam are consistent with the currents observed at the analyser plates.

Southon *et al.* (1977) have reduced the effect to a negligible level by carefully cleaning the analyser plates in turn with acetone, alcohol and distilled water, and then filling the electrostatic analyser with argon at $\sim$ 50 Pa (0.4 Torr) and establishing an rf discharge between the plates. However the systematic error increased slowly to $\sim$ 40 ppm in the course of a few weeks. Repeated use of the rf discharge would again temporarily reduce or eliminate the error, but with progressively less effectiveness.

The mechanism and behaviour appear to preclude the possibility of eliminating this systematic error permanently. Accordingly, the practice has been adopted of determining the spacing of a well known, wide doublet (such as the one shown in equation (8.14)) to assess the magnitude of the effect and then using this information to apply a correction to the narrow doublet under study.

A misalignment of the principal slit may have a much more serious effect than a simple broadening of the mass spectral line. Depending upon the source arrangement, some types of ions may illuminate a longer section of the slit than others. Ewald (1953) has shown, for example, that the focusing properties of a gas discharge source may differ for different ions. Under certain conditions a beam of oxygen ions reaching the slit from this source was several millimetres long while the CH_4 ion beam arriving at the same time was only 0.1 mm in length. Under these conditions one peak may not only be considerably wider than the other (which should lead to its discard), but may also be displaced from its true position relative to the other. A particularly vicious situation is one in which ion beams of equal length illuminate different portions of a misaligned slit. The matched appearance of the resulting doublet gives no hint that all is not well.

In addition, Ewald has shown that ion groups may emerge from a source with different angular widths. Thus, if the angular aperture, 2α, of the collimating slit be too large, the ions corresponding to one doublet member may not completely fill it, whilst those belonging to the other doublet member may more than do so. In some instruments, the latter group may be subject to a greater second order direction focusing aberration, and a consequent displacement.

Systematic errors may also arise when the two members of the doublet are formed in the ion source with different energy distributions. Such a situation may occur if dissociation energies differ or if the two species are ionized at different locations in the ion source. In either case, the effect of residual velocity focusing aberrations is to displace one peak with respect to the other.

This last effect may be minimized in mass spectrometers which possess an intermediate direction focus (e.g. §§5.8, 5.12). In this case the instrument is set up for visual peak matching of the given doublet and the ion accelerating potential V_a, is varied slowly so that the ion beam is intercepted by one of the edges of the slit S_β, which is located at the direction focus of the electrostatic analyser. When ΔV_a is adjusted to the value at which the two matched peaks disappear simultaneously, the condition required for the application of Bleakney's Theorem, namely, that the trajectories for the ions M, M' be identical, is achieved (Nier, 1957; Barber *et al.*, 1964, 1971).

8.8 Peak matching in the rf mass spectrometer

The technique of peak matching was first applied to the determination of atomic mass differences by Smith & Damm (1953*a*, *b*) working with the 'mass synchrometer' after its conversion to rf operation (§6.9). As described previously, a given ion group may be scanned across the detector by a saw tooth modulation of the frequency of the rf applied to the modulator. The detector was a Faraday collector connected through a preamplifier to an oscilloscope which was swept at the same rate as the rf modulation, so that a peak was displayed on the oscilloscope screen.

The peak matching in this case is based on an extension of the Swann (1931)–Bleakney (1936) theorem, as suggested by Snyder and given by Smith & Damm (1956). The ion orbit is determined by the Lorentz force equation:

$$M\frac{d^2\mathbf{r}}{dt^2} = e\left[\mathbf{E} + \frac{d\mathbf{r}}{dt} \times \mathbf{B}\right]. \tag{8.15}$$

From this we see that, for constant $\mathbf{B}$, if the electric field $\mathbf{E}$ is changed to $\mathbf{E}' = k\mathbf{E}$ and the time scale is altered by $t' = t/k$, then an ion of mass $M' = M/k$ having the same initial conditions will describe the same orbit. Accordingly, if all dc voltages and both the amplitudes and frequencies of all rf voltages are changed so that:

$$MV = M'V' \tag{8.16}$$

and

$$Mf = M'f' \tag{8.17}$$

or

$$\frac{\Delta M}{M'} = \frac{\Delta V}{V} = \frac{\Delta f}{f}, \tag{8.18}$$

then the two ion groups follow identical trajectories through the instrument. During one sweep of the oscilloscope the peak for mass M is displayed; on the next sweep, all of the frequencies and voltages are altered as required by (8.18) and the peak for mass M' is displayed. In the arrangement described by Smith & Damm, one of the traces was split vertically by a small 12 kHz square wave so as to give a double trace. The position of the other peak was then matched between the components of the double trace. Inasmuch as the mass difference is sensitive to the applied frequency, the frequencies corresponding to the matched condition are measured and (8.17) or (8.18) used to obtain the mass difference. Clearly the method benefits from the relative ease and high precision with which frequency measurements can be made.

With the early instrument (Smith & Damm, 1953*a*, *b* 1956) the precision of locating a peak was approximately 1/1000 of the width at half-maximum and an important series of measurements was made for nuclides with $A < 40$, with a precision of better than one part in 10^7. With the later improved instrument (Smith, 1960, 1972) which used higher frequencies and fields and further benefited from improved rf apparatus, Smith achieved a resolving power at half-maximum of $2–4 \times 10^5$ and showed that the standard setting error for a single reading was $\sim 1/2500$ of the peak width at half-maximum. Although the frequency ratio could be determined with a precision of a few parts in 10^{10}, this was usually somewhat reduced by systematic effects (Smith, 1972; Smith & Wapstra, 1975), especially those arising from surface charges. Following Smith's death in 1972, this instrument was moved from Princeton to the University of Technology in Delft. In subsequent work (Koets, 1976; Koets, Kramer, Nonhebel & LePoole, 1980) some modifications have been made, including an increase in the resolving power to 10^7 (FWHM), although the expected precision in locating a peak has not as yet been realized.

8.9 Standard for atomic mass

By the mid-1950s, the suitability of ^{16}O as the primary standard of atomic mass was seriously questioned. There was a growing discontent with the historical, but illogical, situation in which ^{16}O served as the physicists' standard of mass, whereas the mixture of ^{16}O, ^{17}O and ^{18}O, as found in nature, was the basis for chemical atomic weights. The situation would have been more tolerable if the isotopic constitution of naturally occurring oxygen had been invariant, as the two mass scales would then have been related by a fixed constant. However natural variations in the oxygen isotopes do exist, and correspond to a variation in such a conversion factor of 1.000 268–1.000 278 (Nier, 1950).

Several proposals were made to unify the two scales and remove the imprecision in the definition of the chemical scale. For example, the imprecision could have been removed by defining the isotopic abundances of 'natural' oxygen or by defining the chemical scale in terms of the $^{16}O = 16$ scale by a particular constant. However, these proposals did nothing to unify the two scales, and suffered somewhat from their arbitrary nature. On the other hand, a change by the chemists to $^{16}O = 16$ would have required a shift of 275 ppm in all chemical quantities whose values depend on the size of the mole. The extensive tabular revision which such a modification would have involved made this proposal unattractive to the chemists.

Other standards of mass were then considered which would require abandoning both the old physical and chemical scales, namely, $^{19}F = 19$, $^{18}O = 18$ and $^{12}C = 12$. In the case of ^{19}F (Wichers, 1956) only a small adjustment from the chemical scale would have been necessary. Further it is anisotopic (existing only as ^{19}F) and forms fluorocarbons which are useful as comparison masses. However the latter are not nearly as extensive as are hydrocarbon fragment ions. A similar objection holds for ^{18}O, plus the disadvantage that it is a rare ($\sim 0.2\%$) isotope.

A. O. Nier and A. Olander (Kohman *et al.*, 1958) independently suggested that a unified scale be based on $^{12}C = 12$. This required a relatively small change in the chemical scale (~ 43 ppm), and removed the ambiguity in it. From the physicists' viewpoint it introduced a more desirable standard inasmuch as many atomic masses had been determined against hydrocarbon comparison fragments. The introduction of ^{12}C as a standard thus removed the uncertainty involved in the ^{12}C to ^{16}O mass ratio.

Accordingly, in 1963, coordinated action was taken by both the International Union of Pure and Applied Physics (IUPAP) and the International Union of Pure and Applied Chemistry (IUPAC) to unify the two scales on the basis of

$$^{12}C = 12\,\text{u, exactly,} \tag{8.19}$$

where 'u' stands for the *unified* atomic mass unit.

Although they have no official status, several nuclides have been studied with very high precision with the object of using them as secondary standards. These are ^{1}H, ^{2}D, and the difference $^{37}Cl - {}^{35}Cl$ in particular, and to a lesser extent ^{13}C, ^{14}N, ^{16}O, ^{35}Cl and ^{37}Cl (Wapstra & Bos, 1977*a*, *b*; Williams & Duckworth, 1972; Smith & Wapstra, 1975). A molecular fragment based on carbon and involving the above secondary standards may then be used as a comparison in a doublet (given by equation (8.2)) from which an 'absolute' mass for the other member may be calculated.

8.10 Mass differences from nuclear physics

A nuclear reaction may be expressed in the general form

$$a + X \rightarrow Y + b + Q, \tag{8.20}$$

where a is the bombarding particle, b is the emitted particle, Q is the net energy release, and X and Y are the initial and final isolated atoms in their nuclear and atomic ground states, respectively. The Q value may be obtained by determining the energies of the particles a and b, and their directions relative to one another. Thus, from equation (8.20), one may derive the mass difference $X - Y$.

In the case of charged particles, the most accurate values are found by deflecting the particles in magnetic or electric fields of known strength. However, in some circumstances solid state detectors for such particles provide sufficient advantages in size, efficiency or simplicity that they are used. The very important determination of γ-ray energies is now carried out almost exclusively with carefully calibrated Ge(Li) detector systems (Helmer, Greenwood & Gehrke, 1976; Helmer, van Assche & van der Leun, 1979; Greenwood, Helmer, Gehrke & Chrien, 1980).

Most of the possible combinations of proton, neutron, deuteron, triton, ^{3}He, α particle or γ-ray are represented in the compilation of reactions by Wapstra & Bos (1977*b*) for which significantly precise Q values have been determined.

Of special interest are the (n, γ) reactions involving the capture of thermal neutrons. At 160 of the 200 mass numbers from $9 \leqslant A \leqslant 208$, there is at least one (n, γ) value which has a precision better than 2 keV and for 20 of these cases, there are two. Of these values, 75% have errors of less than 1 keV while a few have errors as small as 0.2 keV. There are 60 (p, γ) reactions which have a precision of 3 keV or less, and 55 (p, n) threshold energies which have comparable errors. Generally, the remaining charged particle reactions carry somewhat lower precision, in the range of 1–15 keV.

Nuclear decays constitute special cases of equation (8.20). For example, determinations of the energies of α particles may be used to establish the mass difference between parent and daughter nuclides. Thus the Q value is calculated by taking into consideration the conservation of linear momentum in the decay. Where high precision is involved, the requirement that parent and daughter be in their atomic ground states must also be taken into account (Wapstra & Bos, 1977*b*). The precision for this type of data is as good as 0.05 keV (Rytz, Grennberg & Gorman, 1972), although it is more commonly in the range 1–6 keV.

Similarly, determinations of the end points for β^- decay yield the atomic mass difference between parent and daughter. In the case of positron emission, the parent–daughter difference is given by:

$$Q_{\beta^+} = E + 1022\,\text{keV}, \qquad (8.21)$$

where E is the end point energy and 1022 keV is twice the rest energy of the electron. For decays where Q_{β^+} is less than 1022 keV, the decay must proceed by electron capture. The precision of β decay Q values is very high;

in cases where β end point energies are determined, 'good' precision is in the range 1–13 keV. A discussion of the factors affecting the determination of the end point energy including atomic effects, has been given by Bergkvist (1972).

8.11 Atomic mass evaluation

As is evident from the preceding sections, there is now a very large body of data from mass spectroscopic work and from the studies of nuclear reaction and decay Q values. These two kinds of data are related through the conversion relationship:

$$1\,\text{u} = 931\,501.2 \pm 0.3\,\text{keV}, \tag{8.22}$$

given by Cohen & Wapstra (1983). Exhaustive compilations of such data have been made by Wapstra & Bos (1977*b*) and Wapstra & Audi (1985).†

In many cases, the data involving a particular nuclide will relate its mass to the masses of other nuclides in more than one way. That is, from a mathematical viewpoint, we may regard each datum as giving a mass difference in the form of an equation expressed in terms of parameters (the atomic masses) which we wish to find. The number of equations greatly exceeds the number of parameters and so the solution is overdetermined. Therefore, the method of least squares (Bearden & Thomsen, 1957; Mattauch, 1960; Taylor, Parker & Langenberg, 1969) is appropriate for the derivation of self-consistent 'best' values for these parameters.

For some nuclides, only a single datum links the mass to the main body of data. Clearly, such a nuclide has a mass which is not overdetermined and so does not enter into the main least squares calculation. It is termed a 'secondary' nuclide since its mass is calculated relative to the main body of data after the least squares calculation is completed.

Such a major evaluation of all of the atomic mass data has been undertaken several times with the encouragement of the IUPAP Commission on Atomic Masses and Fundamental Constants, for example the compilation of Wapstra & Bos (1977*a*, *b*) which initially involved 4418 data, plus 608 estimates from nuclear systematics. Of these, 1293 of the 3829 nuclear Q values and 405 of the 1197 mass doublets were not used for a variety of reasons. For example, more recent data may have shown a value to be clearly in error or a value may be superseded by a more precise value which greatly outweighs the earlier one. When several values determine a single mass difference, a single value is calculated and used in the least squares evaluation. Thus, 442 reaction values are replaced by 187 average

† Note added in proof: Wapstra and Audi (1985) appeared after the writing of this section.

values. Finally, 1286 of these values link 'secondary' nuclides (including the 608 estimates from nuclear systematics) which are not overdetermined. Thus, the core least squares evaluation involves 1787 equations in 702 parameters.

It should be noted that the mass spectroscopic data are of two types. In one type, the mass doublet involves a well known hydrocarbon member so that the mass difference gives the 'absolute' mass of the other member. In the second type, a doublet such as that given by equation (8.11) yields mass difference information (there between ${}^{A+2}X$ and ${}^{A}Y$) which involves nuclides usually nearby each other in the mass table.

The presence of a very large number of precise doublets of this second type, coupled with the complementary precise (n, γ), (p, γ) and (p, n) data has led to the formation of a relatively rigid 'backbone' along the line of β stability. Since very precise 'absolute' mass values are few in number, the high precision of *mass differences* along the backbone may introduce the possibility that, in some fairly extended regions, the 'absolute' masses may drift systematically high or low by amounts well outside the assigned errors. It has been suggested (Wapstra & Bos, 1977*b*; Sharma *et al.*, 1977; Kozier *et al.*, 1979, 1980) that the reliability could be much improved by the determination of about two-dozen 'absolute' masses for nuclides at strategic locations about 10–20 u apart.

Fig. 8.11 gives a survey of results of the 1977 Atomic Mass Evaluation (Wapstra & Bos, 1977*a*, *b*) by showing the standard deviations associated with the most precisely known absolute atomic mass for each value of $40 < A < 240$. At intervals of five mass numbers are points which give the differences between the 1971 values (Wapstra & Gove, 1971) and the 1977 values (Wapstra & Bos, 1977*a*). The error bars are the standard deviations given in the 1971 table only. The changes suggest that the absolute mass values in one table and/or the other have indeed been affected somewhat by systematic errors, especially in the region $A > 140$.

As described at the beginning of this chapter (§8.1), the interest in atomic masses from the perspective of nuclear physics is centred, not on the absolute values, but rather on the manner in which the binding energy of the nucleus varies with the numbers of neutrons (N) and protons (Z). Such information reflects primarily the mass *differences*, which are known in absolute terms to higher precision than are the masses.

8.12 Atomic masses of the heavier atoms and nuclear stability

Although Aston (see Aston, 1942) obtained masses for some of the isotopes of chromium, krypton and xenon, most of the pioneer work among the heavier atoms was done by Dempster during the period 1936–38. This consisted of a general survey of the masses of the heavier atoms, and led to his well known version of the packing fraction curve (Dempster, 1938) where the packing fraction, f, is defined by

$$f = \frac{{}^{A}_{Z}M - A}{A}. \tag{8.23}$$

This superseded Aston's earlier version and remained the standard for a dozen years, during which period a number of additional mass determinations were made, but failed to add much to the general picture.

An important result of Dempster's packing fraction work was his discovery that the slope of the packing fraction curve in the region $90 < A < 104$ was twice its value in the region $180 < A < 208$. This could be construed as evidence either for a modest improvement in nuclear stability near $A = 180$ or for a marked deterioration in it near $A = 90$ (Dempster, 1938; Feenberg, 1947). This question, unresolved at the time, was sub-

Fig. 8.11. Precision in atomic mass determinations. The error for the most precisely known mass at each A (1977 mass table, Wapstra & Bos, 1977*a*) is indicated. The points with error bars give the differences between the 1971 and 1977 mass evaluations.

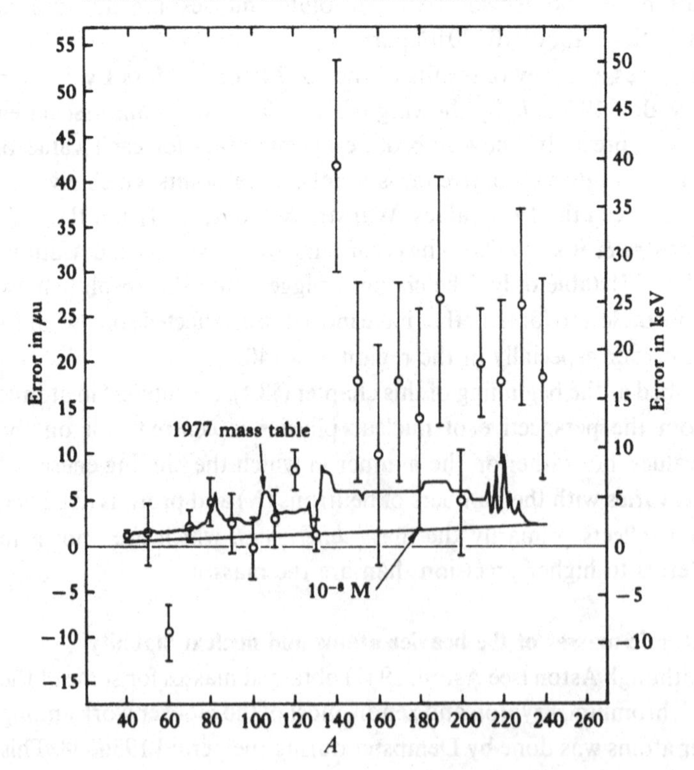

sequently investigated and the latter interpretation found to be correct (Duckworth, Woodcock & Preston, 1950). An explanation of this sudden and rather unexpected change in slope at $A \sim 90$ was conveniently provided by the concept of nuclear shells, proposed shortly before by Maria Mayer (1949, 1950), Haxel, Jensen & Suess (1949), Feenberg & Hammack (1949) and Nordheim (1949).

The shell model hypothesizes a strong spin-orbit coupling, which increases with orbital angular momentum. This hypothesis, when combined with the level scheme for particles in a simple potential well (Elsasser, 1934; Margenau, 1934), gives rise to the prediction that 28, 50, 82 and 126 proton or neutron configurations are particularly stable, in agreement with a compelling array of experimental evidence (Mayer, 1948). The 'magic' numbers mark the spin-orbit splitting of the $f_{7/2}-f_{5/2}, g_{9/2}-g_{7/2}$, $h_{11/2}-h_{9/2}$ and $i_{13/2}-i_{11/2}$ groups respectively.

The break in the packing fraction curve at $A \sim 90$ was attributed to the completion of the 50-neutron shell, and comparable mass effects were subsequently found to be associated with the 28-proton ($A \sim 60$, Duckworth & Preston, 1951), 50-proton ($A \sim 120$, Duckworth & Preston, 1951), 82-neutron ($A \sim 140$, Duckworth, Kegley, Olson & Standford, 1951) and 28-neutron ($A \sim 52$, Collins, Nier & Johnson, 1952) configurations. The 82-proton and 126-neutron shells combine to cause a pronounced effect at ^{208}Pb, marking the threshold of the region of natural radioactivity. These effects are clearly shown in fig. 8.12 which shows the classical packing fraction curve and the more commonly used binding energy per nucleon curve (equation (8.1)), where both are drawn for the odd-A nuclides. Equations (8.1) and (8.23) can be combined to give

$$\text{be/nucleon} = \frac{Z^1\text{H} + (A-Z)n}{A} - (l+f), \tag{8.24}$$

where f represents the packing fraction. Thus the maximum in the be/nucleon curve occurs at roughly the same A as does the minimum in the packing fraction curve, and the nuclear shell effects found in the one are clearly reflected in the other. These effects include, in addition to those mentioned above, evidence for the completion of the $2p$ ($A = 16$) and $3s$ ($A = 40$) shells, and for a broad region of stability between the 82- and 126-neutron shells (Hogg & Duckworth, 1954).

Variations in nuclear binding, which reflect changes in nuclear structure, may be shown more distinctly through the use of a variety of nuclear systematics. In particular, neutron and proton separation and pairing energies and both α and β decay energies may be plotted as functions of either N or Z (for example, see Wapstra & Bos, 1977*a*). One of these quantities, the double neutron separation energy, S_{2n}, is derived directly from doublets of the type given by equation (8.3) (Barber *et al.*, 1963) and is

$$S_{2n} = 2n - (^{37}\text{Cl} - {}^{35}\text{Cl}) - \Delta M_2. \qquad (8.25)$$

In fig. 8.13, S_{2n} is shown for nuclides having even-N in the region near the magic number 82. The dramatic effect at the closed shell is the major feature in this figure. Also of special interest is the behaviour of the S_{2n} curves following the dramatic drop at $N = 82$. Initially these curves exhibit the same shape as those (not shown) which precede the 82- shell closure, but at $N = 88$ there is a clear discontinuity for all of the elements (Barber *et al.*, 1964). The values of S_{2n} for $N = 90$ that would be expected on the basis of the behaviour for $N < 88$ lie much below the measured values. This has been attributed to the onset of nuclear deformation that occurs in this region. Indeed, these curves give a clear indication of the abruptness of this change in nuclear shape. Above $N = 92$ one may interpret the small systematic variations in these curves (Barber *et al.*, 1973) in terms of the variation in spacings between the Nilsson single particle levels (Ogle, Wahlborn, Piepenbring & Fredriksson, 1972) which are calculated for deformed nuclei.

More recently, atomic mass determinations have been extended to include certain nuclides that are well away from the line of β stability. To date,

Fig. 8.12. Packing fraction and binding energy per nucleon curve.

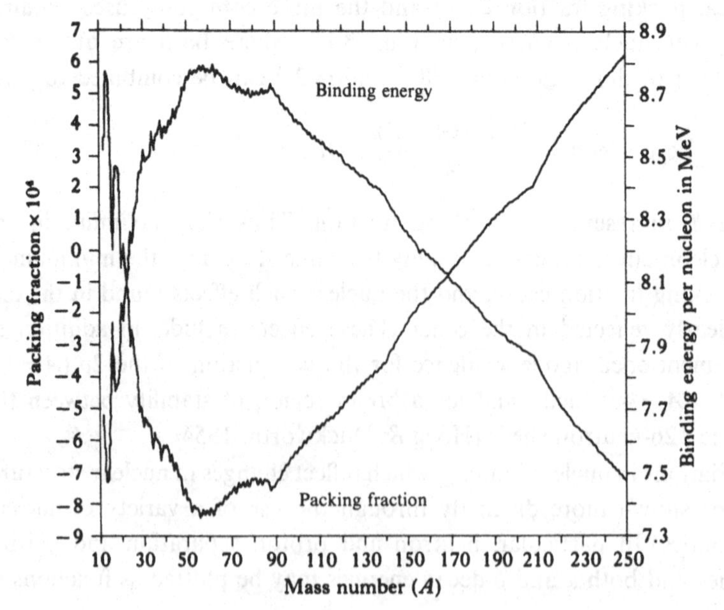

such determinations have been restricted to series of isotopes of elements which are produced with high efficiency in surface ionization sources (§3.2). Atomic masses for Li and Na have been determined by Thibault *et al.* (1975, 1980) using a single focusing instrument (fig. 8.14) with a precision of 80–1100 keV. In other studies in the heavier mass region, masses for Rb, Cs (Epherre *et al.*, 1979; Epherre, Audi, Thibault & Klapisch, 1980; Blair, Halverson, Johnson & Smith, 1980) and Fr (Audi *et al.*, 1980) have been obtained with double focusing instruments with precisions of 24–500 keV.

Initially these experiments were done with the Bernas-type ion source (Bernas, 1970; Thibault *et al.*, 1975, 1980), which also served as the accelerator target. This source consists of a sandwich arrangement of alternate layers of carbon and uranium foils surrounded by a rhenium foil container in the side of which is cut a slit. The accelerator proton beam passes perpendicularly through the foils, causing $U(p,$ fission) reactions to occur, with the fission products recoiling into the carbon. A high current is passed through the rhenium so that the entire assembly is operated at a high

Fig. 8.13. Double neutron separation energy systematics in the region $82 \leqslant N \leqslant 110$.

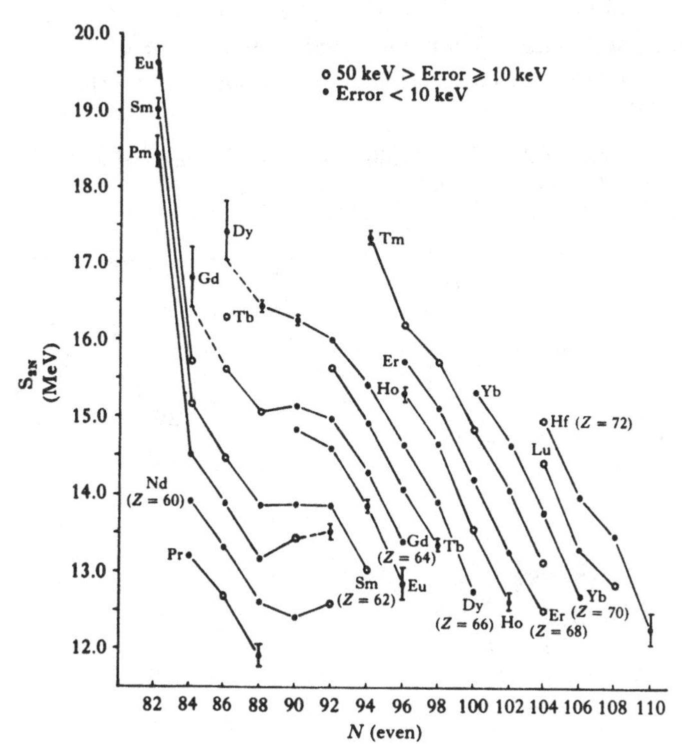

temperature. In particular, alkali metals diffuse rapidly through the carbon to the slit where they are surface ionized with high efficiency.

In more recent experiments (Epherre *et al.*, 1979, 1980; Audi *et al.*, 1980) this arrangement has been replaced by an isotope separator on-line (ISOLDE) where the final collector has become a Re cone operated so as to catch the 60 keV ions and surface ionize the material of interest. The source of the high resolution instrument then discriminates in favour of the elements that surface ionize readily so that unresolved components of the ISOLDE beam are removed. The potential of the Re cone is 9 kV, a more convenient value for the high resolution instrument.

In both cases the mass determinations were made by techniques that are variations of those described previously.

Other experiments directed toward the study of the nuclear mass surface far from the line of beta stability are those in which on-line isotope separators are used to produce a beam of ions having a given A (for example, Rudstam, 1976; see also §5.4 and the references therein to the conferences on electromagnetic isotope separators). These ions are then available for α and β decay studies in which the decay Q value is determined in the conventional way.

Such studies are of great interest in nuclear physics inasmuch as they make possible the investigation of the models used to predict atomic masses in a region far from β stability. Generally the measured masses diverge fairly

Fig. 8.14. Schematic diagram of on-line mass spectrometer and ion source. (After Thibault *et al.*, 1976.)

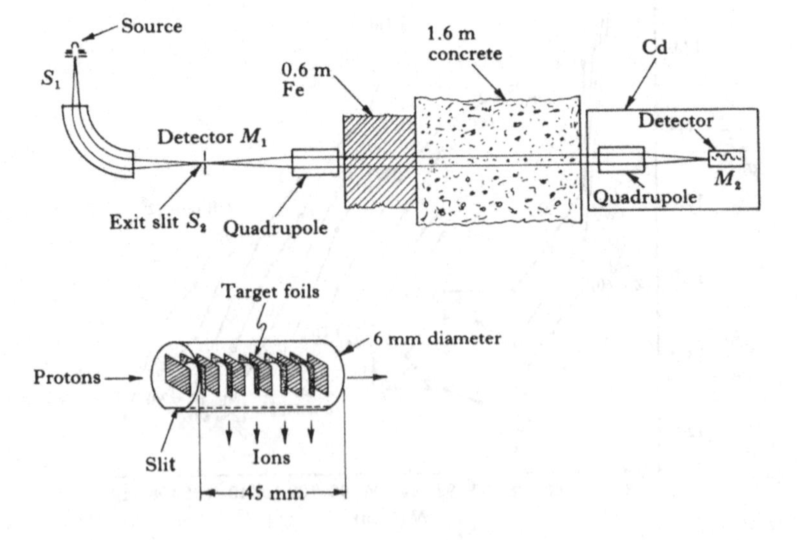

quickly from predicted values as the distance from the line of β stability increases (for example, Epherre *et al.*, 1979). In addition, special features have emerged. In Na, beginning at $N=20$ and extending up to $N=23$, a region of deformation was discovered for such very neutron-rich nuclei, in contrast with the stable nuclides which are spherical and manifest the shell closure at $N=20$ (Thibault *et al.*, 1980). In Cs, the well known deformation at $N=88$, referred to previously, does not occur until $N=90$ (Epherre *et al.*, 1980), while in Rb, a new region of deformation is seen to begin at $N=60$.

The considerable interest in this very active part of nuclear physics is summarized in the recent *International Conferences on Nuclei Far From β-Stability* (Klapisch, 1976; Hansen & Nielsen, 1981) and the *International Conferences on Atomic Masses and Fundamental Constants* (Sanders & Wapstra, 1972, 1976; Nolen & Benenson, 1980; Klepper, 1984).

Note added in proof: The ion cyclotron resonance method (§6.10) provides a means of determining accurately, in terms of frequency, the mass difference between members of a mass spectral doublet. Using this approach Selymas & Lippmaa and Tal'rose have obtained values of 18 588 $\pm$ 2 eV and 18 573 $\pm$ 4 eV, respectively, for the $^3T-^3He$ mass difference (see Tal'rose, 1985). This FT–ICR method will undoubtedly see much wider use in atomic mass determinations.

9

Mass spectroscopy in nuclear research

The role of precise atomic mass determinations in revealing general trends in nuclear stability, as well as indicating the saturation character of nuclear forces, the stability effects associated with shell structure and quadrupole deformations, etc., has been summarized in Chapters 7 and 8. In this chapter, we describe certain other applications of mass spectroscopy to nuclear physics research.

Radioactivity

9.1 Identification of naturally occurring radioactive isotopes

Mattauch's rule (Meitner, 1926; Mattauch, 1934) of nuclear stability forbids the existence of stable isobars whose atomic numbers differ by unity. It is, therefore, to be expected that one member of each of the naturally occurring adjacent isobaric groups will be radioactive. There are ten of these groups, namely, ^{40}Ar ^{40}*K* ^{40}Ca, ^{50}Ti ^{50}V ^{50}Cr, ^{87}*Rb* ^{87}Sr, ^{113}*Cd* ^{113}In, ^{115}*In* ^{115}Sn, ^{123}Sb ^{123}*Te*, ^{138}Ba ^{138}*La* ^{138}Ce, ^{176}Yb ^{176}*Lv* ^{176}Hf, ^{180}Hf ^{180}*Ta* ^{180}W and ^{187}*Re* ^{187}Os. In all but one case radioactivity has been detected and identified with one member: the symbols for the radioactive nuclides are italicized.

Certain general observations can be made regarding such radioactive species. The fact that they have persisted to the present day is proof that their half-lives are at least comparable to the age of the solar system ($\sim 5 \times 10^9$ yr). This also suggests that the radioactive transitions are highly forbidden, involving large spin (and parity) differences between the reluctant parents and their daughters. Furthermore, the total decay energy will generally be small. We give below a brief account of the methods by which the several radioactive nuclides listed above were identified. As will be seen, mass spectroscopy has played a major role in these identifications.

The existence of ^{40}K, which constitutes but 0.0119% of natural potassium, was discovered by Nier (1935) and its radioactivity subsequently observed by Smythe & Hemmendinger (1937) who used the first high intensity mass spectrometer to obtain separated isotopes of potassium. The half-life of ^{40}K is 1.31×10^9 yr. It decays by electron capture to ^{40}Ar (12%) and by β^- decay to ^{40}Ca(88%). The former decay mode forms the basis for the widely used 'potassium–argon' method of dating rocks and minerals (see §13.3).

The radioactivity of ^{87}Rb was established by Mattauch (1937) by isotopic analysis of a sample of strontium extracted (Hahn, Strassmann & Walling, 1937) from an old rubidium-rich mica (lepidolite from Manitoba). The mass spectrum showed a 99.7% abundance of ^{87}Sr, whereas the natural abundance is 7.0%. The unstable ^{87}Rb decays to ^{87}Sr by β^- decay with a half-life of 4.9×10^{10} yr. Its decay forms the basis for the 'rubidium–strontium' method, perhaps the most widely used method for dating minerals and rock formations (see §13.2).

The radioactivity of ^{113}Cd was observed (Greth, Gangadharan & Wolke, 1970) by measuring the β^- activities of a sample of normal cadmium (^{113}Cd–12.26%) and one enriched in ^{113}Cd (96.38%). Atomic mass data had earlier indicated that ^{113}Cd was the unstable member of this isobaric pair. It decays by β^- emission with a half-life of 9.3×10^{15} yr.

In the case of ^{115}In, a searching investigation (Martell & Libby, 1950), involving the use of two isotopically enriched samples of indium in which the $^{115}In/^{113}In$ ratios were 1249 and 0.53, respectively, showed ^{115}In to be the unstable member of the ^{115}In–^{115}Sn pair. The half-life of ^{115}In is 4.4×10^{14} yr and the decay is to ^{115}Sn by β^- emission.

The radioactivity of ^{123}Te was first observed by Watt & Glover (1962), who reported that the decay was by K capture with a half-life of 1.24×10^{13} yr. Separated or enriched isotopes played no part in this work.

The discovery of ^{138}La as a naturally occurring nuclide (0.089% abundance) by Inghram, Hayden & Hess (1947) led rapidly to its investigation for radioactivity, which was observed and elucidated by Pringle, Standil & Roulston (1950), Pringle, Standil, Taylor & Fryer (1951), and Mulholland & Kohman (1952). It decays by electron capture to ^{138}Ba (67%) and by β^- decay to ^{138}Ce (33%) with a half-life of 1.35×10^{11} yr.

The existence in nature of ^{176}Lu (2.6% abundance) was established by Mattauch & Lichtblau (1939) and its radioactivity subsequently observed by Flammersfeld & Mattauch (1943). Although ^{176}Lu is unstable against decay to ^{176}Yb, that mode of decay has not been observed. It decays to ^{176}Hf by β^- decay with a half-life of 3.6×10^{10} yr.

The discovery of naturally occurring ^{180}Ta (0.012% abundance) was made by White, Collins & Rourke (1955). This nuclide had also been known to be radioactive with an 8 h half-life. Recent work (Sharma *et al.*, 1980) has shown that the 8 h state is the ground state which decays by β^- decay to

^{180}W and by K capture to ^{180}Hf. The naturally occurring state has been shown to be an excited state located 77 ± 9 keV above the ground state and having a half-life of greater than 10^{13} yr. This is the only case where a naturally occurring nuclide is not in the ground state.

In investigating the natural radioactivity observed in rhenium, Hintenberger, Herr & Voshage (1954) extracted the osmium from a rhenium-rich molybdenite and found that it comprised 99.5% ^{187}Os. This 'rhenium–osmium' β^- decay has a half-life of 5×10^{10} yr and is sometimes used for dating. The unstable ^{187}Re constitutes 62.6% of ordinary rhenium.

The exceptional isobaric group for which radioactivity has not been observed is that of ^{50}Ti–^{50}V–^{50}Cr. The existence of ^{50}V (0.25% abundance) was established by Hess & Inghram (1949) and Leland (1949). It is known from atomic mass data that this nuclide is energetically unstable against decay to ^{50}Ti and ^{50}Cr by 2.213 MeV and 1.039 MeV, respectively. A lower limit on the half-life has been established as 4×10^{16} yr.

A special case – not in the category of the above – is the mass identification of the known α radioactivity of samarium. In the first experiment to be undertaken Dempster & Wilkins (1938) deposited the isotopes of samarium on a photographic plate which was then placed face-to-face with an unexposed plate for several weeks. When then developed, the first plate showed a line for each isotope whilst the second, or 'transfer', plate showed only one line caused by α radiation from the first. In this way Rasmussen, Reynolds, Thompson & Ghiorso (1950) showed ^{147}Sm to be the α emitter, a result confirmed by themselves and by Weaver (1950) using separated isotopes of samarium.

9.2 Study of branching ratios

Existing in nature are about 60 stable even–even isobaric pairs, with atomic numbers in each case differing by two. Between the members of each pair is located an odd–odd nuclide, of intermediate atomic number, which is unstable with respect to both of its even–even neighbours. As a rule, because of the energies available, or because of complexities in the decay schemes, either the $Z \rightarrow Z+1$ (β^- emission) or the $Z \rightarrow Z-1$ (β^+ emission or electron capture) mode of decay predominates. However, in a number of cases, branching has been observed. If the decay modes are limited to β^- and β^+ emission the percentage of the decays following each route can be determined by radioactive counting. However, electron capture is a different matter. Here, capture by the nucleus of the orbital electron is followed by a rearrangement of the orbital electrons, resulting in X-ray or Auger electron emission. Counting of these emissions is difficult as it is usually done in the presence of other emissions.

The mass spectrometer was first used to study branching ratios by Hayden, Reynolds & Inghram (1949). In this work, which involved the addition to neutron-irradiated europium of known amounts of normal samarium, europium and gadolinium (stable isotope dilution), the relative daughter abundances (^{152}Sm and ^{152}Gd) formed in the branching decay of ^{152}Eu were determined to an accuracy of about 15%. Reynolds (1950) significantly improved this precision in his study of ^{64}Cu, formed by the reaction ^{63}Cu + n. After a time that is long compared to the 12.8 h half-life of ^{64}Cu, the copper was dissolved and to it were added known amounts of electromagnetically separated ^{58}Ni and ^{68}Zn. Nickel and zinc were then extracted from the mixture and analysed in the mass spectrometer to determine the ^{64}Ni/^{58}Ni and ^{64}Zn/^{68}Zn ratios, from which the branching ratio ^{64}Ni/^{64}Zn was calculated. This value of 1.62 ± 0.11, when combined with the known β^-/β^+ ratio, gave the final percentages $\beta^- = 38.2 \pm 1.6\%$, $\beta^+ = 18.6 \pm 1.2\%$ and K capture $= 43.2 \pm 2.0\%$.

In later work (Crocker, Werner & Cherrin, 1968), the branching ratio for ^{170}Tm was found to be K capture/$\beta^- = (1.44 \pm 0.03) \times 10^{-3}$. In this work ^{169}Tm was neutron irradiated for 198 d and then allowed to 'cool' for 779 d, prior to the addition of accurately known amounts of electromagnetically enriched ^{166}Er (73%) and ^{173}Yb(85%). The three elements were separated from one another by ion exchange chromatography prior to mass spectrometric analysis. In this case, the $Z \rightarrow (Z-1)$ branch contained no β^+ decay, as the energy available was insufficient for that mode.

9.3 Double *β* decay

In double β decay, which consists of the simultaneous emission of two β^- (or β^+) particles, a nuclide is transformed directly into one whose atomic number is greater (or less) by two. In this process an energetically inaccessible intermediate nucleus is by-passed. The double β decay process is studied in the hope that it will shed some light on the properties of neutrinos.

At one time, when it was assumed that neutrinos had zero mass, two forms of decay were postulated. The first was 'neutrino-less' in the sense that the virtual neutrino emitted with the first β particle was absorbed during the emission of the second. The second form of decay involved the emission of two neutrinos along with the two β particles. As this involved the formation of four particles, rather than two, it was much less likely than the first form of decay. Thus, it was hoped that the study of double β decay would differentiate between the two postulates ($T_{1/2} \sim 10^{14}$–10^{16} yr and

$\sim 10^{22}$–10^{24} yr respectively). Mass spectroscopic work bearing on this matter took two forms: the determination of the energies available for double β decay (for example, ^{96}Zr–^{96}Mo, Geiger *et al.*, 1953; ^{130}Te–^{130}Xe, Halsted, 1952; and ^{150}Nd–^{150}Sm, Johnson & Nier, 1957; McLatchie, Barber, Duckworth & van Rookhuyzen, 1964) and the search for the daughter products of double β decay (for example, ^{130}Xe in very old tellurium ore, Hayden & Inghram, 1953; Ogata, Okano & Takaoka, 1966; Kirsten *et al.*, 1968). The latter experiments suggested that the half-life of ^{130}Te was of the order of 10^{21} yr, thus apparently disproving the postulate of neutrino-less decay and unholding the conservation of leptons.

Subsequent to that work, however, it has emerged that two distinct kinds of neutrino exist, the electron–neutrino and the muon–neutrino. Furthermore, (a) the electron–neutrino may have a non-zero mass and (b) double β decay involving such a neutrino may be neutrino-less. These possibilities have revived interest in the experimental investigation of double β decay. Thus for ^{76}Ge–^{76}Se decay, assuming an electron–neutrino mass of 35eV, the neutrino-less decay lifetime is calculated to be 10^{21} yr (Haxton *et al.*, 1981). Simpson, Campbell & Mahm (1982) are attempting to observe this decay by detecting (in a very large, hyperpure germanium detector) the peak corresponding to the sum of the two β particles. Ellis *et al.* (1984) have ascertained that the total energy available for the decay is 2040.71 ± 0.52keV.

9.4 Half-life determinations

The usual method of determining the half-life of a radioactive material is to measure its activity as a function of time. An equally fundamental approach is to determine either the rate of disappearance of the parent isotope or the rate of formation of its daughter. Nier (1939) employed the daughter growth method to determine the half-life of ^{235}U, by observing the amount of ^{207}Pb (the end product of the ^{235}U series) in uranium minerals of different ages. By this mass spectroscopic method a value of 7.13×10^{8} yr was obtained, a figure which was subsequently confirmed by precise counting determinations (Fleming, Ghiorso & Cunningham, 1951; Sayag, 1951). Later work has used isotope dilution to determine daughter concentrations. In this way half-lives have been obtained for ^{137}Cs by Rider, Peterson & Ruiz (1963) and for ^{87}Rb by McMullen, Fritze & Tomlinson (1966).

The other approach – to follow the disappearance of a parent isotope – was first employed by Thode & Graham (1947), who discovered long-lived ^{85}Kr as a fission product (see §9.5) and periodically measured its abundance relative to the stable isotopes of krypton. The original sample was studied over a 7-yr period to obtain a value for the half-life of 10.27 ± 0.18 yr (Wanless & Thode, 1953).

For long half-lives the daughter-growth method is more accurate since even a small number of disintegrations can cause an appreciable fractional change in the daughter abundance. Similarly, for short half-lives, the parent decay is the more accurate.

Mass spectroscopy and nuclear fission

The discovery of the slow-neutron fission of uranium (Hahn & Strassmann, 1939) raised the question of the identity of the fissile isotope. This was predicted by Bohr & Wheeler (1939), on the basis of semiempirical arguments, to be ^{235}U (0.715%), rather than the predominant ^{238}U (99.28%). This prediction was verified by Nier, Booth, Dunning & Grosse (1940), and confirmed by Kingdon, Pollock, Booth & Dunning (1940), by neutron bombardment of samples of ^{235}U and ^{238}U which had been separated in a mass spectrometer. Nier secured uranium ions by electron bombardment of UBr_4 and, in his most successful separation, obtained 3.1 μg of ^{238}U, and a corresponding amount of ^{235}U plus ^{234}U (0.0058%). The bombardment, performed at Columbia University, showed ^{235}U, and possibly ^{234}U, to be fissile with slow neutrons, but ^{238}U to undergo fast fission only. Substantially the same results were obtained with the Kingdon-separated samples.

Subsequently, huge separators (calutrons, see §5.4) were developed for the separation of macroscopic quantities of the uranium isotopes for the study of the fission process in detail. In addition, as we shall see below, mass spectroscopy has contributed directly to the elucidation of several aspects of the fission process.

9.5 Determination of fission yields

The 'heavy' and 'light' products of neutron-induced fission do not possess unique masses, but are distributed over two rather broad mass ranges which, for the slow-neutron fission of ^{235}U, are centred at $A \sim 95$ and $A \sim 140$ respectively. The general shape of this double-humped fission yield curve was first established by radiochemical methods (*Plutonium Project Report*, 1946). At high excitation energies (~ 50 MeV) the two humps merge, indicating that the fission is 'symmetric', that is, the fission products are approximately equal in mass.

In 1947 Thode & Graham reported the analysis of a sample of fission-product rare gases which had been extracted by Arrol, Chackett & Epstein

(1947). The isotopic constitution of the krypton and xenon in this sample differed vastly from that found in nature, as shown for krypton in fig. 9.1. This analysis established the efficacy of the mass spectrometric approach in the study of fission yields, an approach which was extensively exploited at both McMaster University and the University of Chicago.

Primary fission products are neutron-rich and highly unstable. They move towards stability by β^- emission (usually) or neutron emission (rarely). A series of successive β^- decays leading from primary to stable end product is known as a fission chain. Radioactive members of such a chain are studied by radiochemical methods, whilst the total chain yield is found by mass spectrometric determination of the stable end product. Together these two approaches have done much to elucidate the fission process and to improve devices based upon it.

The mass spectrometric approach involves the analysis of very small quantities of material. Notwithstanding this handicap and the fact that a single type of ion source is not appropriate to all elements, Petruska, Thode & Tomlinson (1955) quantitatively assayed fission-product krypton, rubidium, strontium, xenon, cesium, cerium, neodymium and samarium to obtain absolute fission yields for 28 mass chains. This and subsequent work in various laboratories has provided much information on the neutron-induced fission of ^{232}Th, ^{233}U, ^{235}U, ^{238}U and ^{239}Pu, as well as on the

Fig. 9.1. Isotopic constitution of normal (on the left) compared with fission product krypton (on the right). (After Thode & Graham, 1947.)

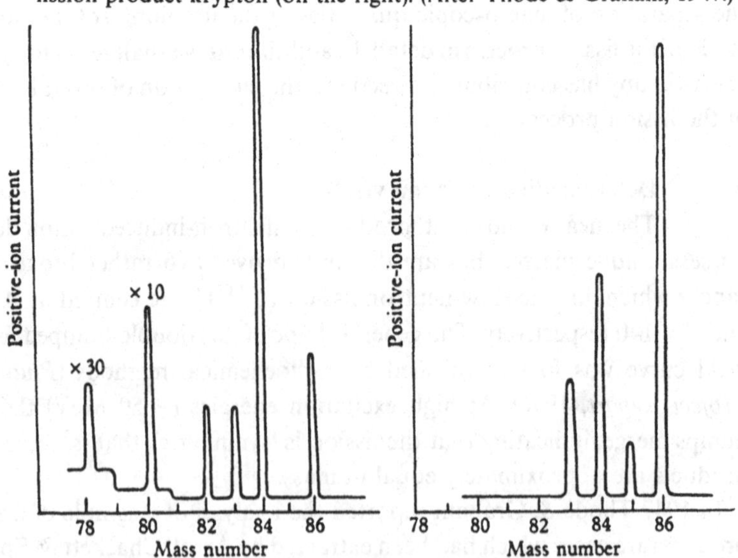

spontaneous fission of ^{235}U and ^{238}U. Amongst other things, this work has revealed the effect of the 50- and 82-neutron configurations on the fission process. The data for neutron-induced fission are periodically compiled, together with the corresponding data from radiochemical sources (see, for example, Crouch, 1977). Recent work on stable end products (for example, Rosman, De Laeter, Boldeman & Thode, 1983), involving nanogram amounts of material, has focused on low-yield products (cadmium, tin and tellurium etc.) corresponding to symmetric fission.

The development of 'on-line' mass spectrometry has made possible the use of mass spectrometers for the study of unstable fission products, a pasture previously reserved for radiochemistry. This subject is elaborated below in §§9.8 and 9.9.

9.6 Reactor operation

Fuel burn-up

Mass spectrometers are regularly used to ascertain the extent of 'burn-up' of the fuel in nuclear reactors. For example, in a reactor based on the fission of ^{235}U, the ratio $^{235}U/^{238}U$ decreases with continued operation of the reactor. This ratio controls the neutron flux and also determines the capture (by ^{238}U)-to-fission ratio for neutrons. As de Bièvre (1978) has described, several accurately known synthetic mixtures have been prepared by Shields and coworkers at the National Bureau of Standards, Washington with $^{235}U/^{238}U$ ratios ranging from 0.02% to 15%. These can be used to calibrate the instruments used to ascertain this ratio, upon which the economics of most nuclear reactors critically depends. It appears that absolute errors in the ratio of 0.01% can be achieved, using a surface ionization source of ions plus a double collector or a system of switched beams.

Neutron absorption cross sections

Emergency control of nuclear reactors can be provided by introducing nuclides with high neutron absorption cross sections. The first such stable nuclide to be identified was ^{113}Cd. In 1947 Dempster found, in a neutron-irradiated sample of cadmium, that the abundance of ^{113}Cd had decreased from 12.26% to 1.2%, whilst the abundance of ^{114}Cd had increased from 28.9% to 39.5%. In the same way Lapp, Van Horn & Dempster (1947) observed the large neutron absorption cross sections of ^{149}Sm, ^{155}Gd and ^{157}Gd. During neutron irradiation of a sample, the change in number of atoms (N) of a neutron absorbing isotope is given by

$$\frac{dN}{dt} = -\sigma NF, \qquad (9.1)$$

where σ = neutron absorption cross section and F = neutron flux. Hence, during time T the number of atoms of the isotope drops to

$$N = N_0 e^{-\sigma FT}. \qquad (9.2)$$

Using this relationship the value of σ can be determined. Uncertainties in the integrated flux (FT) can be eliminated by simultaneously exposing samples of BF_3, and using the well known cross section of ^{10}B (Petruska, Melaika & Tomlinson, 1955). Typical of accurate work of this sort is the determination of σs of ^{169}Tm, ^{170}Tm and ^{171}Tm (Crocker, Werner & Cherrin, 1968).

Neutron flux determinations

The inversion of the method described above obviously provides a means of determining the neutron flux in a reactor. For such determinations the activations of ^{147}Sm and ^{157}Gd are usually employed. The accuracy of such determinations increases with the strength of the flux. As small samples can be used, very local flux densities can be ascertained.

Capture/fission ratios

In the case of ^{235}U, this requires the determination of the ratio $^{236}U/^{235}U$ which, for small reactor burn-up, is not favourable for measurement (2×10^{-3}). Such exacting work has been done using a three-stage tandem mass spectrometer (White, Sheffield & Rourke, 1958; see §5.3).

9.7 Detection of natural nuclear reactors

In June 1972, in the course of routine mass spectrometric analyses of uranium being produced for use in nuclear reactors, Bodu, Bouziques, Morin & Pfiffelmann (1972) encountered samples in which the ^{235}U abundance, normally 0.72%, was as low as 0.44%. These samples had come from a uranium mine at Oklo in the Haut–Ogoné region of the Republic of Gabon and it was suggested that they had either undergone isotopic fractionation or been involved in a chain reaction. Very shortly thereafter Neuilly *et al.* (1972) studied certain isotopic abundances for cerium, neodymium, samarium and europium obtained from the Oklo deposits, and found clear evidence for fission-produced rare earth elements. Thus, the abnormally low amounts of ^{235}U were attributed to a natural fission reaction that had taken place in ancient times.

Subsequent studies revealed ^{235}U levels as low as 0.29% and established that the natural nuclear reactor had functioned about 2×10^9 yr ago for a period of about 10^5 yr. The *enhancement* of ^{235}U was also observed, due to the production of ^{239}Pu (and its subsequent decay to ^{235}U) in the very high neutron flux that had obtained at that time – about 10^{21} neutrons/cm^2/s (see de Bièvre, 1978).

As Roth *et al.* (1975) have suggested, a study of the abundance in nature of certain isotopes (for example, ^{82}Se, ^{99}Ru, ^{105}Pd, ^{130}Te), which are normally rare but which are important products of ^{235}U-neutron fission, could provide evidence for other Oklo-type reactors. Devillers, Lecompte, Lucas & Hagemann (1978) have studied ruthenium from this point of view, but without positive results.

On-line mass spectrometry

Amongst the most important applications to nuclear physics are those in which reaction products produced in fission or under bombardment by accelerated beams are analysed *in situ* by a mass spectrometer or are transported rapidly to a remote instrument for analysis.

9.8 General requirements for on-line mass spectrometers

As the reaction products to be studied and/or separated are invariably in short supply and are subject to radioactive decay, certain general considerations apply to the design of mass spectrometers used in this work. The first desideratum is that of high transmission, which is normally achieved by designing the analyser to accept a wide angle beam in both the horizontal and vertical planes. The design of such an instrument would be based on the focusing properties described in Chapter 2, including those associated with higher order focusing and/or associated with non-homogeneous magnetic fields. The second design objective is a highly efficient ion source, and certain of those described in Chapter 3 (for example, Bernas-Nier, Nielsen and Forced Electron Beam Induced Arc Discharge (FEBIAD)) were designed with this aspect in mind. The source may also be required to differentiate between a *mélange* of reaction products possessing different chemical characteristics. Thus, a heated filament could yield ions of the alkali elements in the presence of other elements. The third aim is to reduce 'hold-up' times to a minimum, in order to accomplish the analysis before the decay of the reaction product of interest. If the mass spectrometer is literally 'on-line' there is little problem of this sort. In experiments, however, in which the reaction products are transported rapidly to the ion source from their place of formation, the usual requirement is that the hold-up time be kept to less than 1 s. Common transport mechanisms are (a) fast-moving tapes and (b) a helium jet which can carry aerosols and their adsorbed reaction products over a distance of several metres with little loss of the reaction products en route. Experiments involving these types of transport, of course, are not 'on-line', but they frequently achieve the same purpose.

9.9 Representative on-line mass spectrometers

The first on-line mass spectrometer was constructed by Kofoed-Hansen & Nielsen (1951) who bombarded uranium with neutrons from the Copenhagen cyclotron and swept the fission-product krypton into the ion source of an isotope separator. This instrument did not lead directly to a second-generation instrument. Later, in the early 1960s, R. Bernas and his colleagues at the University of Paris (e.g. Klapisch *et al.*, 1966) exposed a filament from a triple-filament ion source (§3.2) to a high energy beam of protons (40 MeV–25 GeV) with a view to studying the lithium that was produced in the spallation of C, V, Ta and Pt. When the irradiated filament was replaced in the ion source, $^{7}Li/^{6}Li$ ratios were found to lie between 1.4 and 3.12 (*cf.* the natural ratio of 12.33) depending upon proton energy and the target nucleus. These dramatic results led to the development of on-line techniques for the determination of the masses of unstable nuclides of the alkali elements as described in §8.12.

As a result of the work of Bernas and others, mass separators were developed that were on-line with the targets of charged particle accelerators. These instruments, known as ISOL (Isotope Separator On Line), separated the mass components of the reaction-product beam in order that the nuclear properties of the nuclides belonging to a given mass number could be studied. The principal reactions produced by protons in the GeV energy range are (in order of frequency) spallation, fragmentation and fission. The reaction products are both neutron-poor and neutron-rich, and are often far from the region of β stability. It was to separate these nuclides into their various isobaric groups that an on-line separator known as 'ISOLDE' as built at the CERN 600 MeV proton synchrocyclotron and put into service in 1967. The key elements in this and similar on-line mass separators are the target matrix and the ion source. The combination of bombarding particle and target determines the nature of the reaction products, which are either released directly in the ion source or transferred there continuously. Conditions of operation of the ion source determine the composition of the ion beam to be separated. The beam is usually maximized for singly charged ions and often for those of one element only.

The proper combination of target and ion source also ensures that product nuclei under study pass through the separator in times that are short in comparison to their half-lives. For some systems the elapsed time can be as short as 10 ms. These considerations as they were applied to a modified ISOLDE (fig. 9.2) and to other similar instruments are described by Ravn (1979).

A second family of reactions is that produced by the beams from heavy ion accelerators. These reactions can also lead to neutron-rich and neutron-poor products but, as for the proton accelerator, the main use of on-line mass separators has been to provide separated beams of neutron-poor varieties that lie far from stability. Because of the short range of heavy ions, only thin targets can be used. In a typical ion source the reaction products

Fig. 9.2. The on-line isotope separator ISOLDE-II shown with related experiments in 1978. The 600-MeV proton beam (1) is focused on the target and the ion source unit (2), whilst the 60-keV ions are mass analysed by the magnet (3). Individual mass groups are selected in the electrostatic switchyard (4) and distributed through external beam lines (5) to the experiments, namely, alpha and proton spectroscopy (6), high resolution mass spectrometry (7), beta-gamma spectrometry (8 and 9), range measurements of ions in gases (10), optical pumping and laser spectroscopy on mercury (11), atomic beam magnetic resonance (12), collection of radioactive sources for off-line work (13, 14, 15), beta-decay Q-values determined by coincidences with magnetic 'orange' spectrometer (16) and spectroscopy of β-delayed neutrons (17). (After Hansen, 1979.)

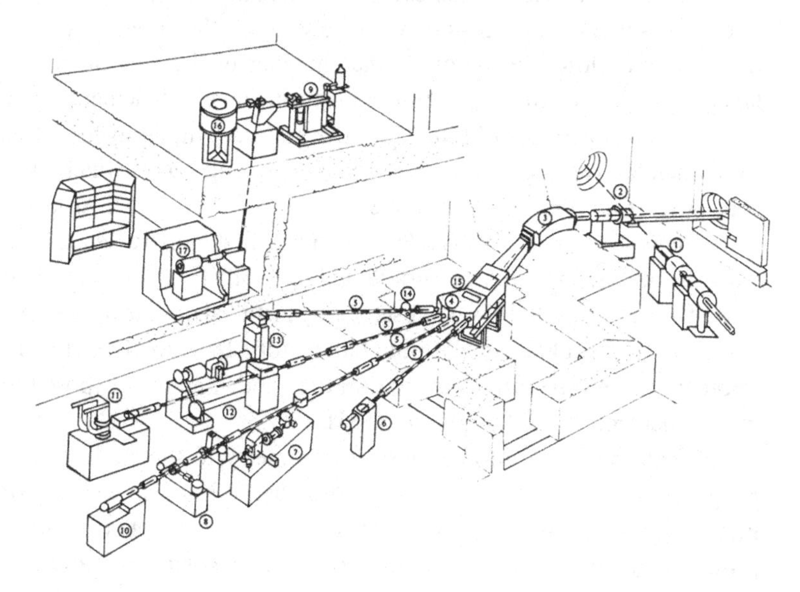

recoil through a thin tungsten window into the ion source which may be a surface ionization type (Bogdanov *et al.*, 1976) or plasma type (Burkard & Roeckl, 1976). The targets and ion sources used in this application have limited lifetimes because of the beam heating and sputtering to which they are subjected.

On-line mass spectrometers have also been developed for the direct and immediate study of fission products. A very thin sample of fissionable material is placed in the neutron beam from a reactor and the resulting fission products emerge from the thin layer with their original energy and in very high charge states. In 1956 Cohen, Cohen & Coley deflected a portion of this fission-product beam with a large sector magnetic field in the course of determining the energy distribution for fission products at mass 97. Subsequently Ewald, Konecny, Opower & Rosler (1964) fed the fission-product beam to a double focusing mass spectrograph (similar to the Mattauch–Herzog geometry – see §5.10), but the work was hampered by low intensity. It led, however, to the construction of a form of positive-ray parabola apparatus (see §1.2) which was installed on-line with the high-neutron flux reactor at Grenoble. In this instrument, known as 'LOHENGRIN' (Armbruster *et al.*, 1976), a 45° magnetic sector field (radius = 4 m) is followed by a 35.35° radial electric field (radius = 5.6 m). The fission products are focused on parabolae characteristic of mass and charge, whilst the energy distribution extends along the parabolae. Another pioneer on-line mass spectrometer for the study of short-lived fission products was 'OSIRIS', which came in service in 1968 (Borg *et al.*, 1971). In this two-directional focusing magnetic analyser the uranium to undergo fission was enclosed in the discharge chamber of the ion source as thin layers of U_3O_8. In this way the ion beam included all elements that are volatile at a temperature of 1500 °C and, in addition, the decay loss of short-lived nuclides was significantly reduced over arrangements which required transport from the reaction zone to the ion source. A comparable instrument, known as 'TRISTAN', which was first constructed for off-line work at Ames Laboratory of Iowa State University (McConnell & Talbert, 1975; Talbert *et al.*, 1976), was subsequently installed on-line to the high flux reactor of the Brookhaven National Laboratory. In it, up to 8 g of ^{235}U was impregnated into a graphite matrix on the anode and then exposed to the intense neutron beam (Gill *et al.*, 1981).

The importance of on-line mass spectrometry to the study of nuclear properties cannot be overemphasized. Further, it is a field which assumes increasing importance as new accelerators make available for study new groups of far-from-stability nuclides. Reviews by Ravn (1979) and Hansen

(1979) provide excellent overviews of the technical aspects of on-line mass spectrometers and the range of investigations which they make possible. In addition, the periodic conferences on *Electromagnetic Isotope Separators* (see §5.4 and related references) and on *Nuclei Far From Stability* (Klapisch, 1976; Hansen & Nielsen, 1981) and on *Atomic Masses and Fundamental Constants* (see §8.12 and related references) provide detailed information on more recent developments.

10

Physical inorganic aspects of mass spectrometry

Ionization and dissociation of molecules

The study of collisions between electrons and molecules began with Lenard in 1902. In 1913 Franck & Hertz demonstrated that electrons require a certain minimum energy in order to cause ionization in a gas, and that this minimum energy, or 'ionization potential', depends upon the nature of the gas. This led to many investigations in which the ion current was plotted as a function of electron energy. Here, the ionization potential corresponds to the energy at which ion current is first detected. These experiments suffered from the limitation that, except for monatomic gases and metallic vapours, the nature of the ions so formed was either not known or, at best, surmised.

In the meantime gaseous ions had been studied extensively by Thomson, Wien, Aston & Dempster, using the methods of positive-ray analysis. In particular, Dempster (1916) had investigated the relative numbers of H_3^+,H_2^+ and H_1^+ ions produced at different pressures by 800eV electrons.

In 1922 Smyth combined these two types of experiments. In his work, ions were created by the impact of electrons whose energy was variable and known, and were then subjected to positive-ray analysis. Within the next ten years this general method underwent many refinements and improvements, notably by Smyth (1925), Hogness (Hogness & Lunn, 1925), Kallmann (Kallmann & Bredig, 1925), Ditchburn & Arnot (1929), Tate & Bleakney (Bleakney, 1930*a*), and their coworkers. During this period the modern arrangement of transverse electron beam was developed in Tate's laboratory, and modified by Bleakney (1929) for operation at low pressures. Since then, the study of electron–molecule collisions has proceeded in many directions, and has revealed much information concerning ionization and dissociation processes.

In this chapter, frequent use will be made of the term 'appearance potential'. This is defined as the minimum energy which a bombarding electron must possess in order to produce a particular ion from a particular molecule – it is the potential at which the ion in question makes its first appearance. The ionization potential is thus one of the appearance potentials.

10.1 Determination of ionization potentials by electron bombardment

When the ion current reaching the collector of a mass spectrometer is plotted against the energy of the ionizing electrons (using an electron bombardment source as in fig. 3.3), a curve of type *A* in fig. 10.1 is obtained.

As anticipated from the Franck & Hertz experiment, the ion current does not begin until the electron energy exceeds the ionization potential, after which it rises rapidly. In this way, the energy needed to remove a single electron from the molecule, and the corresponding energies for two, three or more electrons, may be determined. Such experiments not only substantiate spectroscopic values of the ionization potential but, in the case of multiply charged ions, frequently provide information not hitherto obtained from atomic or molecular spectroscopy.

Fig. 10.1. Ionization curves: *A*, obtained conventionally; *B*, obtained by the method of Fox, Hickam, Kjeldaas & Grove (1951).

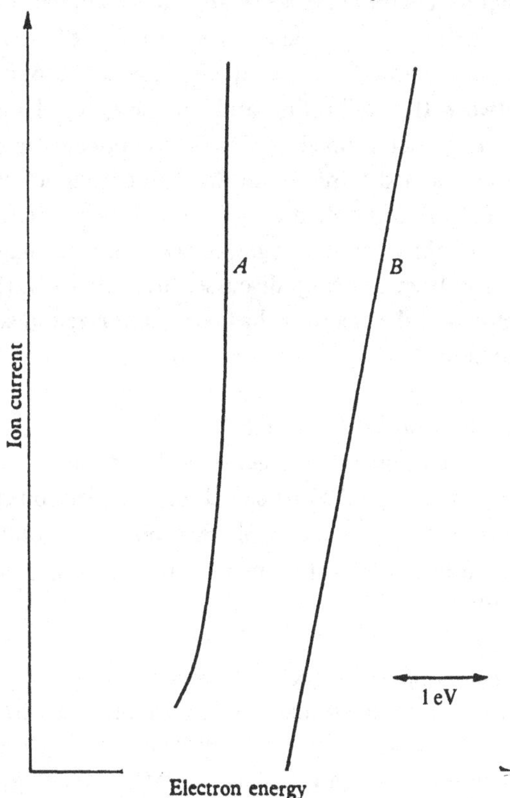

Unfortunately, as fig. 10.1 demonstrates, curve A tails off near the ionization potential over a region of about 1.5 eV. Most of this tailing is the result of an energy spread in the electron beam caused by (a) the emission of electrons from the filament with initial energies of this order, (b) a voltage drop along the emitting length of the filament, (c) the presence of space charges and contact potentials, and (d) the drawing-out field, which removes the ions which are formed. Under these conditions it is difficult to locate the true ionization potential with any degree of certainty. It has, therefore, become customary to make use of a calibrating gas, usually argon, whose ionization potential is well known from spectroscopy. Since the contact potential is critically dependent upon the source conditions, including the source pressure (Waldron & Wood, 1952), the calibrating and sample gases must be introduced into the mass spectrometer as a mixture. This gives rise to two curves of type A, one for each component.

These curves have been analysed in two principal ways. The 'vanishing current' method identifies the ionization potential with the potential at which the ion current is first detectable, while in the other method, 'linear extrapolation', the linear part of the curve is extrapolated to cut the electron energy axis. The former method is regarded as the more reliable (Mariner & Bleakney, 1947), since it eliminates differences in the shape of the ionization efficiency function between the calibrating and sample gases. Thus, the vanishing current point is determined for each component, and the difference between these assumed to represent the difference in ionization potential. The values of ionization potential so found are accurate to 0.1–0.2 eV. The limitations of this method, together with certain suggested empirical variations, have been critically discussed by Robertson (1955). The effect of energy spread of the electrons has been minimized in several ways. These are outlined below.

Monoenergetic electron beam method

Energy selection of electrons was achieved by the use of a 127° electrostatic analyser by Marmet and Morrison (1962) and electron energy spreads of 0.05 eV were obtained. A hemispherical electron monochromator has been used by other workers (Simpson & Kuyatt, 1963; Maeda, Semchuk & Lossing, 1968).

Retarding potential difference (RPD) technique

Here electrons from the electron gun encounter a retarding potential V_R, which removes the lower energy electrons and guarantees that the least energetic of those transmitted have, in fact, zero energy. The transmitted electrons then fall through an accurately known potential, V, following which they create positive ions which are observed in a magnetic analyser. A reduction of ΔV_R in the retarding potential results in an increase in the observed ion current caused by those electrons whose energies lie between eV and $e(V + \Delta V_R)$. Thus, by varying V and recording, for each value of V, the difference in the ion current corresponding to retarding potentials of V_R and $V_R - \Delta V_R$, an ionization probability curve is obtained which is essentially identical with one obtained with monoenergetic electrons. This technique was first introduced by Fox, Hickam, Kjeldass & Grove (1951, 1955) and has been used by several others.

The energy-distribution difference (EDD) method

What virtually amounts to a variation of the RPD method was developed by Winters & Collins (1966) who observed the difference in ion currents produced by two slightly different ionizing electron distributions. In this method the energy spread amongst the ionizing electrons is accepted as a fact of life, but different energy distributions are achieved by varying the potential through which the electron beam falls. In this way, results comparable in accuracy to the RPD method were achieved without modifying the ion source of the mass spectrometer. This EDD method has subsequently been used by others in conjunction with computer data acquisition systems (for example, Johnstone & McMaster, 1974).

Electronic deconvolution techniques

Electronic deconvolution techniques using a conventional electron bombardment source, yielding derivatives of ion current with respect to electron energy, were introduced by Morrison and his coworkers to obtain appearance potentials (for example, Morrison, 1953*a*, *b*, 1954; Dromey, Morrison & Traeger, 1971). These lend themselves admirably to the use of microcomputer techniques.

10.2 Ionization potentials by photoionization

Single photon ionization

In this method, the gas is irradiated with monochromatic radiation and ions produced after ejection of photoelectrons are analysed by a mass spectrometer. The optical part consists of a monochromator equipped with a concave grating and a light source. For example, the Seya–Namioka type of vacuum monochromator may be used at near-normal incidence (70°). In this case, the wavelength resolution is 0.2–0.3 nm, corresponding to an energy resolution of $\sim 0.03\,eV$ at $\sim 12\,eV$. Hydrogen and helium discharges are used as sources. For the single ionization process, the cross section for photoionization has a finite value at threshold, which results in a sharp rise (almost step-function) in the ion current.

In an important development known as photoelectron–photoion coincidence spectroscopy (Brehm & von Puttkamer, 1968), cases may be studied in which the ion is created in excited states. Here a photon of energy $h\nu$ produces a molecular ion A^+ plus a photoelectron whose kinetic energy E is measured (for example, by an electrostatic analyser). The ionization potential for the event is then $I = (h\nu - E)$ and the excitation energy $E^* = I - I_0$, where I_0 is the lowest ionization potential. The photoelectrons and the corresponding ions can be detected by a coincidence spectrometer in which the electron signal is delayed by a suitable interval.

Multiphoton ionization

Ionization of a gas atom/molecule can be caused by the focusing of a powerful laser beam, even when the photon energy (for example, 1.78 eV for a ruby laser) is less than the ionization potential of the gas ($\sim 10\,eV$). This is due to the simultaneous absorption of several quanta by the atom/molecule via virtual states of lifetime $\tau < 1/\nu$, where ν is the laser photon frequency. These intermediate states can be studied by a tunable (dye) laser, the ionization current showing discrete peaks corresponding to two-photon, three-photon, etc., absorption on scanning the frequency of the laser. The population in a selected quantum state can also be studied without interference from other molecules or states and the sensitivity considerably enhanced by this 'resonance ionization' technique.

10.3 Interpretation of the ionization curves for atoms

The ionization threshold laws were examined by Wigner (1948) and Geltmann (1956) who showed that under certain restrictions the cross section $\sigma(E) = C(E - E_0)^{n-1}$, where E_0 is the threshold energy and n is the number of outgoing electrons. This leads, for the case of electron impact ionization to a linear law for single ionization, a quadratic law for double ionization, and so on. Similarly for photoionization, a step function is indicated for single ionization, a linear law for double ionization, and so on. Actual experiments show step functions for the photoionization of Ar, Kr, and Xe, and finite slopes and more complicated ionization curves for other gases. For electron impact, a linear threshold law for single ionization and a quadratic law for double ionization appear to exist in some cases, with breaks in slope that were interpreted at the time in terms of excited states (Marr, 1967; Marchant, Pacquet & Marmet, 1969).

The structure of the ionization curves is profoundly influenced by the effects of autoionization (Fano & Cooper, 1965). This process represents the spontaneous ejection of an electron from states that lie at energies above the lowest ionization energy of the atom. The higher members of the Rydberg series lie above the first ionization limit and can thus decay by autoionization into the lowest ionization continuum. This effect is seen strongly in the threshold region, imparting a rich structure to the ionization curve that is not easily resolved.

10.4 Determination of ionization cross sections

Bleakney (1930*a*, *b*), working in Tate's laboratory at the University of Minnesota, was the first to use the mass spectrometer to study the probability of multiply charged ion formation as a function of electron energy. He obtained ionization efficiency curves for Ar^+, Ar^{2+}, Ar^{3+} and Ar^{4+}, as shown in fig. 3.4, and also for Ne^+, Ne^{2+}, Ne^{3+}, Hg^+, Hg^{2+}, Hg^{3+}, Hg^{4+} and Hg^{5+}.

Although the ionization efficiency curves have the same general form for all gases, there are significant differences in the absolute values of their maximum ionization cross sections. This is shown clearly in results presented by Barnard (1953) for He, Ne, N_2, CO and Ar, which bear out the expectation that multielectron molecules are more vulnerable to ionization that those with small total nuclear charge.

10.5 Determination of energies of excited states of molecules

The ions formed by electron impact may be formed in excited states, as well as in the ground state. Consider the case of the nitrogen molecule, whose ground state is known to be (Mulliken, 1932)

$$KK(\sigma_g 2s)^2(\sigma_u 2s)^2(\pi_u 2p)^4(\sigma_g 2p)^2, {}^1\Sigma_g^+.$$

The first ionization potential of nitrogen corresponds to the removal of an electron from the $(\sigma_g 2p)$ orbital, and is known, from spectroscopic evidence (Worley, 1943), to be 15.577 eV. The second ionization potential represents the removal of an electron from the $(\pi_u 2p)$ orbital, and has the value 16.93 eV (Worley, 1953).

The appearance potential of the N_2^+ ion has been determined frequently by mass spectroscopic methods, a representative value being 15.62 eV (Katzenstein & Friedland, 1955). This measurement clearly refers to the first ionization potential. Presumably, the technique of Fox *et al.* (1951, 1955) is capable of revealing, by a break in the ionization efficiency curve, the second ionization potential and, possibly, others, including a third known by spectroscopists to occur at 18.733 eV (Worley & Jenkins, 1938). There may be evidence (McDowell, 1954) for the 16.93 eV state (that is, ~ 1.2 eV above ground) in Hagstrum's (1951) conventionally obtained N_2^+ curve.

In other cases, for example with oxygen (Tate & Smith, 1932) and ammonia (Mann, Hustrulid & Tate, 1940), breaks in the ionization efficiency curve corresponding to excited states of the molecular ions were clearly observed. The work of Fox *et al.* provided a compelling example of such observation. Here, the doublet ground state of krypton, with a separation of 0.66 eV, was successfully resolved.

10.6 Determination of molecular bond dissociation energies

Frequently ionization, as a result of electron impact, is accompanied by dissociation of the target molecule. In general, in order to understand this dissociation completely, one must know (a) the specific charge and state of excitation of each of the dissociation products, (b) their initial kinetic energies and (c) the appearance potential for the dissociation products in question. In early mass spectroscopic investigations only (a) and (c) were studied. By use of retarding potentials, Lozier (1931, 1933, 1934) and Hanson (1937) elucidated (b) and (c). Later, Hagstrum (1951) determined (a), (b) and (c) for each ion observed in the dissociation of several diatomic molecules.

The fact that dissociation products may possess zero or non-zero kinetic energy may be understood by reference to fig. 10.2. This diagram illustrates, for a diatomic molecule AB, the transitions from the ground state (curve 1) which are permitted by the Franck–Condon principle. These transitions, it will be recalled, are represented by vertical lines on this diagram, and these should lie between the vertical dotted lines, which represent the classical turning points of the ground state vibrational motion of the molecule. Curve 2 represents a stable ionic state, while curve 3 represents a repulsive one. A transition to the latter inevitably leads to dissociation products with non-zero kinetic energy as, for example, the transition GP, where the dissociation products share the kinetic energy T''.

Dissociation may also occur in a small fraction of transitions to the stable ionic state of curve 2, with zero kinetic energy for the transition GM, and with non-zero kinetic energy for the transition GN. In the latter case dissociation takes place with release of kinetic energy T'.

When it is possible for the dissociation products to possess zero kinetic energy, as in transitions to curve 2, the appearance potential may be used to deduce the bond dissociation energy. Thus

$$AB + e^- + I_{A^+,B} \rightarrow A^+ + B + 2e^-, \tag{10.1}$$

where $I_{A^+,B}$ is the minimum kinetic energy which the bombarding electron must possess in order to create from the molecule AB, the ion A^+ and the neutral atom B. That is, $I_{A^+,B}$ is the appearance potential for A^+ from the molecule AB. The ionization potential of A, I_A, will presumably be known from other studies, whence the energy, E_D, required to break the molecular bond is

$$E_D = I_{A^+,B} - I_A. \tag{10.2}$$

In the case of hydrogen, for example, $I_{H^+,H}$ is known experimentally to

Fig. 10.2. Potential energy vs nuclear separation for a typical diatomic molecule AB.

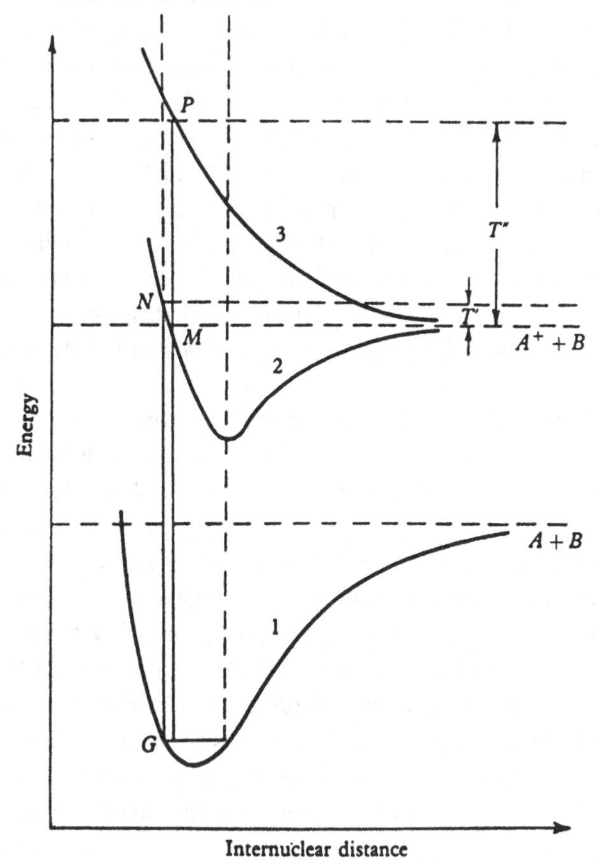

be 18.0 ± 0.2 eV (Hagstrum & Tate, 1941), and $I_H = 13.57$ eV. These results lead to a value of 4.43 ± 0.2 eV for the strength of the hydrogen molecule bond, in excellent agreement with the spectroscopic value given by Herzberg (1950). Many bond energies have been obtained in this manner, including those of hydrocarbons and other compounds (Stevenson, 1942, 1943, 1950, 1951; Hipple & Stevenson, 1943; Dibeler, 1947; and many others).

The first attempt to study the initial kinetic energy of dissociation products was made by Bleakney (1930*b*) who used retarding potentials in the source of the mass spectrometer to discriminate between ions with zero and non-zero energy. Later, Hagstrum & Tate (1941) studied initial kinetic energies by their effect upon the shapes of the mass spectral peaks. Retarding potentials following mass analysis (§10.7) were employed by Hipple, Fox & Condon (1946) in their study of metastable ions, and were suggested by Fox & Hipple (1948) as a method of determining initial kinetic energies. In the subsequent work of Hagstrum (1951), retardation following mass analysis was used for this purpose. Here, the specific charge and appearance potential of each of the ions observed was also determined.

In these refined experiments, concerned with the ionization of certain diatomic molecules by electron impact, Hagstrum learned much concerning both simple and dissociative ionization of CO, N_2, NO and O_2. In the course of this work he also deduced a number of characteristic energies, namely, the dissociation energies of CO, N_2, and NO, the sublimation energy of C, the electron affinity of O, and the energy of the excited state of O^-. The relation between many of these results and their spectroscopic counterparts was discussed by Hagstrum, by McDowell (1954) and by Eyring (1955).

As already stated, the direct photoionizing or electron impact transitions obey the Franck–Condon principle, so that the positions and momenta are unchanged by the transition and the accessible vibrational levels depend on the relative positions and shapes of the potential energy curves. If the curves are similar, only the ground vibrational state of the ion is accessible, while if the ion and molecular distances are different the higher vibrational levels become accessible and the maximum transition probability shifts to a higher vibrational level. Thus a study of electron (or photon) impact thresholds should give information about the relative probabilities of transitions to the vibrational states (the Franck–Condon factors). One thus expects a 'staircase' structure with step-heights proportional to Franck–Condon factors for photoionization, and for electron impact ionization a series of straight-line segments, the onset of each segment representing an excited state or higher vibrational level. A large number of studies have been made, and the results summarized in a review by Rosenstock (1976). The structure of the ionization curves is complex due to the contributions of direct and autoionization. Here, however, the actual smoothed-out threshold curves have rather similar shapes for various molecules, so that, with a rare gas curve for calibration, ionization potentials can be determined to within 50–100 mV. For photoelectron studies the use of photoelectron–photoion coincidence spectroscopy, as described earlier, is particularly effective. Here, one identifies the ejected electrons corresponding to each transition to a set of vibrational levels, and obtains (for each transition) the different vibrational progressions as a 'bar-graph' of the Franck–Condon factors. Detailed studies have been made for NO, N_2, O_2, H_2O, NH_4, CH_4, C_2H_2, CO_2 and CO, as well as for methyl, benzyl, etc., radicals and fragmentation processes have been studied in detail in N_2O, C_2H_2, etc. (Rosenstock, 1976).

The photoelectron–photoion coincidence technique is also effective in the study of the kinetic energy released upon fragmentation of molecular ions (Stockbauer, 1977). Here the photon pulse traverses the ionizing region and is detected. Some photons create photoions and photoelectrons. The photoelectrons are directed towards an energy analyser (say, a 127° cylindrical analyser followed by an electron multiplier). In the meantime some of the photoions fragment. When the electron energy analyser records a photoelectron with zero initial electron energy any fragment ions associated with that event are accelerated into a time-of-flight mass spectrometer to determine their masses and their initial kinetic energies (the latter from the peak widths). In this way, for example, the kinetic energy of the fragment CH_3^+, formed from CH_4^+, was observed as a function of photon energy. A similar technique has been used by Niwa, Nishimura, Nozoye & Tsuchiya (1979). An overall review of work done using the technique of photoelectron–photoion coincidence spectroscopy has been given by Eland (1980).

The fragmentation of large molecules is fairly well described by the quasi-equilibrium theory of mass spectra developed by Eyring and coworkers (Rosenstock, Wallstein, Wahrhaftig & Eyring 1952). Rosenstock (1968) has summarized the essential features and developments of this theory. In this theory, the fragmentation of the molecular ion is considered as separate from the ionization by the impacting electron and the distribution of the internal excitation energy. It is assumed that between the ionizing event and the decomposition, there is ample time for the randomization of the excitation energy – the parent molecular ion with its many low-lying electronic states transferring the electronic energy into vibrational energy. The excess energy is thus not retained in the bond from which the electron is removed, but distributed over all internal degrees of freedom. The subsequent fragmentation process (which takes place when enough vibrational energy is concentrated in some particular bond) is thus described as a series of competing, consecutive, unimolecular reactions. The rate constants for these unimolecular reactions can be calculated by the activated complex theory.

The theory has been applied to a number of molecules, listing possible modes of decomposition and calculating reaction rates using activation energies evaluated, in turn, from appearance and ionization potential measurements. The calculations show good agreement with measured fragment ion abundances.

10.7 Study of metastable ions

Hipple, Fox & Condon (1946) demonstrated that some of the primary ions formed in hydrocarbon gases by electron impact are metastable, with a half-life of $\sim 10^{-6}$ s. These ions are sufficiently stable to be drawn in quantity from the ion source and accelerated, but they spontaneously dissociate during their passage through the analyser section of the mass spectrometer. Their existence was suspected from the occurrence in mass spectra of certain broad peaks at positions corresponding to non-integral masses.

As an example, consider what happens when dissociation occurs at the boundary between accelerator and analyser. The metastable ion, of mass M_0, will acquire, during acceleration, kinetic energy $T_{M_0} = \frac{1}{2}M_0 v^2 = Ve$. If this ion dissociates into an ion of mass M, and a neutral fragment of mass $M_0 - M$, the new ion will possess kinetic energy $T_M \sim (M/M_0)Ve$. It will continue through the analyser in accordance with the usual relation $r = \sqrt{(2MT)}/eB$. It will, therefore, appear at the collector for the same combination of V and B as does a normal ion of mass

$$M^* = M^2/M_0. \tag{10.3}$$

In general, dissociation may occur at any stage during acceleration or analysis, and there is the release of internal energy upon dissociation to be taken account of, so that conditions are complex. The final result, however, will be a broad band in the mass spectrum located between M_0 and M^*, and a broader than normal peak at M^*.

The ions resulting from the dissociation of these metastable systems were identified by Hipple, Fox & Condon by exploiting the fact that their kinetic energies, as mentioned, are less than those of normal ions. A retarding field, inserted between the analyser and the collector, provided the means of determining the kinetic energies and, consequently, the values of M and M_0. Throughout the work the vacuum was such that there was no question of dissociation by collision, an effect which had given rise to 'bands' in several early mass spectrographs (Mattauch & Lichtblau, 1939).

In the study of *n*-butane, for example, Hipple, Fox & Condon showed that three prominent peaks arising from metastable ions are

$$C_4H_{10}^+ \rightarrow C_3H_7^+ + CH_3, \quad M^* = 31.9;$$
$$C_4H_{10}^+ \rightarrow C_3H_6^+ + CH_4, \quad M^* = 30.4;$$
$$C_4H_7^+ \rightarrow C_3H_5^+ + H_2, \quad M^* = 39.2.$$

Several other hydrocarbons were similarly studied.

Ion–molecule reactions

The study of ion–molecule reactions dates back to J. J. Thomson who observed 'secondary rays' from collision processes in his parabola spectrograph. Aston discovered that these form well defined bands (Aston bands) which are related to metastable ions. In view of their significance in basic chemistry, ion–molecule reactions have been studied intensively and the field has received much impetus from space studies. These studies yield useful thermochemical data and valuable insights into the mechanisms of reactions that occur in plasmas, in the upper atmosphere and elsewhere. Techniques which have been developed for the study of ion–molecule reactions are briefly summarized below. The subject is extensively treated by Stevenson (1963), McDaniel *et al.* (1970), Franklin (1972) and Durup (1974).

10.8 Ion–molecule reactions observed in conventional mass spectrometers

Early studies were made with an electron bombardment source run at pressures $> 10^{-3}$ Pa ($> 10^{-5}$ Torr) and with differential pumping of the analyser. Secondary ions were distinguished from primary ions by their pressure dependence (p^n for a reaction of order n), and by varying the repeller voltage (which varies the energy of the primaries).

10.9 Čermák–Herman technique

This simple and useful technique (Čermák & Herman, 1961) permits collection of secondary ions with no contamination by primaries. The bombarding electrons pass through what is normally the ionization region of the source with an energy that is less than the ionization potential. Thus, no primary ions are produced in that region. As they approach the trap or collector, however, these same electrons acquire sufficient energy to create ions. Some of the ions so-produced are accelerated backwards into the ionization region where, if they undergo no collisions, they strike the walls of the chamber. Others initiate ion–molecule reactions. The secondary ions so-produced have little kinetic energy and are drawn out of the source of the mass spectrometer.

10.10 Ion–molecule reactions produced in pulsed ion sources

Here (Tal'rose & Frankevich, 1960), one uses a pulsed beam of electrons followed by a variable time delay during which no voltages are applied inside the ionization chamber. This is followed by an ion extraction pulse that draws both primaries and secondaries into the acceleration region en route to the analyser.

10.11 Afterglow techniques for study of ion–molecule reactions

While laboratory discharges provide a suitable environment for the study of ion–molecule reactions, electron excitation processes complicate the situation. Hence, quantitative studies of ion–molecule reactions may be prudently confined to the afterglow regime, that is, following the input energy pulse. In this case, a measurement of the decay rate of the relevant ions allows the rate constant for a reaction to determined. The general procedure is to initiate the discharge by a pulsed rf oscillator and sample the ions through a pinhole which leads into a mass spectrometer. The decay of the ion current is measured using an electron multiplier. A variation is the 'flowing afterglow technique' (Fehsenfeld, Ferguson & Schmeltekopf, 1966) in which the recombining afterglow plasma is spatially separated from the active discharge by a rapidly pumped flow of the discharge gas (see fig. 10.3). In this way spatial resolution is substituted for time resolution. In a typical case, the flow velocity of the gas is $\sim 10^2$ m/s, with the gas being exhausted by a Roots-type blower backed by a fast forepump. The ion composition of the plasma is monitored at the end of the reaction zone by a quadrupole mass spectrometer in which both the primary and product ions are observed as a function of time. Another advantage of this technique is the chemical manipulation of the glows that is made possible by introducing neutral reactants into the afterglow. A great deal of circumstantial information relating to the technique has been compiled by McDaniel *et al.* (1970).

10.12 Use of drift tubes for study of ion–molecule reactions

Drift tubes in which ions produced by a source drift and diffuse down a tube under an applied electrostatic field have also been used to study ion–molecule reactions. In this work, which requires knowledge of the diffusion properties of ions, the ion population arriving at the remote end of the drift tube is determined with a mass spectrometer. In a more elaborate version, the drift tube is placed between two mass spectrometers, the first serving as the source of ions whilst the second serves as the analyser. By introducing appropriate materials into the drift tube, the reactions $Ar^+ + O_2$, $Ar^+ + N_2$, $Ar^+ + NO$ and $Ar^+ + CO_2$ have been studied in the energy range less than 1 eV (Thomas, Barassin & Burke, 1978).

10.13 Use of ion-cyclotron resonance for study of ion–molecule reactions

This technique is useful in the study of slow ion–molecule collisions. As described in §6.10 an ion moving in a uniform magnetic field B describes a helical path between successive collisions, the 'cyclotron frequency' being $f = Be/2\pi M$ (equation (6.1)). If, in addition, an rf field is applied normally to B, there is an absorption of rf energy when the exciting frequency matches the cyclotron frequency. The ionic species can be identified by the frequency at which absorption occurs and the ionic density can be calculated from the magnitude of the absorption. To give an example, an N^+ ion moving in a magnetic field of 0.15 T has a cyclotron frequency of 1.7×10^5 Hz.

In one form of apparatus (fig. 10.4) ions are formed in an electron bombardment source (region 1) and travel down the ion-cyclotron (ICR) resonance cell, which is the analyser section (region 2), to the total ion monitor (region 3). Ions may also be formed by laser desorption and by the 'soft ionization' methods described in §3.10. 'Trapping voltages' are applied

Fig. 10.3. Flowing afterglow technique for study of ion–molecule reactions. (After Fehsenfeld, Ferguson & Schmeltekopf, 1966.)

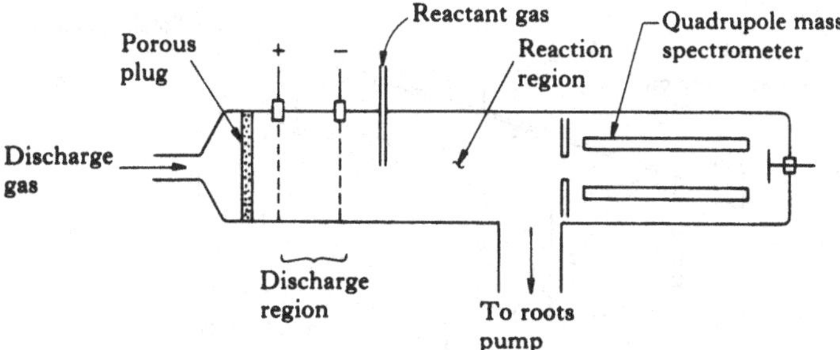

to the side plates of the ICR cell to constrain the ion motion to the central region, thus enhancing the reaction kinetics. The ICR cell is located in a very homogeneous magnetic field, and a marginal oscillator detector system monitors the resonance (as in a nuclear magnetic or electron spin resonance apparatus). The ICR cell is used as the capacitive element in a resonance circuit and a phase-sensitive detector senses the energy level changes in the oscillator.

Double resonance techniques can also be used in ion-cyclotron resonance. Thus, in studying the reaction $A^+ + B \rightarrow C^+ + D$, the main rf field is set for the detection of C^+. If now, an auxiliary rf field at the resonance frequency of A^+ is applied, a significant change in intensity of C^+ is observed, showing that A^+ is the reactant ion. Detection of couplings between ions in complex reactions is also possible. In addition, under low pressure conditions, ICR has the advantages of high resolution and sensitivity. Excellent reviews exist of this technique and its applications (Gross & Wilkins, 1971; Baldeschwieler & Woodgate, 1971; Marshall, 1985).

The ion-cyclotron resonance technique has been widely used for the study of ion–molecule reactions with both inorganic and organic vapours. Thus, for example, Davidson, Powers, Sue & Sue (1977) have studied equilibrium constants of proton-transfer reactions of the type $AH^+ + B \rightleftharpoons BH^+ + A$, whilst gas-phase reactions of alkyl nitrates, involving mainly hydride abstraction and NO^+ transfer, have been studied by Fand & McMahon, (1978). Ions trapped in an electron space charge, in order to enhance the rate of ion–molecule reactions (Bourne & Danby, 1968; Herod & Harrison, 1970), have also been used with ICR (Freiser, 1978).

Fig. 10.4. Ion cyclotron resonance with drift region for ion–molecule reactions. The electron beam and the magnetic field are collinear. (After Baldeschwieler, Benz & Llewellyn, 1966.)

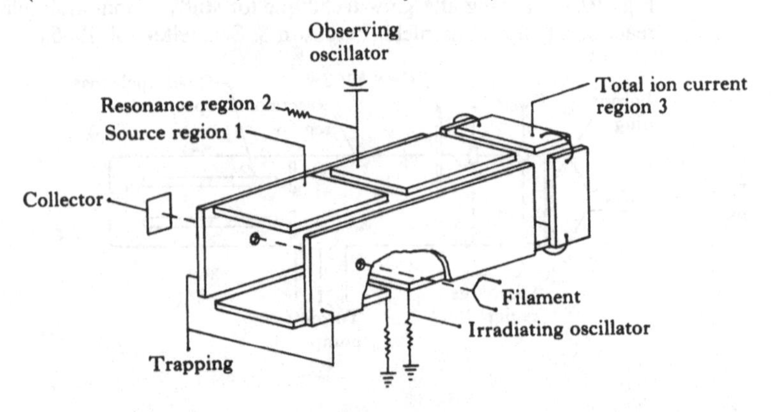

10.14 Ion beam techniques for study of ion–molecule reactions

A beam of monoenergetic ions of selected mass is either (a) transmitted through a static low-pressure target or (b) allowed to intersect a molecular beam. The primary and secondary ions emerging after collision are analysed with respect to energy and mass. If desired, the post-collision analyser can be rotated about the scattering centre for angular distribution analysis. The wide separation of the source and reaction regions ensures that the reactant ions have decayed to ground states, and enables one to achieve higher densities in the reaction region and in the primary ion source. Also, the energies of the primaries can be closely controlled. Various configurations have been used by different workers. Thus, Lindholm (1954), as also Tal'rose *et al.* (1966), used two magnetic analysers, one in a vertical plane and the other in a horizontal plane with the reaction region between them. Giese & Maier (1966) used a 2.54 cm (1.00″), 90° analyser to form a primary beam at 2–100 V in tandem with a 30.5 cm (12.0″), 90° analyser following the reaction chamber. A more sophisticated arrangement was used by Futrell and coworkers (Futrell & Mille, 1968; Futrell & Tiernan, 1972), comprising two double focusing analysers in tandem, with a deceleration lens and collision chamber separating them. Typical arrangements for angular distribution studies are those of Turner, Fineman & Stebbings (1965) and Champion, Doverspike & Bailey (1966) who used rotatable quadrupole mass filters as post-collision analysers. Herman & Wolfgang (1972) used an apparatus capable of both angular and energy distribution studies. A later variant is the 'merging beam' technique where the problem of low beam intensities at low energies is circumvented by making the ionic and molecular beams collinear. For beams of equal masses and energies E and $(E-\Delta E)$, the centre-of-mass energy $T=(\Delta E)^2/8E$, so that for $E=3\,\text{keV}$, $\Delta E=100\,\text{eV}$, $T=0.3\,\text{eV}$. Here the primary ion beam and the molecular beam (obtained again from an ion beam traversing a charge-exchange cell) are merged with the aid of a merging magnet, and the product ions analysed.

10.15 Some general comments on results

The literature on ion–molecular reactions is vast and has been reviewed comprehensively in the books by McDaniel *et al.* (1970), Franklin (1972) and elsewhere by Ferguson (1975), Jennings (1978) and others. The principal observations relate to cross sections and reaction rates. If N is the density of molecules in the reaction chamber, and x is the length of interaction, the primary and secondary beam intensities are related by

$$I_S = I_P[1-e^{-Nqx}] \approx I_P N q x, \tag{10.4}$$

where q is the cross section involved. Thus, q can be calculated from measurements of I_S and I_P. Also, the rate constant of the reaction, k, is defined by

$$dN_S/dt = kN_PN, \tag{10.5}$$

where N_S, N_P and N represent the number densities of the primary and secondary ions and the neutral vapour. As

$$I_S = e(dN_S/dt) \tag{10.6}$$

and

$$I_P = eN_P\bar{v}_P/x, \tag{10.7}$$

$$k = q\bar{v}_P. \tag{10.8}$$

With a single mass spectrometer, the energy of the primaries is determined by the repeller potential E_r, and it can be seen that

$$\bar{v}_P = \sqrt{(eE_rx/2M_P)}, \tag{10.9}$$

so that the reaction rate is $q\sqrt{(eE_rx/2M_P)}$.

A large body of data is extant on cross sections and reaction rates – in fact, most of the known simple reactions of ions with chemically neutral gases have been studied, and there are some measurements with O, N, H and other free radicals. Negative ion reactions have also been studied (Franklin & Harland, 1974). In addition, the kinetic energy dependencies of the rate constants have also been investigated. The particular studies that relate to ion–molecule reactions in the upper atmosphere and ionosphere are summarized in Chapter 14.

A special type of ion–molecule reaction – charge transfer (for example, $Ar + NO^+ \rightarrow Ar^+ + NO$) – has also been extensively studied. Energy resonance appears to be a requirement in this case and the manifold of vibrational levels for molecules enhances this possibility. An additional requirement is a favourable Franck–Condon overlap between the ion and neutral states of reactant.

Another important reaction that has been studied with negative ions is that of associative detachment,

$$A^- + B \rightarrow AB + e^-.$$

Here the energy required to detach an electron is provided by the formation of a bond between the reactants. A necessary condition for this reaction is that the reactants approach on an attractive potential curve into the autodetaching region of AB^-.

Considerable insight into reaction mechanisms has been obtained by angular distribution measurements, as has been summarized by Massey & Burhop (1972).

High temperature mass spectrometry

The mass spectrometer is an invaluable aid to the study of the chemistry of systems at high temperatures as it affords a precise means of determining the concentrations (partial vapour pressures) of all the species in vapour phase. First used by Inghram (1953) and Honig (1954), it has become a routine technique for investigating the thermodynamics of solid–vapour equilibria at high temperatures, and several reviews of this field are available (Goldfinger, 1965; Grimley, 1967).

In this technique, a molecular beam that is representative of the vapour in thermodynamic equilibrium is obtained from a Knudsen effusion cell into which the sample material has been introduced. The Knudsen cell is a black-body cavity with an effusion orifice much smaller than the sample area and is heated to the desired temperature by a filament heater or by electron bombardment. Both single and double cells are employed (fig. 10.5). The molecular beam effusing from the cell enters the ion source of the mass spectrometer, which is usually of the electron bombardment type. As outgassing from the heated walls of the cell results in a background, a movable shutter is employed to distinguish the effusing species from the background. The shutter also isolates the ion source. Sector field instruments, time-of-flight analysers and quadrupole analysers are used.

The mass spectrum contains the contributions of the parent and fragment ions of all the effusing species. From this one must determine the

Fig. 10.5. Knudsen cell assembly and electron bombardment ion source.

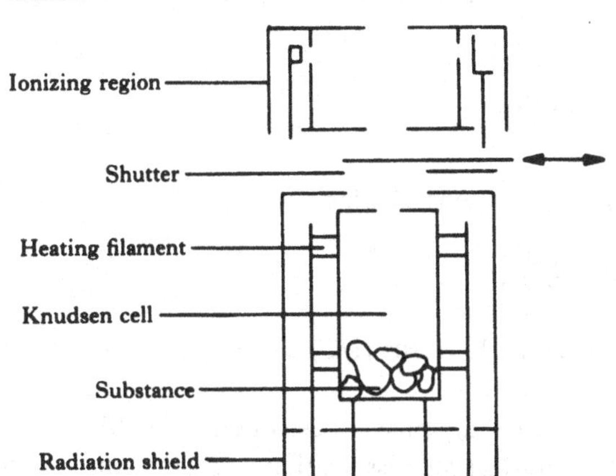

partial pressures of all the species present, as the various thermodynamic data are calculated from the partial pressures. The first step is to identify all the ionic species from the masses and the expected isotopic abundances. The next is the identification of the neutral vapour species by associating with each ion, one or more of the neutral species – for example, with alkali halides, MX, one obtains ions M^+, MX^+, M_2X^+, $M_2X_2^+$ and smaller fractions of others. An associated measurement is the ionization efficiency curve, that is, the ion current as a function of the ionizing electron energy. The appearance potential of an ion indicates whether it is formed by simple ionization or by dissociative ionization (in the latter case, the appearance potential (AP) exceeds the ionization potential (IP) by the dissociation energy in the simplest case).

To relate the ion intensity to the partial pressure, P, one notes that the rate of effusion

$$v = P\bar{c}/4RT. \tag{10.10}$$

A linear relationship of the ion current (I^+), with excess electron energy is assumed, so that one has an effective cross section

$$\langle\sigma\rangle = \frac{\sigma \cdot \Delta E}{E - AP}. \tag{10.11}$$

This yields

$$I^+ = \text{const}\, v\langle\sigma\rangle\Delta E/\bar{c} \tag{10.12}$$

$$P = KI^+\, T/\Delta E. \tag{10.13}$$

The constant can be determined by calibration with a standard (for example, Ag).

The partial pressures ($\approx$ fugacities) can yield a variety of thermodynamic data. Thus, equilibrium constants can be calculated using Van't Hoff's equation:

$$\frac{\partial \ln K}{\partial T} = \Delta H^0/kT^2. \tag{10.14}$$

Rearranging this equation, one obtains

$$\frac{\partial \ln K}{\partial(1/T)} = \frac{-\Delta H^0}{R}. \tag{10.15}$$

Thus, the slope of the ln K versus $(1/T)$ curves yields ΔH^0, the reaction enthalpy.

A variety of reaction enthalpies have been obtained in this way – heats of sublimation and vapourization, dissociation energies, heats of formation and reaction, etc. A large number of elements, compounds and alloy systems have been studied and the applications reviewed (Drowart & Goldfinger, 1967; Grimley, 1967; Drowart, 1985). The mass spectrometer has a particular advantage at high temperatures where complex systems with several gas–solid and gas–gas equilibria are encountered. A typical case is the vapourization of carbon, where the equilibrium vapour involves species C, C_2, C_3, C_4 and C_5. To cite another example, heats of dissociation can be directly determined when molecular species and dissociation products can both be measured, for example Cr_2O_3 which yields Cr, O, CrO, CrO_2. When the dissociation products are not all measureable, one can still calculate heats of dissociation in conjunction with the thermochemical data.

Another aspect of 'high temperature mass spectroscopy' is the growing application of mass spectrometry to plasma diagnostics. This involves the continuous monitoring of M/e species and their abundances in plasma formed in gas discharges, afterglows, flames, etc. Usually, the ions are extracted from the plasma region through an orifice in a plate of adjustable bias and mass analysed, for example in a quadrupole instrument. The influence of the Debye sheath formed when the plasma touches the metal orifice plate, the sheath length-to-mean free path ratio, and other factors to ensure correspondence between observed ion currents and the ion densities in the plasma have been discussed by various authors (for example, Hasted, 1974). Diagnostic procedures for the study of gas discharges have been discussed by Studniarz & Franklin (1968) and Vasile & Smolinsky (1974), whilst Fristrom (1975) has reviewed the sampling of flame processes in a rocket chamber, in spark ignition, etc.

11

Applications of mass spectroscopy to organic chemistry

11.1 Historical development

The application of mass spectrometry to the quantitative analysis of hydrocarbon and other mixtures became fairly routine by the early 1950s. In this work, mass spectroscopists observed regularities which suggested that chemical properties influenced the fragmentation behaviour of organic compounds subject to low energy (~ 70 eV) electron impact. It was realized that this could prove a powerful tool for structural analysis, and the elucidation of organic structures in terms of fragmentation patterns was pursued extensively with low resolving power (< 500) instruments.

The next development was the use of high resolving power instruments ($> 5 \times 10^4$ at 10% of peak height) to determine with greater precision the masses of ions resulting from fragmentation. Since precise masses of all possible combinations of C, H, N, O are readily calculable, an unambiguous determination of the chemical composition of a molecular or fragment ion may be made from its mass. In this way, the instrumental capability for structural analysis of complex compounds was improved, especially for natural products and molecules of biochemical interest. The advent of computerized accumulation and analysis of spectra greatly enhanced the scope of this type of study. In recent years, the development of 'softer' techniques of ionization, namely, field and chemical ionization, with their simpler fragmentation behaviour and large abundance of parent (or parent $\pm$ 1) peaks, has further extended this technique.

Finally, a dramatic advance occurred with the coupling of the gas chromatograph to the mass spectrometer with an associated data acquisition and analysis system (a system designated as 'GC–MS/computer'). In such a system, the gas chromatograph separates the components of an organic mixture and delivers them, one by one, to the mass spectrometer which, in turn, delivers a fast-scan output to the computer. In such a system, nanogram quantities of mixtures of complex biochemical compounds can be analysed with high speed and accuracy.

Thus, mass spectroscopy has come to play a key role in organic chemistry, both as an analytical tool for the identification of compounds and, along with infrared, ultraviolet, and nuclear magnetic resonance spectroscopy, as a tool for the elucidation of molecular structure. Whilst this is not the place to describe in detail these specialized and rapidly expanding areas, some general descriptions will be given below. For a more extensive survey of these matters, with emphasis on the organic chemistry involved, the interested reader is directed to the *Specialist Periodical Reports* prepared through the Royal Society of Chemistry (see, for example, Johnstone, 1981) and to the review issues of *Analytical Chemistry*. Attention is also drawn to the recommendations of the IUPAC Commission on Molecular Structure and Spectroscopy for symbols and nomenclature (Beynon, 1978).

11.2 Hydrocarbon analysis

The earliest significant application of the mass spectrometer to organic chemistry, that of hydrocarbon analysis, was an outgrowth of the study of ionization by electron impact. While this approach is now largely of historical interest, it illustrates features that are incorporated in current practice.

In this work, the sample was introduced as a gas into the source of the mass spectrometer, where it was subjected to electron bombardment. Under such bombardment, both ionization and dissociation of the hydrocarbon molecules take place, each hydrocarbon constituent giving rise to a characteristic mass spectrum or 'cracking pattern', as it was frequently called. Fig. 11.1 shows a portion of the mass spectrum resulting when *n*-butane (C_4H_{10}) is introduced into the source. The strong peaks at $(m/z)^{\dagger} = 39$, 41, and 43 are due to the $C_3H_3^+$, $C_3H_5^+$ and $C_3H_7^+$ fragments, respectively, while the peak at $m/z = 58$ is from the undissociated parent molecule. The small peak at $m/z = 59$ represents such a molecule with one atom of ^{13}C in place of the predominant ^{12}C.

Suppose one has a mixture of the five hydrocarbons listed in table 11.1, the amount of each constituent being unknown, but desired. The first step is to determine the cracking pattern of each constituent by introducing the pure gases by turn into the source of the mass spectrometer and obtaining graphs of the sort shown in fig. 11.1 for *n*-butane. Thus, for methane, the peak at $m/z = 15$ is almost 85% as intense as the base peak at mass number 16, which corresponds to the parent or undissociated molecule. Peaks are also present at mass numbers 1, 2, 12, 13, and 14, but are not included here

† Here, following the convention in chemistry, z is the charge number and m/z the mass to charge ratio.

Fig. 11.1. Partial mass spectrum of *n*-butane.

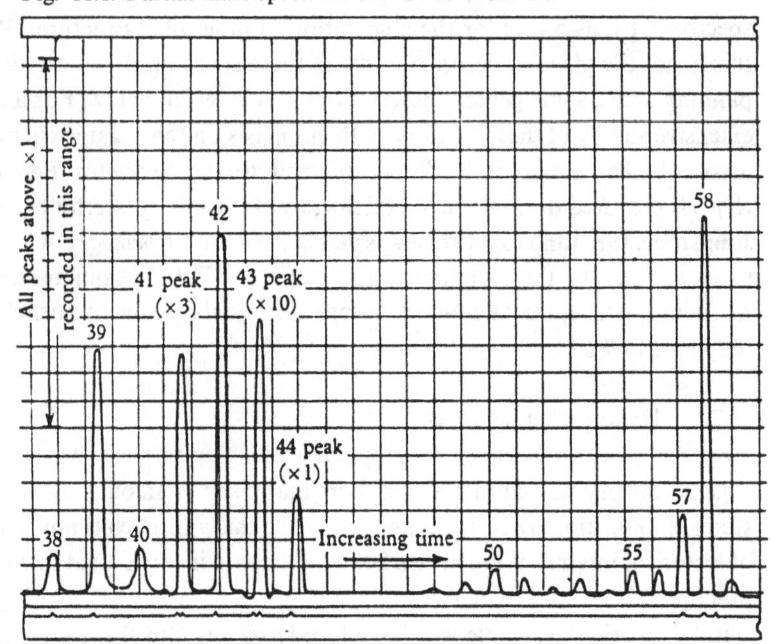

Table 11.1. *Cracking patterns of some hydrocarbons*

Hydrocarbon		Mass number 15	16	27	28	29	30
Methane	CH_4	84.8	100	—	—	—	—
Ethane	C_2H_6	4.42	0.07	33.2	100	21.5	26.2
Propane	C_3H_8	5.90	0.16	40.3	59.5	100	2.19
n-Butane	C_4H_{10}	4.83	0.11	37.9	33.3	44.8	0.98
Iso-butane	C_4H_{10}	5.55	0.16	27.9	2.51	6.37	0.14
Mixture		16.73	13.44	36.16	77.78	59.31	13.00

Hydrocarbon		Mass number 41	42	43	44	57	58
Methane	CH_4	—	—	—	—	—	—
Ethane	C_2H_6	—	—	—	—	—	—
Propane	C_3H_8	12.9	5.86	23.1	29.2	—	—
n-Butane	C_4H_{10}	28.4	12.4	100	3.36	2.46	12.6
Iso-butane	C_4H_{10}	37.7	33.9	100	3.32	3.01	2.75
Mixture		9.41	4.99	20.74	14.53	0.25	0.50

since they are of smaller intensity and are not essential for the analysis. Similar data are given in table 11.1 for the other four constituents, ethane having its base peak at $m/z = 28$, propane at 29, while *n*-butane and isobutane (2-methylpropane) show maximum peaks at 43. Extensive compilations of such mass spectral data for numerous chemical compounds have been given in book form by Stenhagen, Abrahamson & McLafferty (1969, 1974), for example.

In addition to the cracking patterns, it is necessary to know the sensitivity of the mass spectrometer for each constituent. Thus, for methane, it had been ascertained earlier that, under agreed-upon conditions, a principal peak height of 52.2 corresponded to a pressure in the source 10^{-3} Torr of methane.

The unknown mixture is now introduced into the source and gives rise to a mass spectrum such as that given in the bottom line of table 11.1. Each peak for the unknown is the algebraic sum of up to five contributions. As long as there are five suitable peaks under observation, five simultaneous equations may be written down, in which the variables are the five unknown percentages. Here, since there are many more than five peaks, the problem is greatly overdetermined. In this case, the answers resulting from the solution of the five simultaneous equations may be checked at some of the unused mass numbers. The accuracy with which such a mass spectrometric analysis may be made is typically about 1%.

It should be emphasized that the above example is a simple one involving, as it does, light hydrocarbons for which pure calibration materials are available, and for which the interpretation of the spectra is not difficult. The severe limitations of this technique in analysis of more complicated mixtures of organic compounds have been drastically mitigated by the use of chromatographic separation of the sample before it is introduced to the mass spectrometer ion source, as described in the following section.

11.3 Gas chromatography–mass spectrometry

The analysis of organic mixtures, referred to in §11.2, has been greatly simplified and extended by the addition of the gas chromatograph to the mass spectrometer (Holmes & Morrell, 1957; Fenselau, 1977; Gudzinowicz, Gudzinowicz & Martin, 1976), an arrangement commonly identified as 'GC–MS'. As is well known, the gas chromatograph efficiently separates the components of volatile mixtures according to their transit times through the GC column. With specific column packings and carrier gases, it can be applied to a variety of complex compounds such as steroids or amino acids. In principle, the separated components can then be delivered one by one to the mass spectrometer for identification and analysis.

In practice, however, there are two main problems to overcome. First, there is a large pressure difference between the gas chromatograph and the mass spectrometer ion source. Gas chromatographs operate at pressures slightly above atmospheric pressure, whereas electron bombardment and field ionization sources operate at $< 10^{-2}$ Pa ($< 10^{-4}$ Torr) and chemical ionization sources operate at $\sim 10^2$ Pa (~ 1 Torr). Second, the amount of effluent is small in comparison with the carrier gas. As described below, a variety of interfacing arrangements have been developed to overcome these difficulties.

Direct coupled interface

Here a capillary tube is used to transfer the GC effluent. The required pressure drop occurs along the capillary (splitter valves and restrictions may be introduced) as a result of sufficiently high speed pumping at the ion source end.

Effusive interface

Here the effluent undergoes molecular flow along a porous glass fritted tube (Watson–Biemann type) with a constriction at the inlet and outlet. In this arrangement the lighter carrier is selectively removed through the pores of the fritted tube.

Jet orifice interface

Here the effluent flows as an expanding jet from a nozzle through an orifice. The arrangement functions as a skimmer in which the lighter carrier gas diverges more than the heavy sample component and is removed by high speed pumping. The higher momentum heavy components are selectively transmitted through the orifice.

Permeable membrane interface

Here an elastomer membrane selectively transmits the organic components by a solution–diffusion process.

Generally, in GC–MS combinations, the entry of any particular component from the GC column is detected by a monitor of the total ion current at the exit of the ion source. The monitor is used to initiate a fast scan so that the mass spectrum of that particular component is obtained. In this way, mass spectra of all of the components in the mixture are recorded in sequence. The enormous amount of data generated by such a GC–MS system makes desirable the use of a computer for the automated acquisition, normalization and quantification of such spectra.

As the multicomponent mixture is being eluted in the GC, spectra are scanned repetitively and the ion current is recorded as a function of both mass and time. When a computer is used, the information can be retrieved and total-ion-current mass chromatograms reconstructed. Selective ion monitoring may also be used in which the intensities of preselected ions are recorded as a function of time.

A feature of the GC–MS is its extremely high sensitivity (10^{-6} to 10^{-10} g for the identification of compounds and 10^{-9}–10^{-12} g for selected ions), which makes it attractive for the study of trace metabolites from natural compounds, environmental contaminants, etc.

Computerization of mass spectra obtained in GC–MS analyses has made available large libraries of mass spectra ($\sim$ 40 000 entries), and several methods have also been developed for identification and interpretation of unknown mass spectra by comparison with these known ones (McLafferty & Venkataraghavan, 1979; Henneberg, 1980; Milne, Heller, Heller & Martinsen, 1980). In these library-search retrieval methods, both 'forward'– (the compound represented by a given spectrum) and 'reverse' – (the mass spectrum that could arise from a given compound) search techniques are used, by matching of a definite number of peaks. Pattern recognition methods are also employed, using the sample and unknown data sets. These are valuable in analysing complex mixtures such as those encountered in environmental studies, waste gases from industry, medical analyses, etc.

11.4 Liquid chromatography–mass spectrometry

While less common than the GC–MS systems described in §11.3 the liquid chromatograph has also been used for the separation of involatile materials followed by a mass spectrometer (Arpino & Guiochon, 1979). The different components that are separated along with the elutriant solvent from a liquid chromatographic column are introduced into the ion source (for example, via a moving belt or wire) and ionized. The principal differences between experimental arrangements lie in the specific details of the apparatus coupling the chromatograph to the ion source.

One approach has made use of the chemical ionization source with the LC solvent serving as the ionizing reagent gas (for example, Melera, 1980). In another approach (Blakley, McAdams & Vestal, 1980), the solvent and sample are vapourized rapidly in a laser beam at a jet nozzle. The expanding jet is sampled through a small orifice which leads to a chemical ionization source (§3.4).

11.5 Organic structure analysis – basic instrumentation

The mass spectrometer has become an important tool for the elucidation of molecular structure. Some of the common features of the instrumentation used for this work may be summarized in terms of the major components of the system.

Sample inlet system

A gas or a liquid with a low boiling point can be introduced by a leak from a reservoir at a pressure of 10^2 Pa (1 Torr). For a liquid with a higher boiling point, a heated reservoir–inlet system in used. A compound with very low volatility can be analysed by using a 'direct-insertion' sample inlet in which the sample, placed at the end of a long rod, is inserted into the ion source (of the electron bombardment type) through a suitable vacuum lock assembly. The sample, situated close to the electron beam, is then heated to yield appreciable vapour pressure below its decomposition temperature. The same arrangement can be adapted for the field ionization source, with the heated sample being near the field tip. For field desorption, the sample is deposited on the field tip (by evaporating a solution having a volatile solvent) prior to its insertion into the source.

Choice of ion source

The standard electron bombardment (EI) source is by far the most extensively used for producing ions of the organic samples, although the techniques of field ionization (FI), field desorption (FD), and chemical ionization (CI) are assuming increasing importance. Many commercial instruments may be fitted with combined, or interchangeable, EI–CI–FD (FI) sources.

Analyser for separation of ionic masses and/or energies

The direction focusing sector magnetic field analyser was the workhorse for many years, but the higher resolution and versatility of double focusing instruments (especially the Nier–Johnson geometry, §5.8) has led to their widespread use in recent years. The sensitivity of either single or double focusing instruments can be improved with axial focusing (see §§2.6, 2.8, 5.4, 5.10–5.14). The quadrupole mass filter has enjoyed wide use in applications where rapid scanning is desired, in particular in GC–MS systems where there is a rapid variation in the sample arriving at the source.

Detection and data analysis system

In simple systems, the output of an electron multiplier, after suitable amplification, may be fed to an electrometer and the spectra recorded on UV-sensitive paper. The Faraday cup and electrometer or the photoplate are also used to a limited extent – the last-named for precise mass measurements of parent and fragment ions.

More commonly, however, the detection of ions is accomplished by arrangements in which the signal information is converted to a digital form so as to be compatible with computer-based data recording. Where comparatively high currents are involved, the average current is detected in analog form and converted to digital form (§4.8). For example, the current could be detected by a vibrating reed electrometer whose output is measured by a digital voltmeter. More often the mass spectrometer signal (after attenuation if necessary) is fed to a sample-and-hold amplifier actuated by an internal clock and, in the hold mode, the signal is digitized by an ADC, whose output is either directly entered into the computer or stored (for example, on magnetic tape) for processing later. Where comparatively low (§4.8) currents are being studied, an electron multiplier or Daly detector (§§4.5, 4.6, and 4.7) followed by a pulse counting system is used.

Both ADC and particle detector arrangements have good sensitivity and speed of response and are inherently compatable with computer systems. In such systems, digital information on line positions and intensities is recorded in the computer which may then be used to reconstruct the mass spectrum and to give the ion intensity and the mass value of each line. For low resolution work, the mass values are nominal integers; for higher precision, the mass scale may be calibrated.

Progress in the field of data acquisition in mass spectrometry has been reviewed by Ward (1971, 1973), Mellon (1975, 1977, 1979), and Daly (1978*a*).

11.6 Types of ions involved in structure analysis

We now consider the types of ions formed by the ion source, with special reference to the electron bombardment source, and the information they provide concerning the molecular structure of the sample.

Of foremost importance is the *molecular ion* formed by the removal of one electron from the molecule. It is present if the molecule vapourizes without decomposition and if all the bonds are strong enough that the probability of bond cleavage is low. Its nominal (integral) mass gives the molecular weight of the compound, while its precise mass (at a precision typically 1 ppm of the mass and determined by peak matching as described in §§8.2–8.7) yields the elemental composition of the molecule. Further, the rules of valence ensure an even number for the normal mass of the molecular ion in general. An important exception to this rule occurs for molecules having an *odd* number of nitrogens, since N is the only common even-mass element with odd valence.

The molecular ion also enables recognition of several polyisotopic elements (such as C, Cl, Br, S, Si) by the pattern which reflects the masses and abundances of the constituent isotopes. The use of compounds enriched with heavy isotopes such as ^{2}D, ^{13}C, may be exploited to give additional information.

Finally, the decomposition of the molecular ion between the source and analyser yields 'metastable peaks', which will be subsequently discussed.

Fragment ions contribute the majority of the peaks in the mass spectrum and arise from the cleavage of specific bonds in the molecular ion, and from rearrangements of the atoms. These electron-impact-induced reactions are governed by chemical properties whose understanding permits the interpretation of the fragmentation pattern in terms of molecular structure. The general feature of these reactions is the localization at specific centres of the positive charge formed during the ionization of the molecule. Loci of preferred charge localization are heteroatoms, π-electron systems, etc. Ions in which the charge is localized at one or other functional group are formed in large abundance due to the cleavage of weaker bonds in the vicinity of the charge centre.

Electron 'book-keeping' is useful in keeping track of the electron shifts which, among other things, determine which fragment is charged. Molecular ions that have lost one electron from their normal complement of paired electrons, are represented by the symbol ($^{+\cdot}$). One also symbolically distinguishes between (a) *heterolytic cleavage* (↷) of a bond in which the pair of bonding electrons is transferred to one partner, and (b) *homolytic cleavage* (↷) where a single electron is transferred along the fish-hook.

As is quickly apparent, the gamut of mass spectral reactions in the fragmentation of organic molecules is very wide. In the following, we attempt to summarize some of the essential features involved in these studies. For more extensive treatments of these matters, with emphasis on the organic chemistry involved, the interested reader is referred to specialized references such as Budzikiewicz, Djerassi & Williams (1967), Johnstone (1972), Schrader (1974) and McLafferty (1980).

It is useful to classify the mass spectral reactions into: (a) simple cleavage reactions involving the cleavage of a single bond and (b) multicentre fragmentations which involve fragmentation of some bonds and formation of others, resulting in molecular rearrangements.

As examples of the former, one has the fission of a weak carbon–iodine bond, viz.

$$CH_3-CH_2-CH_2-CH_2-I]^{+\cdot} \rightarrow C_4H_9^+ + I \qquad (11.1)$$

and the cleavage of the benzylic bond with the rearrangement of the benzyl ion, $[C_7H_7]^+$, viz.,

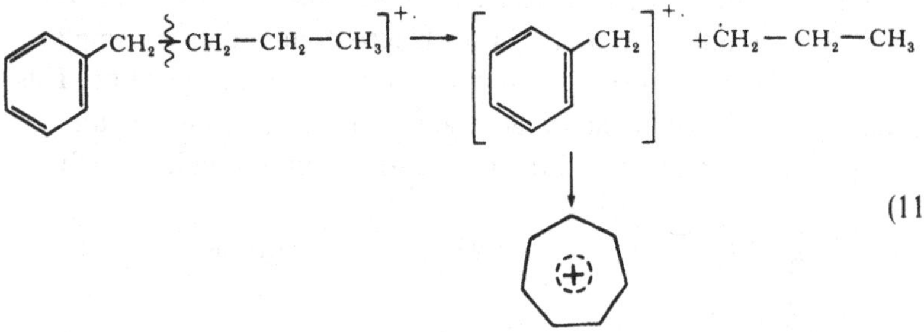

$$[C_6H_5-CH_2 \wr CH_2-CH_2-CH_3]^{+\cdot} \longrightarrow [C_6H_5-CH_2]^{+\cdot} + \dot{C}H_2-CH_2-CH_3$$

$$\downarrow$$

$$\text{(tropylium ion, } C_7H_7^+\text{)} \qquad (11.2)$$

Another example of the *α-cleavage* reaction (breakage of bond α to the carbon atom bearing a heteroatom) is one in which the bond fission is triggered by the odd non-bonding electron on the heteroatom. In this reaction the ion is stabilized by delocalization of the positive charge and the energy required for bond fission is largely compensated by π-bond formation:

$$R-CH_2-CH_2-Q^{+\cdot} \rightarrow R\dot{C}H_2 + CH_2=Q^+, \qquad (11.3)$$

where $Q = OH$, NH_2, SH, etc. This type of positive charge stabilization controls the fragmentation of a variety of organic compounds – for example, primary alcohols to yield $CH_2=OH^+$, primary amines to yield $CH_2=NH_2^+$, aliphatic ketones to yield $R-C=O^+$, and benzoyl compounds to yield

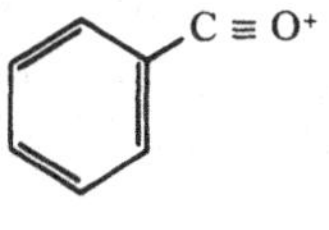

$$C_6H_5-C\equiv O^+$$

Typical of multicentre reactions is the *elimination* fragmentation where a hydrogen atom is abstracted from a carbon atom removed from the heteroatom; for example, loss of H_2O from alcohols:

$$R\text{-}CH(H)\cdots O^{+\cdot}(H)\text{-}CH_2\text{-}CH_2\text{-}CH_2 \longrightarrow R\text{-}\dot{C}H\cdots H\text{-}O^{+}(H)\text{-}CH_2\text{-}CH_2\text{-}CH_2 \tag{11.4}$$

$$\longrightarrow H_2O + R\text{-}\dot{C}H\text{-}CH_2\text{-}CH_2\text{-}CH_2^{+}$$

An extensively studied reaction is the *McLafferty* rearrangement reaction, whose requisites are a polarized double bond and a hydrogen in the γ position. Here, if ionization occurs at the heteroatom, the latter completes its octet by picking up an H atom from a CH group at the γ position. This transfer, in a six-membered cyclic transition state, results in an ion containing the heteroatom and the loss of an olefin molecule, namely,

$$R\text{-}C(=\dot{O}^{+})\text{-}CH_2\text{-}CH_2\text{-}CH(H)\text{-}R' \longrightarrow R\text{-}C(\text{-}\dot{O}^{+}\text{-}H)=CH_2 + CH_2=CH\text{-}R' \tag{11.5}$$

Thus, 2-pentanone

$$CH_3-\overset{\overset{\displaystyle O}{\|}}{C}-CH_2-\overset{\overset{\displaystyle H}{|}}{CH}-CH_3$$

loses C_2H_4 and yields a peak at $m/z = 58$,

$$\left[CH_3-\overset{\overset{\displaystyle OH}{|}}{C}=CH_2\right]^{+\cdot},$$

by this reaction. Another well known case is the peak at $m/z = 74$,

$$\left[CH_2=\overset{\overset{\displaystyle OH}{|}}{C}-OCH_3\right]^{+\cdot},$$

in methyl esters of fatty acids.

Another type of fragmentation is the retro-Diels–Alder reaction which occurs when cyclic olefins expel a neutral olefinic fragment. Thus with tetralin,

$$[\text{tetralin}]^{+\cdot} \longrightarrow [C_6H_4(CH_2)_2]^{+\cdot} + C_2H_4 \tag{11.6}$$

11.7 Interpretation of the mass spectrum

The principles outlined above provide the basis for structure elucidation of organic molecules. Various step-by-step procedures have been suggested in the literature (for example, Schrader, 1974). Thus, the appearance of the mass spectrum provides information on type structure. For example, alkanes have low intensity molecular ions and fragment ions separated by $\Delta M = 14$ (CH_2), aromatic hydrocarbons have intense molecular ion peaks (as the π-electron systems easily accommodate the loss of an electron), whilst a molecular ion with heteroatoms fragments at this bond or at the alpha bond. When the molecular ion peak is known, the fragment peaks can be used to reconstruct the molecular structure.

One either starts from the molecular peak and looks for fragments expelled by simple cleavage or rearrangement reactions, or starts from the low-mass fragment peaks and looks for characteristic groups that are present in the molecule. The latter technique should be applied with caution since low mass ions may have been formed by rearrangement and may not bear a direct relationship to the groups originally present. The mass spectral data are supplemented by UV, IR, and NMR spectral information in addition to chemical (elemental) analysis.

The development of *high resolution mass spectrometry* has been a valuable aid inasmuch as precise mass measurements enable the elemental composition of the ions to be determined. This is possible because the mass defects differ for compounds of nominally the same isobaric mass and, as Beynon (1959) has shown, a mass resolution 10^4–10^5 is adequate for unambiguous determination of elemental composition. For example, acetone C_3H_6O, has $M = 58.041\,866$, and butane, C_4H_{10}, has $M = 58.078\,252$. Compilations of exact masses corresponding to various elemental compositions of all possible combinations of C, H, N and O have been given by Beynon & Williams (1963). A similar but more extensive table, including 16 elements, has been provided by Tunnicliff, Wadsworth & Schissler (1965). The mass measurements are made by scanning the magnetic field or, when precise values are required, by the peak matching technique described in §§8.3–8.5. It is common practice to use computer techniques to match the experimentally measured masses with those in the computed list of values. A direct determination of parent and fragment peak composition together with the fragmentation pattern usually enables one to arrive at a unique solution for the structure that would rationalize the observed fragmentation pattern. However, isomers are difficult to distinguish from one another without reference spectra, and may be difficult even with such data.

Computer-aided interpretative schemes based on the above considerations have been extensively applied to natural products, amino-acid sequencing in peptides and other problems in biochemical and biomedical mass spectrometry.

11.8 Negative ion mass spectroscopy

Negative ions were not initially employed for analytical work or structure analysis because their production rate is much lower under electron impact than is the rate for positive ions. Moreover, the formation of negative ions depends strongly on the energy of the electrons, reflecting the presence of a variety of processes (for example, dissociative and resonance capture, ion-pair formation) that are involved in negative ion production. The sensitivity for negative ions can, however, be enhanced considerably by the proper choice of bombarding electron energy (a few eV). For a variety of polar compounds (for example, amino-acids) the cleavages for negative ions are different from those for positive ions. With such materials rearrangement ions are absent and fragmentation can be minimized, so that molecular structure correlation may be obtained.

The structure of various bioorganic molecules has been studied and the presence of air pollutants (N_2O, NO_2) has been monitored by such negative ion techniques. Negative surface ionization has also been used for Cl^-, and may prove to be applicable to some organic molecules. Increasing use is being made of negative ions for structural and analytical applications. The technique is applicable to compounds with high electron affinities and a large number of compound classes has been studied (Bowie, 1975).

11.9 Application of soft ionization techniques

We now consider the application of the 'soft' methods of ionization (field ionization and field desorption, described in §3.5, chemical ionization, described in §3.4, and particle-induced desorption of organic ions, described in §3.10), which attempt to overcome the main drawbacks of the conventional electron bombardment source, namely,

(a) the absence of molecular ions in many cases, and hence the absence of information needed to derive the molecular weight and formula;

(b) the presence of rearrangement reactions that complicate the interpretation of the structure; and

(c) the inability to distinguish between stereochemical isomers in many cases.

As the term 'soft' implies, the transfer of surplus energy to the molecule during the process is much less than for electron impact ionization. Typically it is less than a few tenths of an eV, whereas in electron impact ionization the corresponding energy is of the order of a few eV. Thus fragmentation of the molecule is minimized and strong peaks characteristic of the intact molecule are produced, so that the determination of the molecular weight and constitution is permitted.

The related techniques of field ionization (FI) and field desorption (FD), described in §3.5, in which the sample is subjected to an intense electric field at a sharp tip or a wire anode, are enjoying an expanding role in organic chemistry. Combined electron bombardment and FI–FD sources are available with most commercial machines, and heated field desorption emitters are used as direct-insertion probes. The main features of FI and FD of importance in organic chemistry are now outlined.

Intense peaks related to the original molecule are obtained even in those cases where electron bombardment yields no molecular ion. For example, carbohydrates yield abundant molecular FI ions, although they are difficult to evaporate without thermal decomposition. When an instrument having high resolving power is used, composition assignments can be made from precise mass measurements.

The fragment ion spectrum in FI and FD is produced by the polarization of the molecules in the strong electric field and differs from the fragment spectrum resulting from electron bombardment. Further, since the spectrum primarily comprises major fragments, its components are complementary to a certain extent. In addition, products due to degradation–pyrolysis yield their own parent ions.

FI and FD provide an important method for the study of the mechanics of the decomposition of organic ions over very short (10^{-12}–10^{-5} s) time intervals. As shown in fig. 11.2, the positive ion formed at the field tip is accelerated to the cathode by the field established by the tip. If this ion decomposes at a distance Δx from the tip, a fragment ion may result which

Fig. 11.2. Schematic diagram of the equipotential lines for a field ionization (desorption) source. (After Beckey, 1977.)

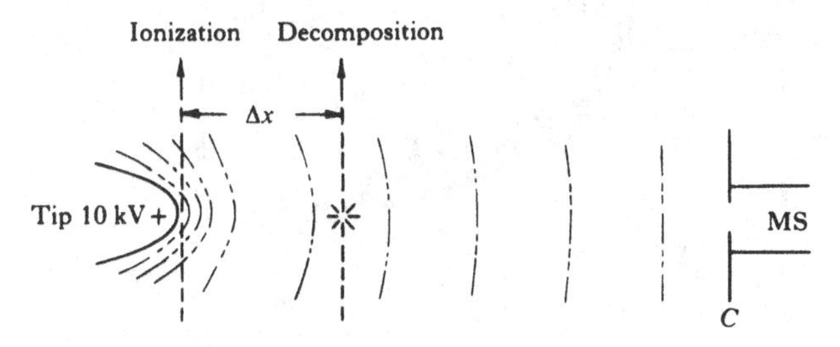

is formed at a potential $V - \Delta V$, where V is the tip potential, so that it arrives at the cathode with a smaller kinetic energy than do those formed at the tip. The energy deficit yields the potential at the point of origin of the fragment and hence its location. From this information the times of flight until decomposition can be calculated and the rate of reaction derived.

The determination of the time scale is made through a determination of the energy loss, which causes, in a single focusing instrument, an apparent shift in the mass scale and a peak shape such as that shown in fig. 11.3 (Beckey, Knoppel, Metzinger & Schulze, 1966). Considerable information on fast, field-induced rearrangement reactions (where the time intervals are short because the fields are strong) has been obtained. A recent example of the application of field ion kinetics to the study of methanol as a condensed layer has been described by Gierlich & Röllgen (1979).

The technique of field desorption (§3.5) is an offshoot of the field ionization method that has proved to be even more valuable for the

Fig. 11.3. Section of field ionization spectrum of diethyl ether, showing the transition between field induced and metastable decomposition. The metastable fragment $C_2H_5OCH_2$ ($M = 59$) appears at $m/z = (59^2/74) = 47.04$. (After Beckey *et al.*, 1966.)

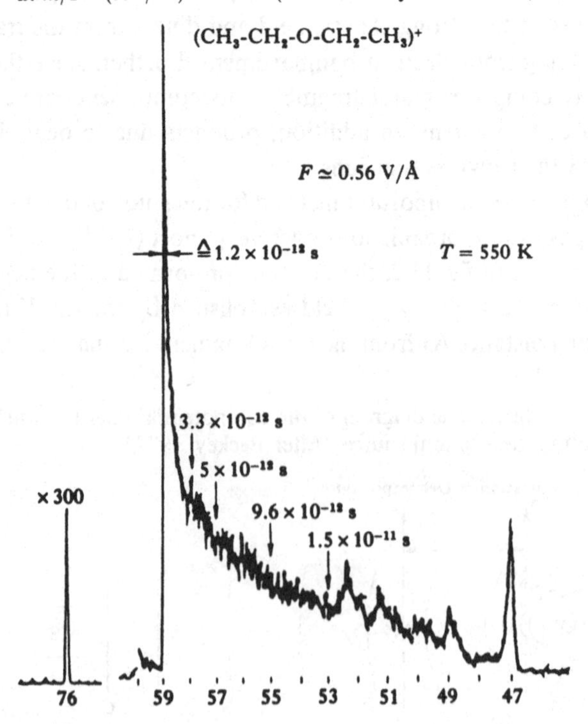

elucidation of organic structures. The field desorption source is operated by loading the sample on the surface of the emitter, for example by dipping the wire into a solution. It is particularly suited to the study of organic molecules that are involatile. The method uses electrical repulsion, rather than vaporization, to transfer the molecules into the gas phase, the compounds (or fragments) being field desorbed in an ionized state.

FD spectra show $(M + 1)^+$ peaks when the energy for the protonation reaction is lower than that for ionization in organic compounds. Doubly charged molecular ions are also found in FD spectra of molecules with conjugated electron systems.

The mass spectra that are observed are determined, in part, by the temperature of the emitter. Depending on the volatility of the sample, the emitter may be cooled below or heated above ambient temperature. The heating of the wire in the presence of a high field causes the desorption and ionization of a variety of ions that are subsequently analysed. The 'best anode temperature' (BAT) is defined as the temperature at which the maximum molecular ion current and minimum fragmentation occurs. Above the BAT, thermal and field-induced fragmentation may be used for structure studies. At temperatures even further above the BAT, the fast, controlled thermal degradation of complex molecules may be investigated.

Generally the spectrum is influenced by the emitter temperature, the morphology of the emitter needles, the solvent, the analysis substances, etc. Considerable work has been done on the preparation of reproducible emitters on the wire and in determining the optimum temperature of the wire for field desorption for each sample (McCormick, 1977, 1979, 1981). In some cases, mixture of the analysis substance with an alkali solution produces polymolecular ions that incorporate, Na, K, etc.

For reviews of field ionization and field desorption studies one is referred to Block (1968), Derrick (1978) and Wilson (1971, 1973, 1975, 1977). A comprehensive review of developments in FD has been given by Schulten (1978) and Wood (1982).

In chemical ionization (CI), the ionization of molecules is effected by reactions between these molecules and ions of a chosen reactant gas (most commonly CH_4 or isobutane). Typical CI sources have been described in §3.4. While CI also involves relatively small amounts of energy, there is a transfer of protons (or hydroxide, alkyl or carbonium ions, etc.) so that CI spectra are markedly different from electron impact or FI spectra. In general, CI spectra are simple, having enhanced peaks of mass numbers close to the molecular ion, and provide evidence of different aspects of structure. Flexibility in changing the spectrum is available by the use of different reactant gases (for example, Ar, NH_3, H_2O have been used in order to alter the reactions by which the ions are formed – Wilson, 1971, 1977). The method has the limitation that it must be possible to volatilize the sample. The simplicity of the mass spectrum also makes the method amenable to GC–MS analysis of mixtures, particularly the CI method where the GC carrier gas can also be the primary ionizing gas of the source. In addition, the CI spectra contain a number of fragment ions which result from processes occurring in the protonated $[M + H]^+$ or alkylated $[M + C_2H_5]^+$ molecules. Thus the spectra of stereoisomers can often be distinguished from each other.

The major components in positive ion CI spectra are the $[M + 1]^+$ ions formed by protonation, the $[M - 1]^+$ ions formed by hydride abstraction from the parent molecule, and fragment ions produced predominantly from the decomposition of the $[M + 1]^+$ and $[M - 1]^+$ ions. In addition one finds fragment ions, for example, as the attack of the reactant ions is concentrated at polar carboxyl groups. Structural information has been obtained by the proper choice of the reactant gas (for example, H_2 for the study of the amino acids, NH_3 for triglycerides). Negative-ion CI spectra often show enhanced sensitivity and reflect stereo chemical effects that are not present in electron impact fragmentation patterns.

Early work in chemical ionization has been summarized by Field (1968); the more recent development of this area has been reviewed in detail by several authors (for example, Wilson, 1971, 1973, 1975, 1977).

For the purpose of comparison, typical spectra obtained by electron impact, CI, FI, and FD methods for the same substance are shown in fig. 11.4 (Fales *et al.*, 1975).

It should be noted that FD is inherently incompatible with a gas chromatograph. However, FI may be used with GC so that molecular mass data may be found, while the use of CI with GC, especially with different reactant gases, may be used to deduce structure.

Less generally used to date, but of rapidly growing interest, are the various techniques of particle-induced desorption of large organic molecules described in §3.10 (^{252}Cf–PDMS, HIIDMS, SIMS, and FAB) and laser-induced desorption. In all of these techniques the spectra are similar to the other soft ionization methods inasmuch as the parent peak is clearly evident but differ in that the fragmentation peaks are much more significant. This latter property makes these methods very promising for the study of structure, especially for large bioorganic molecules (Benninghoven, Jaspers & Sichterman, 1979; Macfarlane, 1982; Macfarlane, McNeal & Hunt, 1980) that are involatile and thermally labile. In particular the

Fig. 11.4. Comparison of the mass spectra of phenobarbitol by electron impact, chemical ionization, field ionization and field desorption. (After Fales *et al.*, 1975.)

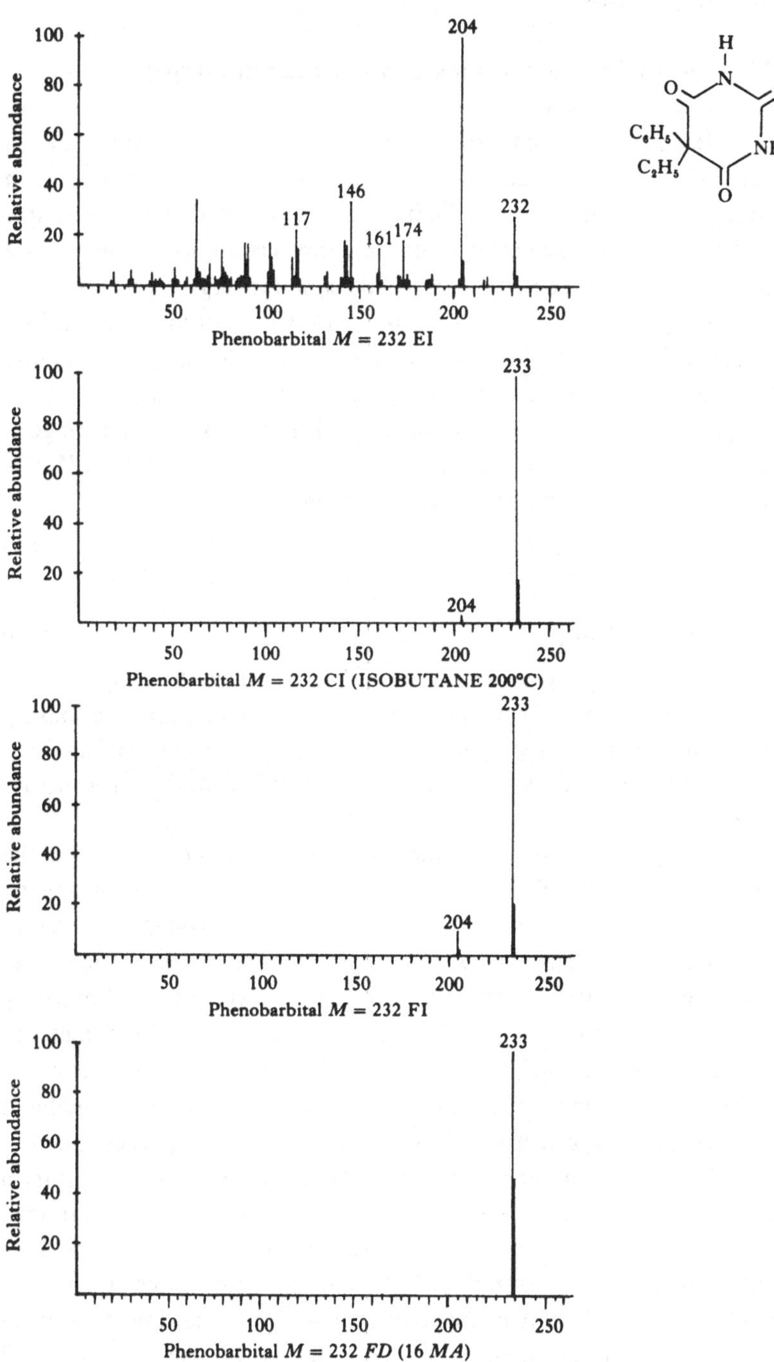

fragmentation takes place in such a way that the sequencing of large molecules such as peptides and oligonucleotides may be deduced and isomers may be distinguished.

11.10 Metastable peaks and mass spectrometer–mass spectrometer (MS–MS) method

In an electron bombardment source, some of the ions are formed in metastable excited states and, as they reside for only about 1 μs in the ion source, those that have longer lifetimes are accelerated from the source before de-excitation. The excited ion may dissociate into an ionic and a neutral fragment, with a partitioning of the kinetic energy according to the masses of the fragments as required by conservation of momentum. In a magnetic field instrument, when the dissociation occurs in the region between the source and analyser, the fragment ion which is left with a lower energy appears at the detector as a broad peak (corresponding to a range of kinetic energies released in the decomposition) at an effective mass M^*. It is easily shown (§10.7) that for a decomposition

$$A^+ \rightarrow B^+ + C, \tag{11.7}$$

then

$$M^* = M_B^2/M_A. \tag{11.8}$$

Such *metastable* mass peaks have been observed since the early work of Aston ('Aston bands'). They offer evidence of the structural relationship between the parent A^+ and daughter B^+ ions, which are observed as sharp peaks, and have been used extensively in studies of molecular structure (Beynon, 1968).

With the application of double focusing spectrometers (Chapter 5) to the investigation of such collisions, it became possible to determine the kinetic energies of metastable ions by varying the accelerating voltage at constant electric sector field. Inasmuch as it is desirable to have the kinetic energy distribution combined with mass analysis, such instruments are frequently operated in 'reversed geometry', that is, the instrument has the geometry in the order: ion source–magnetic analyser–electrostatic analyser–detector. With such an experimental arrangement (sometimes called 'mass analysed ion kinetic energy spectrometer – 'MIKES'; or 'direct analysis of daughter ions' – 'DADI'; but, in more recent work, referred to by the general term 'mass spectrometer–mass spectrometer'–'MS–MS'), a particular ion may be selected and all reactions leading *from* this ion studied (McLafferty *et al.*, 1980). This may be compared with the conventional geometry which is suitable for the study of reactions that lead *to* a particular ion. A collision chamber, into which any collision gas can be leaked, may also be interposed between the magnetic and electric analysers and collision-induced decompositions studied ('collision activation spectroscopy') (McLafferty *et al.*, 1978).

Typically the mass resolution of the magnetic analyser is moderate ($\sim$500) but the electrostatic analyser provides very high energy resolution ($\sim$0.1eV for 8keV ions) so that the fine structures in the kinetic energy spectrum, corresponding to different decomposition modes, can be studied. With a collision chamber, the 'MIKE' spectrum reveals all the neutral fragments lost on collision-induced dissociation. This can be used for direct analysis of mixtures, or for structural elucidation of individual compounds.

These collisional activation spectra are being employed increasingly for characterization of ion structures. The masses and intensities of the peaks are characteristic of the structure of the parent ion, and differ for differing isomeric structures. This is so, particularly, if these structures contain heteroatoms. Thus, while high resolution mass spectrometry yields the elemental compositions of these isomers (by precise mass measurement), the collisional activation spectra enable the structures of the isomers to be deduced. For example, a precursor ion of $m/z = 57$ can arise from different structures – $[CH_2=CHCH\text{-}OH]^+$, $[CH_3CH_2C=0]^+$, etc. These show different collisional activation spectra, and the precursor peak can thereby be characterized. This can facilitate the identification of the parent molecule that produced this precursor ion, expecially if other precursors are also studied.

The analyser components of MS–MS arrangements (or 'tandem mass spectrometers') may be designated as B (magnetic), E (electric) and Q (quadrupole). The use of QQQ, EBEB, BEEB and BEQQ systems, with intervening reaction region(s), to study the details of consecutive collision-induced reactions, has been described by McLafferty (1983, 1985) and Cooks *et al.* (1985). FT–ICR (§6.10) can also be used.

11.11 Biomedical applications of mass spectroscopy

Three particular aspects of mass spectrometry have emerged as significant tools in biomedical investigations, namely, (a) use of stable isotopes, (b) the applications of the GC–MS systems and (c) the study of molecular structures, including the sequencing of polypeptides, by the 'soft' ionization methods outlined in §§3.4, 3.5, 3.10 and 11.9.

The use of 'labelled' compounds (for example, those containing ^{2}H, ^{13}C, ^{15}N and ^{18}O) for the elucidation of fragmentation mechanisms for ordinary organic molecules has been exploited for some time (Holmes, 1975). Here the relative distributions of the labelled ions in the metastable peaks reveals information on the atoms or functional groups involved. The isotope effects also reflect the precursors and degree of randomization in the fragmentations.

The availability of isotopically labelled materials has also made possible a large number of both analytical and metabolic studies which are relevant to human metabolic pathways. Use of isotopically labelled drugs and metabolites provides pico-mole sensitivities in studies of materials such as estrogens in pregnancy urine or specific proteins with ^{15}N-glycine. The inherent sensitivity of GC–MS has also been exploited, for example, for prostaglandins. Another application is that of CO_2 breath tests, where the variations in isotope abundance ($\delta^{13}C$, see §13.9) are measured. A recent review of these diverse applications has been given by Krahmer & McCloskey (1978).

The extreme sensitivity of the GC–MS method has led to its widespread use in the quantification of biological materials. The complexity of these materials makes necessary the optimisation of conditions for the packed and capillary chromatographic columns. EI, FI and CI sources have been used. The precision can be enhanced by the use of the isotope dilution method (§7.9) and computer methods for signal processing. Typical examples of these investigations include determination of prostaglandins (involved in fertility) in human semen, ergotamine and hallucinatory drugs in blood, etc. The method also finds extensive use in food science (Horman, 1979) where the compounds responsible for flavour and aroma are studied and quantified. (A coffee constituent heated to 180 °C produced 60 volatile, aroma-forming compounds!)

Structural studies have also been carried out with a large number of natural compounds (Games, 1979). In addition to the conventional electron bombardment source, the soft ionization methods (CI, FI, FD) provide considerable complementary information. Various derivatives of the natural and biochemical products are used for this purpose. As described in §11.9, stereoisomers can be distinguished by CI spectra, and special techniques in FD such as increasing the emitter temperature or incorporating alkali ions may be helpful in structure studies (for example, for oligosaccharides and glycolipids – Puzo & Prome, 1978). The newer MS–MS techniques (§11.10) are of increasing importance for this type of work.

The sequencing of amino-acid residues of an oligopeptide (peptide sequencing) is of great importance in the light of its relevance to genetic information. The usefulness of the GC–MS technique for this purpose has been demonstrated. Here suitable peptide derivatives may be separated by gas chromatography and their amino-acid sequence determined by mass spectrometry with electron impact and CI sources (Biemann, 1972; Daly 1978*b*). As noted previously (11.9) particle-induced desorption techniques have also been demonstrated for this work (Macfarlane, 1982).

11.12 Environmental applications

Mass spectrometry plays a major role in the identification and quantification of trace impurities in the environment. While the presence of inorganic metallic compounds may be determined at low levels of concentration by spark source mass spectroscopy (§§3.6, 12.2), we are concerned here with organic micropollutants.

Methods of organic mass spectrometry are employed because they provide very high sensitivity, although some form of preconcentration from the usually enormous quantities of sampled material is involved. The advantage of the mass spectrometer is that it is not only a sensitive detector, but also a tool for structure confirmation via the spectra obtained. Hence, the mass spectrometer is widely employed for the detection of environmental pollutants and study of their metabolism and degradation.

The analytical procedure involves (a) sampling, (b) pretreatment and separation, (c) mass spectral analysis. Sampling techniques for obtaining large and representative samples are available (for example, for the study of aerosols, preliminary filtration over millipore filters followed by chemical methods of extraction of the organic compounds) and separation typically involves capillary gas chromatography columns, although liquid chromatography may also be appropriate. Magnetic and quadrupole instruments with a resolving power of ~ 1000 are employed, usually with an electron bombardment ion source, although FI and CI sources are also used. The mass spectra must be scanned at high speed, as the chromatographic peaks are sometimes narrow (that is, in time). Further, each chromatogram yields a number of mass spectra, so that it is imperative that data acquisition systems with computer storage be employed.

The usual procedure for the identification of the mass spectra is to separate the eluate from the GC column into two parts, one coupled to the mass spectrometer ion source and the other to a specific detector. (When aroma and flavour are involved, the latter sensitive detector is the human nose). The mass spectra may be compared with known spectra (§11.3) for the identification of the pollutant.

A spectacular example of such a pollutant is that of the universally present polychlorinated biphenyls ('PCBs') which are found in river waters at ultra-low concentrations ($\sim 50 \times 10^{-12}$ or 50 ng/l). These accumulate dramatically in successive steps in the food chains (for example, water –

zooplankton – fish – seal) so that in seals, PCBs constitute a few *per cent* of the total fat content. This is an example of an organic industrial by-product that remains undegraded in the environment. Other examples are pesticides and their metabolites in agricultural by-products and effluents from smoke-stacks that introduce organic products into the atmosphere. As the characterization of these and other allied compounds solely by GC detectors is vitiated by interference between these and other components, the GC–MS combination is necessary for unambiguous quantification. The same is true of halogenated hydrocarbon pesticides (for example, DDT) which are ubiquitous in the environment. The chemistry, metabolism and degradative mechanisms for these compounds have been studied.

Details of the mass-spectral features and their application to the analysis, structure-elucidation and mechanisms of decomposition of individual molecules have been discussed in *Mass Spectrometry, Specialist Periodical Reports* (Chemical Society, London), under the classification – (a) Organometallic, Co-ordination and Inorganic Compounds and (b) Natural Products. A similar discussion has also been given in *Analytical Chemistry (Reviews)*, particularly by Burlingame, Shackleton, Howe & Chishov (1978).

12

Applications to solid state physics

The electromagnetic and mechanical properties of many solids are critically dependent upon the amounts and types of trace impurities. Amongst the several techniques which have been developed for the identification and/or introduction of such impurities, the mass spectroscopic technique occupies a position of special importance.

The earliest application was that of spark source mass spectrometry (SSMS) to the analysis of trace constituents of solid state materials. This was followed by secondary ion mass spectrometry (SIMS) which was first directed to the analysis of surfaces and later was developed into a microprobe for the study of successive layers located near the surface. Finally, the study of ion penetration of solids led to the development of ion implantation as a versatile tool for semiconductor technology, including the development of new materials.

12.1 Spark source mass spectrometry (SSMS)

Although several analytical techniques have been used to determine trace impurities in solids (for example, optical spectrochemical analysis, X-ray fluorescence and neutron activation analysis), spark source mass spectrometry (SSMS) provides certain special advantages, as will now be described. Figure 12.1 shows typical equipment for SSMS.

In this technique, a vacuum spark is produced by the application of a high ($\sim 50\,\text{kV}$) pulsed rf voltage ($\sim 1\,\text{MHz}$) between the sample electrodes, the ion source being of the type described in §3.6. The vacuum breakdown in the gap between the electrodes evaporates and ionizes the sample material. The ions, which are representative of the sample, have an energy spread of $\sim 1\,\text{keV}$, thus necessitating the use of a double focusing mass spectrometer. For this application an instrument employing the Mattauch–Herzog geometry (§5.7) is usually employed. Here the double focusing condition holds for all masses and the mass spectrum is produced along a plane at the exit boundary of the magnetic analyser. A photographic plate positioned in this plane will integrate the fluctuating ion current from the rf spark and produce the desired mass spectrum. In this arrangement qualitative analysis of impurities is achieved by identifying isotopic lines for the elements of interest whilst quantitative analysis requires densitometry of the lines. Impurity concentrations are calculated from intensity measurements on corresponding isotopic lines for a graded series of 'total charge' exposures. The latter are obtained by collecting a fixed fraction of the total ion beam on a monitor collector and electronically integrating the fluctuating output, the integrated output being used as a beam control switch. In practice, mass spectra are obtained for exposure levels from 10^{-4} to 100 nC, thus covering a range in concentration of $\sim 10^6$. The advantages of spark source analysis are the following:

(a) Complete elemental coverage is obtained over a wide sensitivity range, extending to $\sim$ 10 ppb in favourable cases.
(b) The mass spectra are simple (even with the complications of non-elemental ions, multiply charged lines, etc. as compared to optical or X-ray spectra). For example, the mass spectrum of Fe contains four lines (singly charged) whilst its optical spectrum comprises $\sim$ 3000 lines.
(c) The *relative sensitivity coefficients* (RSC) which relate the measured intensity ratios to actual concentration ratios, vary only by a factor $\sim$ 3 for most elements. A number of studies have confirmed this feature, for example Chakravarty, Venkatasubramanian & Duckworth (1963), Honig (1966), Farrar (1972), Jaworski & Morrison (1974) and Van Hoye, Gijbels & Adams (1980). This is to be compared to RSC variations of 10^4–10^5 for other analytical

Fig. 12.1. Spark source mass spectrometry. The double focusing mass spectrometer is the Mattauch–Herzog type (5.7). Trace elements are detected by the photographic plate whilst the total ion current is monitored at slit S_4, located between the electrostatic and magnetic analysers. In this case provision is also made for electrical monitoring of the major constituent. (After James & Williams, 1959.)

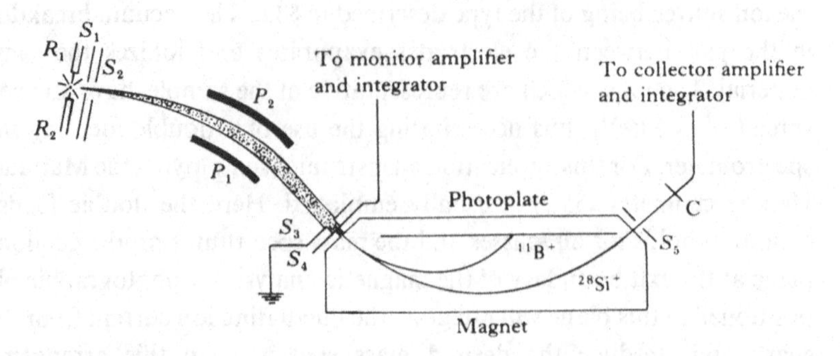

techniques (for example, optical spectrochemical analysis, where large variations are caused by variations in oscillator strengths; or neutron activation analysis, where large variations in neutron absorption cross sections are reflected in the RSCs).

(d) Interelement and matrix effects are minimal in spark source analysis.

(e) Finally, a minimum of sample treatment is required, conducting samples being used as such, and non-conductors press-moulded as small rods with pure graphite or Ag powder.

Electrical detection methods can lead to more accurate analyses, the magnetic field being scanned or switched to bring the resolved ions to an electron multiplier, and the ratio of the collector to monitor (that is, total ion) currents being recorded. Two different techniques have been used. In the 'scanning technique' the outputs of the monitor and resolved ion collector are fed to voltage-to-frequency converters (VFC) whose outputs go to a ratiometer. The ratiometer counter gate is opened to accept a preset number of counts from the monitor VFC output, and the resolved ion output is printed. In the 'peak-switching' techniques, the monitor and ion currents are integrated over larger intervals, the magnetic field or the electrostatic analyser field (plus the acceleration voltage) being switched over the peaks of interest. The latter method appears to yield more reproducible results. Electrical detection for spark source mass spectrometry has been reviewed by Conzemius & Svec (1972) and is illustrated in fig. 12.2.

The measurement of intensities using photoplates has been described in §4.2 and computer techniques are being employed increasingly for the collection and processing of data from plates used in SSMS. Plate densities are scanned at about 1 μm intervals by an optical microdensitometer and these readings, usually digitized (above a preset threshold), together with position information, are fed to the computer, which then calculates ion intensities (using input information relating plate density to ion intensity) as well as peak centroids (Woolston, 1972; Stüwer, 1976).

Besides metals, alloys, and semiconductors, the SSMS technique has been employed for insulators, powders (using suitable briquetting techniques), low melting samples (frozen in the source by cryogenic cooling), reactive and radioactive samples. Analyses of gases in solids have been made under favourable conditions, using cryosorption pumping at the source to reduce blank corrections. Also, trace elements in rocks and minerals have been determined (Taylor, 1971) including the rare earth elements.

In addition to the radio-frequency spark, other types of discharge sources have shown promise of application to the analysis of solids, for example vacuum vibrator (drawn arc) triggered low-voltage discharge (§3.6).

The laser (§3.14) can also be used to evaporate and ionize small samples of material in a *laser microprobe* mass analyser. With a normal Q-switched laser the evaporation rate is $10^{13}-10^{17}$ atoms per flash. This results in a crater $\sim 20\,\mu$m in diameter and 100–500 μm in depth, corresponding to an evaporated sample of $\sim 1\,\mu$g. The ions so produced have energy spreads of 500–1000 eV. As with the spark source – because of the large energy spread – a Mattauch–Herzog instrument is normally used, but an electrostatic analyser followed by a time-of-flight tube is also employed. Impurity levels of 1 ppm can be ascertained. The analytical capabilities of laser probe mass spectrometry have been reviewed by Kovalev, Maksimov, Suchkov & Larin (1978), whilst typical apparatus and applications to materials analysis have been given by Bingham & Salter (1976) and Furstenau, Hillenkamp & Nitsche (1979).

12.2 Secondary ion mass spectrometry (SIMS) and ion microscopy

There is much interest in studying the surface properties of materials. This can be attributed partly to the development of ultra-high vacuum techniques and partly to the development of new methods for the study of the composition and structure of surfaces. One of the powerful

Fig. 12.2. Electrical detection for SSMS, adapted from Conzemius & Svec (1972). With reference to fig. 12.1, the 'beam monitor' is the slit S_4, whilst the 'ion collector' C collects one ion species at a time.

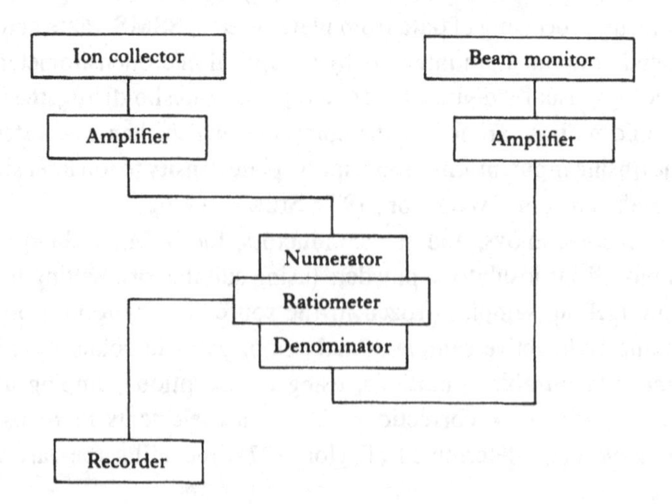

techniques for surface characterization is secondary ion mass spectrometry (SIMS). In this technique, the sample is bombarded with a beam of primary ions of energy 1–10 keV, resulting in the sputtering of secondary ions from the upper atomic layers of the sample (5–50Å in depth). These ions are subsequently analysed by a mass spectrometer and the mass spectrum provides quantitative information on the composition of the sputtered material. The sputtering ion source itself is described in §3.10.

The basic considerations of ion scattering have been described in detail by Honig (1976). A quantitative model for the production of secondary ions was worked out by Anderson (1972) on the basis of local thermodynamical equilibrium between the neutral and charged atomic species in the localized plasma of the bombarded area. This formula, based on the Saha–Eggert equation, has proven to be highly useful in calculating the elemental compositions from the observed secondary positive-ion mass spectra (Anderson & Hinthorne, 1973).

Secondary ion mass spectrometry instrumentation – both experimental and commercial – may be classified according to application, namely, monolayer analysis, ion microprobe and ion microscopy.

Monolayer analysis

Typical of monolayer analysis equipment is that of Benninghoven and coworkers (1971, 1974), (fig. 12.3). Here the sample is located in a stainless steel chamber at very low pressure (10^{-10} Torr), and the mass-analysed primary beam, with energy 1–3 keV, is incident on it. The secondary ions are emitted with initial energy of ~10 eV and are accelerated into a quadrupole mass spectrometer for analysis. The low

Fig. 12.3. Apparatus for the analysis of secondary ions released from a target surface under ion impact. (After Benninghoven & Loebach, 1971.)

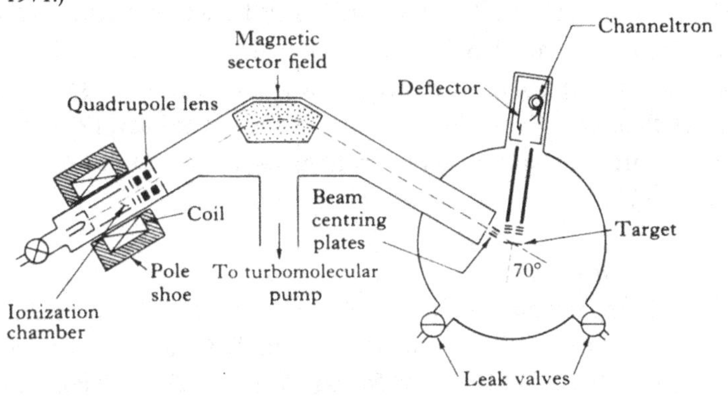

energy of the beam assures that monolayers are being sampled, and the detection sensitivity of this arrangement is extremely high, $\sim 10^{-19}$ g. (This corresponds to a microampere beam of 1 mm diameter, an ionization efficiency of 20% (ion/atom ratio), a mass spectrometer transmission of 10% and detector sensitivity of 10 ions/s.)

The technique has been widely employed for the study of surface reactions, catalysis, surface segregation in alloys, etc. Recent applications of the technique are the analysis of surface monolayers by Kloppel & Seidl (1979) and the analysis of electrosprayed organic layers by Westmore, Ens & Standing (1982). The subject has been reviewed by Benninghoven (1985).

Ion microprobe

The ion microprobe is the analogue of the electron microprobe. In it, the sample is bombarded by a finely focused beam of primary ions – Ar^+ or N^+ for inert ion bombardment or O^+ for reactive ion bombardment. These primary ions are produced in a duoplasmatron ion source (§3.9) and pass through a low resolution magnetic analyser for ion selection. The beam is focused ('demagnified') by two sets of electrostatic lenses (a 'condenser' lens followed by an 'objective' lens) to obtain a 1–2 μm diameter image on the specimen surface, the latter being observed by a long focus microscope. A set of deflecting plates is also used to position the spot in the x and y directions, over a range of about 300 μm. The secondary ions, inhomogeneous in energy, are transmitted into a double focusing analyser for M/e identification. The arrangement is shown schematically in fig. 3.12 and was described by Liebl (1967), Evans & Pensler (1970) and Evans (1971).

The secondary ion output can be obtained from the electron multiplier detector in analog or digital form. Also, to obtain an image of the distribution of a selected element (isotope) over a desired area, the primary ion beam may be swept over the sample surface and the output signal used to z-modulate the intensity of the synchronized beam on a cathode ray screen. Furthermore, the depth profile of the selected isotope may be obtained from variations in the image intensity as successive layers are sputtered away (Pawel, Pensler & Evans, 1972; de Grieve, Figaret & Laty, 1979). The effects that obscure the results of depth profiling have been studied by Schilling & Büger (1978).

Ion microscopy

In the ion microscope direct imaging is achieved by an ion-optical analogue of the optical microscope (Castaing & Slodzian, 1962; Castaing & Hennequin, 1971; Slodzian, 1975; and Morrison & Slodzian, 1975). The method is shown schematically in fig. 12.4.

The primary ion beam is focused to form an image spot ($\sim 10\ \mu m$) on the sample surface and this area constitutes the object. Secondary ions from this area are accelerated into a prism-type mass filter of resolving power ~ 1000. Two-directional focusing is achieved by means of the oblique entrance angle. The mass-resolved beam then enters a *convex electrostatic mirror* with a repeller electrode, a unique feature of the instrument. Here the ions are reflected, energy-filtered and re-enter the prism analyser. The ions emerging from the prism form a one-to-one image of the object which is projected to the cathode of an electron *image converter*. This converted ion-to-electron image strikes a fluorescent screen which gives a visual image of the spatial distribution in the sample area of whatever element (isotope) was selected by the magnetic prism. The instrument may also be used for analysis.

Ion microprobe (and ion microscope) methods have many applications relating to the solid state. Thus, in materials science, compositional correlations of microstructure, segregation along grain boundaries, oxidation, corrosion and diffusion processes have been studied (Morobito & Lewis, 1973). In mineralogy, the study of elemental distribution patterns enlarges considerably the scope of petrography, and studies have been made with meteorites (Hutcheon, Steele, Smith & Clayton, 1978) and with samples from the Oklo 'fossil' fission reactor (Slodzian & Havette, 1976). A novel application is the analysis of small samples of lunar regolith which contain components of 'solar wind' from the sun. This technique has been termed *gas ion probe* (Kiko, Muller & Kirsten, 1979).

12.3 Ion scattering mass spectrometry

This technique is of value in studying the composition of the outermost layer of a solid. It employs the technique of ion scattering in a

Fig. 12.4. Ion microscopy. (After Morrison & Slodzian, 1975.)

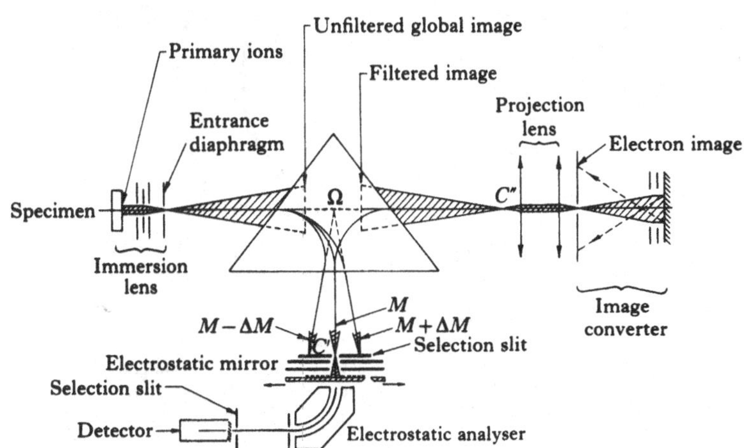

binary elastic collision, that is, an impinging ion is scattered by a surface atom. The energy spectrum of the scattered ion is determined by the mass of the scattering atom. In the simplest arrangement, ions from a monochromatic ion source are accelerated to energies of 1–3keV and bombard the surface of the target, which is located in ultra-high vacuum (fig. 12.5). The scattered ions enter the electrostatic analyser (in this case, of 127° sector angle), where they are energy analysed prior to detection in a channel electron multiplier. The ratio E_s/E_i of scattered and incident ion energies determines the position of the scattered peak. Typical cases of superficial layer composition determined in this way are given by Honig (1976).

12.4 Ion implantation

Atoms – in the form of a beam of specific ions – may be embedded in a solid, thereby modifying its physical and chemical properties. To date, most such ion implantation has been directed towards semiconductors for the fabrication of solid state devices, but the field and potential applications are more extensive. Ion implantation has several inherent advantages: (a) because the ions are injected, the material need not be in thermodynamic equilibrium: thus, new materials can be formed which by-

Fig. 12.5. Ion scattering spectroscopy. (Adapted from Honig, 1976.)

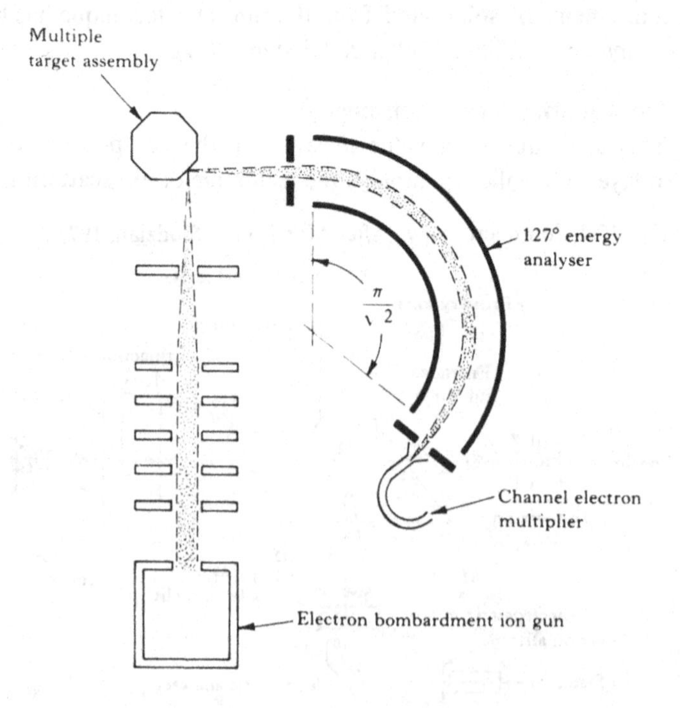

pass solubility rules; (b) their high energies allow the impinging ions to penetrate chemical (for example, oxide) barriers on the substrate; (c) the control over the the dosage, depth profile and uniformity is more precise than in chemical and diffusion methods; (d) the implantation can be done at low temperatures and with vacuum cleanliness. On the debit side, the technique is expensive and causes radiation damage, though this latter can be overcome by annealing techniques.

We now describe briefly the type of equipment used for ion implantation. Inasmuch as the ion energies involved range from a few kilovolts to ~ 0.5 MeV, and the ion beam intensities from microamperes to milli-amperes, mass separator types, rather than nuclear accelerator types, of equipment are usually employed. There are three main components – ion source, mass analyser and focusing system, and the target chamber – used in one of two configurations: (a) the voltage of the ion source corresponds to the required ion energy whilst the analyser and target chamber are at ground potential: thus, there is a free access to the target, but at the cost of a powerful analyser magnet; (b) the source is at a relatively low voltage V and is followed by a conventional mass spectrometer for beam analysis, whilst the emergent beam is accelerated to the target, which is at a high voltage. The high voltage of the target limits experimental flexibility, but intense currents of good beam quality can be obtained with a conventional magnet.

Typical of the first configuration are the systems (fig. 12.6) in which the ion source is enclosed in the high voltage terminal of small Van de Graaff or Cockroft–Walton accelerators whose beam output goes through an analyser and beam scanner plates to the target chamber. For somewhat lower energies (50–80 keV), laboratory isotope separators of the Scandi-navian type (Nielsen, 1957; §3.8) can be used. In this case, the primary acceleration of ions is across a single gap following the ion-source exit, thereby simplifying the focusing ion optics and giving easier access to the source (which yields currents $\sim 10\,\mu$A–1 mA).

Typical of the second configuration are the installations described by Wilson (1967) and Freeman (1967) – (see fig. 12.7). Here one obtains intense beams at low energy, with considerable source flexibility. The mass-resolved ions are then accelerated, focused and made to impinge upon the target. If desired, the energy range can be extended by the use of multiply charged ions.

The ion sources employed are usually of the 'universal type', that is, applicable to the various elements to be used as 'dopants'. The magnetic field arc discharge (§3.7) and plasmatron types (§3.9) satisfy this criterion, and

Fig. 12.6. Ion implantation unit with mass analyser at high voltage. (After Ryssel & Glawischnig, 1982.)

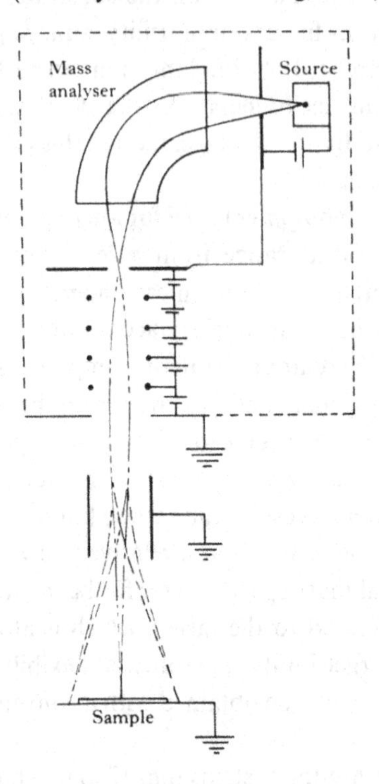

Fig. 12.7. Ion implantation unit with target at high voltage. (After Wilson, 1967.)

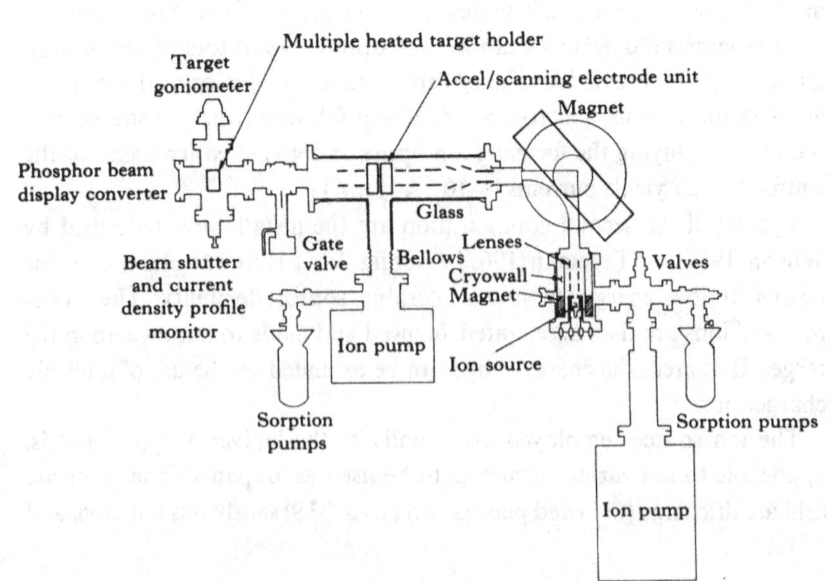

a commonly used version is the Freeman (1967) type which uses a hot-filament arc discharge in a magnetic field (§3.12) with side-extraction. The arc is fed with vapour of the dopant element which is held in a sputtering probe or heated in an oven (fig. 12.8). The hollow-cathode discharge type Nielsen source (§3.8), as well as the duoplasmatron source (§3.9) with gas inlet or vapour feed into the anode expansion cup, are also used, as also is the Penning discharge source (§3.11). For certain specific elements (for example, alkali metals and indium), high intensity surface ionization sources in which the vapour of the element heated in an oven impinges on a heater–ionizer filament are also used (Wilson, 1967).

Beam scanning is necessary for uniform coverage of the implanted area, and can be achieved either by modulating the acceleration voltage with a linear sweep or, for larger areas, by placing a scanning system behind the slit at the exit focus. X–Y scanning is used for beams of circular section. In some systems, the target holder is mechanically scanned through a fixed beam. Accurate target alignment and manipulation are needed (especially in channelling or backscattering measurements) and are achieved by mounting the target on a goniometer arrangement that can be adjusted in vacuum. Measurements of ion beams in a Faraday cage, with suitable precautions to suppress secondary electron emission, are necessary for the estimation of implantation dosage. Detailed descriptions of ion implantation facilities, including target chambers, have been given in Dearnaley, Freeman, Nelson & Stephen, (1973), Wilson & Brewer (1979) and Ryssel & Glawischnig (1982).

When ions enter condensed matter they are subject to 'nuclear' stopping

Fig. 12.8. Source for ion implantation. (After Freeman, 1967.)

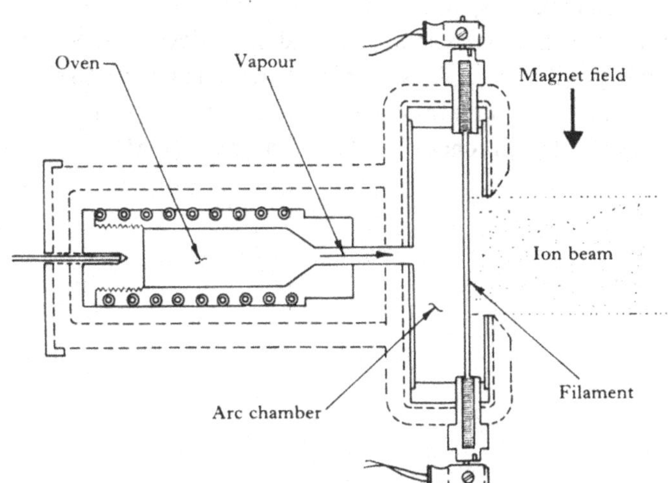

and 'electronic' stopping. The former is an energy loss resulting from elastic collisions between ions and screened nuclei. This effect, which was analysed initially by Bohr (see Bohr, 1948), is important for low energies and/or heavy ions. Electronic stopping was first studied authoritatively by Lindhard (1954) and applies to higher energies and/or lighter ions. For the range of impinging energies 10–500 keV, the experimentally observed 'stopping powers' (that is, dE/dx for various ion-material combinations) are extensive and the underlying theory appears to be in good shape (see Dearnaley *et al.*, 1973). Thus, one can calculate in advance the range of specific ions as they are implanted in various host materials. Not included in the above remarks is the phenomenon of 'channelling'.

Channelling is an important special feature of ion penetration in crystalline materials. It is displayed by ions that are incident along major crystallographic axes or the directions between close-packed planes. For these ions, the trajectories are 'steered' or guided along the close-packed directions by glancing collisions with 'walls' of atoms that provide a smoothed repulsive potential. The energy loss of such 'channelled' ions is controlled by electronic stopping, with the result that the ions have abnormally long ranges, leading to long tails in the range distribution. Although channelling is not much exploited in device fabrication, it is an important tool for the study of lattice location of dopants. The subject of channelling has been reviewed by Dearnaley *et al.* (1973) and Gemell (1974) and the use of channelled ions for lattice studies by Townsend, Kelly & Hartley (1976), Carter & Grant (1977), Chu, Mayer & Nicolet (1978) and Wolicki (1979).

As indicated in the foregoing, the mass spectrometric method of ion implantation provides clear mass resolution, focusing of the ion beam to a small size and precise spatial control of the dopants. For these reasons it is widely used in the manufacture of solid state devices, especially micro-electronic ones (for example, for use in pocket calculators). Useful details of the device fabrication that is made possible by ion implantation are given by Pickar (1975), Hirvonen (1980) and Donnelly (1981).

13

Applications of mass spectrometry to geology and cosmology

There are many natural processes that can alter the isotopic constitution of elements. For this reason mass spectrometric studies of variations in isotope abundance provide a powerful means both of investigating these processes and of ascertaining the time for which terrestrial and extra-terrestrial specimens have been subjected to them. These processes and their effects can be classified as (a) nuclear, where normal isotopic abundances are altered by nuclear processes and (b) non-nuclear, where isotopic fractionation of stable isotopes by various physical and chemical processes leads to variations in isotopic constitution.

Radiogenic isotopes in geology

The concept of isotopy grew out of the study of radioactivity amongst the heavy elements in which early workers (Soddy, 1913; Richards & Lembert, 1914) noticed variations in nature of the atomic weight of lead. The first mass spectroscopic study of lead (Aston, 1929) revealed variations in isotopic abundances and led to early attempts to utilize these variations in geologic dating (Aston, 1933; von Grosse, 1932, 1934). The first systematic study of the variations in nature of the isotopic composition of lead and its correlation with geologic time was made by Nier and his colleagues (Nier, 1938; Nier 1939*a*; Nier, Thompson & Murphey, 1941). During this period Nier (1939*b*) also measured the $^{238}U/^{235}U$ abundance ratio, making possible the calculation of the half-life of ^{235}U. These experiments of Nier and his colleagues were the precursors of many age determinations involving U, Th and Pb. For example, the U–Th–Pb method, especially applied to the commonly found resistate mineral zircon, is one of the important methods of geochronology. Also, the Pb–Pb method, involving only the measurement of isotopic ratios of lead, has been widely applied to lead minerals (for example, galena) for the calculation of 'model' ages (Russell & Farquhar, 1960). Refinements in mass spectrometric techniques have permitted the method to be applied to ordinary rock samples. These matters are elaborated in §13.1.

A number of other naturally occurring radioactive parent–daughter systems are potentially useful for geochronology; however, ^{87}Rb-^{87}Sr and ^{40}K-^{40}Ar are the only ones that are extensively used at present. Although the natural radioactivity of rubidium had long been known (Wood 1904), and Hahn & Walling (1938) had suggested the possibility of Rb–Sr dating, the method was not extensively employed until the 1950s, when the stable isotope dilution technique (§7.9) for the determination of trace concentrations of Rb and Sr in rocks and minerals had been developed. The introduction of the 'isochron' method, relating the ratios ($^{87}Sr/^{86}Sr$) vs ($^{87}Rb/^{86}Sr$) for cogenetic rocks and minerals, provided a more meaningful interpretation of the calculated age values. Improvements in the precision of ($^{87}Sr/^{86}Sr$) measurements also extended the range of applicability of the method, which is one of the most widely used of geochronologic techniques, and which is described in some detail in §13.2.

Also widely used is the K–Ar method. The radioactivity of potassium was known in 1905 (Wood) but the radioactive isotope ^{40}K was not discovered until 1935 (Nier). This led to clarification of the decay scheme (von Weizsäcker, 1937). The applicability of the method, which involves the measurement of ^{40}Ar in K minerals, was demonstrated by Aldrich & Nier (1948). In view of the ubiquitous nature of K in rocks and minerals, the technique has wide applicability. Additionally a variation of the method – the $^{39}Ar/^{40}Ar$ technique – can assist materially in the interpretation of data. The K–Ar method is the subject of §13.3.

The ^{187}Re–^{187}Os pair has also been used in geologic dating, whilst the Sm–Nd pair involving the α decay of ^{147}Sm to ^{143}Nd promises to be of importance in the view of the geochemical coherence of the parent and daughter elements. For the ^{14}C method of dating (Libby, 1955), involving the decay of cosmic- ray- produced ^{14}C in organic samples, and originally carried out by low-level counting techniques, mass spectrometric techniques appear to offer distinct advantages (Muller, 1977; Lutherland *et al.*, 1981; see §7.11). The Re–Os and Sm–Nd methods are touched on in §13.4.

The general principle of dating, using the parent–radiogenic daughter relationship, is quite straightforward. Let us assume that N_i atoms of the parent were initially present ($t = 0$), whilst N_p atoms are present today (time $= t$). Thus,

$$N_p = N_i e^{-\lambda t} \quad \text{or} \quad N_i = N_p e^{\lambda t}, \tag{13.1}$$

where λ is the decay constant. The number of radiogenic daughter atoms produced during this time is

$$D_p - D_i = N_i - N_p = N_p(e^{\lambda t} - 1), \tag{13.2}$$

where D_p and D_i refer to the present and initial concentrations of daughter products. Measuring N_p and D_p (present concentrations) and knowing λ and D_i, the 'age' t can be calculated, if the rock or mineral has remained a 'closed system' (with no loss or gain of parent or daughter) during this period. If the parent is a radioactive mineral (that is, highly enriched in the parent), D_i can be neglected. When this is not the case, if the daughter element has a *stable* isotope of concentration D_0, equation (13.2) can be written

$$[D_p/D_0] - [D_i/D_0] = [N_p/D_0][e^{\lambda t} - 1]. \tag{13.3}$$

A plot of $[D_p/D_0]$ vs $[N_p/D_0]$ for samples of varying $[N_p/D_0]$ ratios is thus linear, the slope being $(e^{\lambda t} - 1)$ and the intercept (D_i/D_0), so that both the age and the initial D_i/D_0 values can be calculated.

13.1 Variations in the isotopic constitution of lead

There are four stable isotopes of lead: ^{204}Pb (1.42%), ^{206}Pb (24.1%), ^{207}Pb (22.1%) and ^{208}Pb (52.4%). The three last-named are the end products of the three naturally occurring radioactive series, namely,

$$^{238}U(T_{1/2} = 4.5 \times 10^9 yr) \rightarrow {}^{206}Pb(RaG) + 8\,{}^4He^{2+},$$
$$^{235}U(T_{1/2} = 7.1 \times 10^8 yr) \rightarrow {}^{207}Pb(AcD) + 7\,{}^4He^{2+},$$
$$^{232}Th(T_{1/2} = 1.4 \times 10^{10} yr) \rightarrow {}^{208}Pb(ThD) + 6\,{}^4He^{2+},$$

and may be present in any sample in abnormal amounts as a result of uranium and/or thorium decay. As the remaining isotope, ^{204}Pb, is non-radiogenic, its presence is evidence of common lead contamination.

The lead–uranium and lead–thorium methods of age determination

These methods require an isotopic analysis of radiogenic lead, together with an analysis of the concentration of U and/or Th. The present amounts of isotopic daughter lead are related to the present parent abundances as follows:

$$\left.\begin{aligned} {}^{206}Pb &= {}^{238}U(e^{\lambda_{238}t} - 1) \\ {}^{207}Pb &= {}^{235}U(e^{\lambda_{235}t} - 1) \\ {}^{208}Pb &= {}^{232}Th(e^{\lambda_{232}t} - 1) \end{aligned}\right\}. \tag{13.4}$$

Three independent age values can thus be obtained. The 'secular' equilibrium conditions that these equations describe, are attained in 10^6, 10^5 and 10^2 yr respectively. The assumption, implicit in these methods, that the mineral has remained a closed system, is frequently invalidated by leaching of lead or uranium and, possibly also, by loss of radon by diffusion. The effect of leaching can be minimized by the so-called 'lead–lead' method in which the ratio ($^{207}Pb/^{206}Pb$) is determined. Dividing the second of the equations (13.4) by the first,

$$\frac{^{207}\mathrm{Pb}}{^{206}\mathrm{Pb}} = \frac{^{235}\mathrm{U}}{^{238}\mathrm{U}} \cdot \frac{e^{\lambda_{235}t} - 1}{e^{\lambda_{238}t} - 1} = \frac{1}{138} \cdot \frac{e^{\lambda_{235}t} - 1}{e^{\lambda_{238}t} - 1}. \tag{13.5}$$

This equation can be solved graphically or by interpolation from a table of values and is less sensitive to loss of uranium or lead from the sample.

The bulk of present-day work on the U–Th–Pb method has been concentrated on zircon, as this is a mineral that is resistate and retentive for uranium and lead and is widely disseminated in a variety of rocks. The uranium and thorium ($\sim$ 1000 ppm) and lead are all measured by isotope dilution (for example, with spikes ^{235}U, ^{230}Th, ^{204}Pb) after hydrothermal decomposition of the mineral in a sealed vessel (Krogh, 1973). A surface ionization source is employed for these measurements, as well as for the lead isotope–ratio determinations. Although a discordancy was noted in the different 'ages' of cogenetic zircons, it was observed that ($^{206}Pb/^{238}U$) vs ($^{207}Pb/^{235}U$) shows a linear relationship. This was explained by Wetherill, Tilton, Davis & Aldrich (1956) on the basis of episodic lead loss of differing extent for different samples. Starting with a 'concordia' curve connecting (206/238) vs (207/235) as a function of age (fig. 13.1), they showed that for all samples of age t_1, that had experienced varying lead loss (or uranium gain) at time t_2, the points fall on a chord connecting t_1 and t_2. One thus plots the points (206/238, 205/237) for various cogenetic zircons, and finds the age of 'provenance' as the upper intercept of the best fit straight line with the

Fig. 13.1. Concordia curve plotting (206/238) vs (207/235) as a function of age. (After Gebauer & Grünenfelder, 1979.)

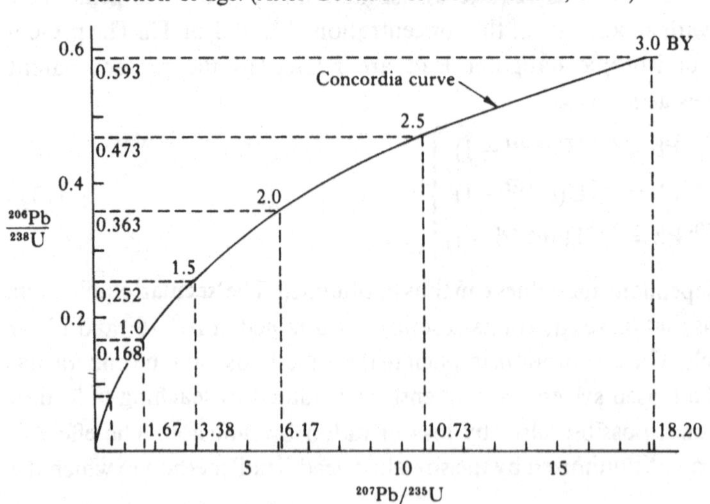

concordia, the lower intercept yielding the age of 'episodic' lead loss. In some cases the interpretation of the lower intercept was problematical; this was explained by Tilton (1960) on the basis of continuous diffusion. Here the loss is a function of age and D/a^2 (where D = diffusion coefficient and a = grain size), and the concordia is linear at low values of D/a^2 but curves toward the origin at large D/a^2, so that only the upper intercept has a physical significance. The problem has also been treated by Wasserburg (1963), whilst Catanzaro (1968) has reviewed the interpretation of zircon ages.

The common lead method

It was first supposed that common lead (for example, as found in lead minerals such as galena) was isotopically invariable; instead, Nier (1938) found large variations amongst common lead samples. This led him to distinguish between 'primeval' lead, or lead as it existed at the time of formation of the earth's crust, and common lead, which consists of primeval lead to which has been added a certain quantity of radiogenic lead. This is the Holmes–Houtermans model (Holmes, 1946; Houtermans, 1951) which assumes that (a) at the early stage of the earth's history, uranium, thorium and lead were uniformly distributed in a homogeneous source, for example the mantle, (b) the solidification of the earth fixed the uranium–thorium–lead constitution for each locality introducing regional differences, (c) thereafter, prior to the mineralization of lead, no changes in uranium–thorium–lead abundances occurred except those due to radioactive decay and (d) at the time of mineralization radiogenic additions to the Pb ceased. The lead mineral can thus be regarded as constituting a permanent record of the isotopic constitution of the lead at the time of the mineral's formation. Thus if T and t denote the ages of the earth and of mineral formation, and μ and W denote the ratios ($^{238}U/^{204}Pb$) and ($^{232}Th/^{204}Pb$), and we denote the present and initial ratios of (206/204) by a and a_0, of (207/204) by b and b_0, and of (208/204) by c and c_0, then

$$\left.\begin{aligned} a &= a_0 + \mu[e^{\lambda_1 T} - e^{\lambda_1 t}] \\ b &= b_0 + \frac{\mu}{137.8}[e^{\lambda_2 T} - e^{\lambda_2 t}] \\ c &= c_0 + W[e^{\lambda_3 T} - e^{\lambda_3 t}] \end{aligned}\right\}. \tag{13.6}$$

We can eliminate the unknown μ, from the first and second equations to obtain

$$\frac{(b - b_0)}{(a - a_0)} = \frac{1}{137.8} \cdot \frac{e^{\lambda_2 T} - e^{\lambda_2 t}}{e^{\lambda_1 T} - e^{\lambda_1 t}}. \tag{13.7}$$

Equation (13.7) can be used (a) to calculate the mineralization age t, if the primeval age T is known or (b) to calculate the crustal age T if the ages of mineralization have been measured, for example by the Rb–Sr method (see §13.2). It is known that meteorites are fragment bodies formed early in the solar system, and the sulphide phases known as troilites, which have been formed in many meteorites after partial melting and differentiation, contain appreciable Pb but little U or Th. The Pb in these meteorites can thus be considered as representative of the composition of 'primeval' Pb. Measurements of the isotopic composition of Pb in troilites were made by Patterson (1955, 1956) and using these values (a_0, b_0) and lead isotopic ratios measured in other stone and iron meteorites (for which $t = 0$), he was able to calculate $T = 4.55 \times 10^9$ yr. This agrees with Rb–Sr ages determined from a large number of meteorites. Further, the isotopic ratio for modern lead present in oceanic sediments (Murthy & Patterson, 1962), falls on the 'geochron', that is, the ($^{207}Pb/^{204}Pb$) vs ($^{206}Pb/^{204}Pb$) plot, thus confirming the close identity in age of the earth and meteorites.

The model developed above also permits the dating of galena (lead ore) samples. For this purpose, one considers the 'growth' curves, which represent $^{207}Pb/^{204}Pb$ vs $^{206}Pb/^{204}Pb$, for different values of $\mu = {}^{238}U/^{204}Pb$, usually in the range 8–10 (see fig. 13.2). Thus, from equation (13.7), where a and b are the measured isotope ratios, one calculates t, using the (currently accepted) values T(age of the crust) $= 4.57 \times 10^9$ yr, $a_0 = 9.307$ and $b_0 = 10.294$. This 'model age' is calculated graphically, or

Fig. 13.2. Single-stage lead growth curve with some 'conformable' lead data. (After Faure, 1977.)

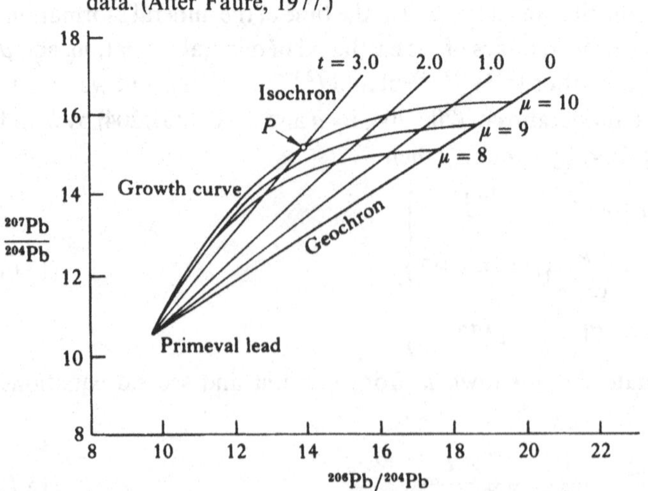

from a table of values using interpolation. One can then calculate $\mu = (a - a_0)/(e^{\lambda_1 T} - e^{\lambda_1 t})$. The Th/U ratio for the region can also be determined, using $^{232}Th/^{204}Pb = W = (c - c_0)/(e^{\lambda_3 T} - e^{\lambda_3 t})$ (where c_0 is known) and $\mu = {}^{238}U/^{204}Pb$. Thus,

$$^{232}Th/^{238}U = W/\mu. \tag{13.8}$$

Russell and coworkers (Russell & Farquhar, 1960; Kanaseiwitch, 1968) showed that a wide range of galenas had isotope ratios that plotted 'conformably' on the growth curve (207/204) vs (206/204) with $\mu \sim 9$. These deposits were proposed by Stanton & Russell (1959) as having a mantle source, having been deposited in sedimentary basins during submarine volcanism. Their 'model ages' then correspond to the time of migration of the ore deposits from the homogeneous source region.

Leads whose isotopic ratios do not fit on the single-stage growth curve (usually from hydrothermal deposits that are genetically related with igneous intrusives) are called *anomalous* leads. Frequently these samples fall on a straight line above the growth curve, intersecting it at some point. This situation has been explained by the addition of radiogenic lead to the common source lead. With certain assumptions it is possible to derive limits to the age of the deposit, as well as to the age of the rocks that supplied the radiogenic lead (Russell, 1972). The other case is that of *multistage* leads, where the lead has been associated at different periods, with environments of differing μ values. Thus, for a two-stage system, where the lead has spent time $T_0 - t_1$ in an environment with $\mu = \mu_1$, and $t_1 - t_2$ in a different host rock with $\mu = \mu_2$,

$$[^{206}Pb/^{204}Pb] = a = a_0 + \mu_1[e^{\lambda_1 T_0} - e^{\lambda_1 t_1}] + \mu_2[e^{\lambda_1 t_1} - e^{\lambda_1 t_2}] \tag{13.9}$$

and similarly for $b = (^{207}Pb/^{204}Pb)$. It can be seen that the plot of b vs a is a straight line of slope $m = (y - b)/(x - a)$ passing through the points, $a_1 = (206/204)t_1$, and $b_1 = (207/204)t_1$ and (a_2, b_2). Some leads have been interpreted in this way (Kanaseiwitch, 1968).

Finally, following revisions to the decay constants of U and Th, the data for the 'conformable' leads did not fit the one-stage evolution curve, the departures being serious for the younger deposits. A two-stage growth curve was proposed (Stacey & Kramers, 1975) with a change in μ value at 3.7×10^9 yr (when the major crust-mantle differentiation occurred). This, and another model which assumes a continuous decrease of μ with time (Cumming & Richards, 1975), accommodate the available data for conformable leads on the revised growth curves. Studies of a number of galenas from Indian ores show a good fit with the Stacey–Kramers curve (Venkatasubramanian, Jayaram, & Subramanian, 1981).

The Pb–Pb method has also been applied to feldspars (Doe, 1967) and to whole rocks. Here the lead is extracted by volatilization or electrochemical techniques, and an HPO_3 gel technique is used to obtain suitably high currents with a thermal ion source. Using this method, an age of 3.7×10^9 yr was obtained for the rocks of Isuo, W. Greenland (Moorbath & Pankhurst, 1976). Much of what is described above, concerning variations of lead isotopes in minerals and rocks, is exposed in greater detail in a review by Doe (1970).

13.2 The rubidium–strontium method of age determination

The isotope ^{87}Sr, constituting 7.02% of common strontium, is the radiogenic daughter of 4.5×10^{10} yr ^{87}Rb. It is formed in the decay $^{87}Rb \rightarrow {}^{87}Sr + \beta^-$. Because of the low activity of ^{87}Rb, the amount of Sr resulting from the decay is not large. However, there are several minerals (Ahrens, 1949), notably lepidolites and biotites, which are at once Rb-rich and Sr-poor and early work was concentrated on these minerals. The amounts of ^{87}Rb and ^{87}Sr were determined by isotope dilution. From equation (13.2) the basic age equation is

$$^{87}Sr_p - {}^{87}Sr_i = {}^{87}Rb_p(e^{\lambda t} - 1). \tag{13.10}$$

Thus, in this early work, the age was computed from the measured ratio $^{87}Sr/^{87}Rb$, as the initial ^{87}Sr could be neglected. With the development of the isochron method, and improved precision in the measurement of $^{87}Sr/^{86}Sr$ ratios, the method began to be applied to common rocks and minerals. The isochron equation is obtained by dividing the age equation by ^{86}Sr, the concentration of the stable isotope, to yield

$$[^{87}Sr/^{86}Sr]_p - [^{87}Sr/^{86}Sr]_i = [^{87}Rb/^{86}Sr][e^{\lambda t} - 1]. \tag{13.11}$$

A plot of ($^{87}Sr/^{86}Sr$) vs ($^{87}Rb/^{86}Sr$) for various cogenetic rocks or minerals thus yields a straight line of slope $m = (e^{\lambda t} - 1)$ and intercept = $(^{87}Sr/^{86}Sr)_i$, the initial ratio. Cogenetic rocks of differing composition covering a wide range of (Rb/Sr) ratios are produced in a rock body by fractional crystallization of a magma, so that the isochron method can be applied to whole rocks. On the other hand, an isochron can also be drawn using a rock sample and minerals separated from it, and sometimes by using density fractions.

The ^{87}Rb and ^{86}Sr values are determined by isotope dilution, with suitable spikes (or calculated from Rb and Sr concentrations measured by X-ray fluorescence). The $^{87}Sr/^{86}Sr$ ratios are determined by using a thermal ionization source on unspiked samples. Isobaric interference between ^{87}Rb and ^{87}Sr necessitates complete removal of Rb from the Sr solution, and this is achieved by eluting the rock solution through an ion exchange column (^{87}Sr does not interfere with Rb analysis as Rb is emitted at much lower temperatures in the surface ionization source). In the simplest case, the mass peaks are recorded by magnetic (or electrical) scanning and the ratio calculated as the average of several scans. Mass fractionation effects can be corrected by using the ($^{86}Sr/^{88}Sr$) ratio $= 0.1194$. More sophisticated systems of the sort described in Chapter 7 are used for high precision measurement of $^{87}Sr/^{86}Sr$ ratios. Standard deviations of $\sim 10^{-4}$ in the (87/86) ratios are achieved in most cases.

The initial time ($t = 0$) for the Rb–Sr clock is the time when the rocks or minerals under study had the same initial ($^{87}Sr/^{86}Sr$) ratio. As diffusion rates in molten magmas are rapid at the time of crystallization or emplacement of the rock body, the Sr isotopes get uniformly distributed among the available lattice sites in the minerals, that is, the ($^{87}Sr/^{86}Sr$) ratio is uniform in the system studied. The subsequent isotopic distribution in the systems with time is as shown in fig. 13.3, so that at any time t, the ($^{87}Sr/^{86}Sr$) ratios plot on the isochron shown in fig. 13.4. Metamorphic events induced by an increase in temperature have thus a profound effect on Sr-isotope ratios. One general pattern is that during the metamorphic episode the Sr migrates among the individual mineral phases so that the 'mineral' isochron yields a slope corresponding to this time t. The whole-

Fig. 13.3. Relationship between $^{87}Sr/^{86}Sr$ and $^{87}Rb/^{86}Sr$. (After Faure, 1977.)

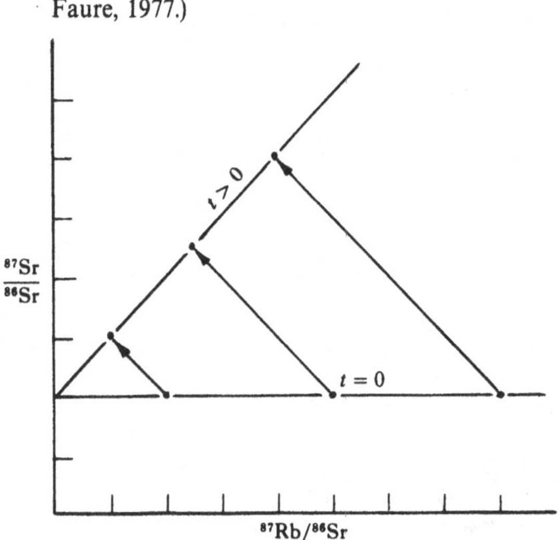

rock 'hand samples' however remain closed systems and the 'whole-rock isochron' yields the older time of emplacement. Sometimes, if the metamorphism is intense, the Sr in whole-rock samples themselves is isotopically homogenized, as the temperature rises above the blocking temperature. Such behaviour exhibited by isotopic systems was investigated by Hart (1964) in a region in Colorado where Precambrian rocks are cut by much younger Tertiary rocks. The possibility of dating sedimentary rocks using anthigenic or detrital minerals or fine-grained rocks (for example, shales) has been discussed (for example, Moorbath, 1969).

Together with the K–Ar method, the Rb–Sr method has made a major contribution to our understanding of the chronology of terrestrial, lunar and meteoritic rock samples. The salient features and main conclusions of these results are discussed in §13.5. Quite apart from this, the ($^{87}Sr/^{86}Sr$) ratios provide relevant information on the geochemical evolution of rock systems. This is because there is good basis for the assumption that the $^{87}Sr/^{86}Sr$ ratio for the material that separated out from the solar nebula had a uniform value, and subsequent variations are due to differing Rb/Sr ratios of the environment of the samples (for example, meteorites). These results are also summarized in §13.5.

Fig. 13.4. Isochron plot for Rb–Sr dating. Slope of the line is $(e^{\lambda t} - 1)$. (After Faure, 1977.)

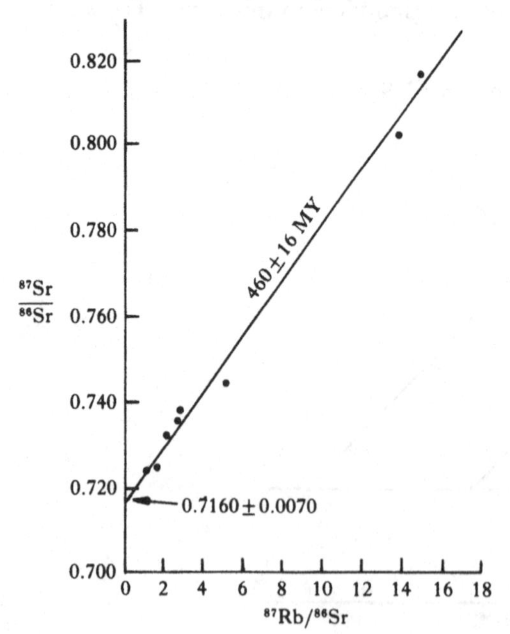

13.3 The potassium–argon method of age determination

There are three stable isotopes of argon ($^{36}Ar = 0.337\%$, $^{38}Ar = 0.063\%$, $^{40}Ar = 99.6\%$), the predominance of ^{40}Ar resulting from the K capture decay of ^{40}K in the earth and the subsequent escape of ^{40}Ar into the atmosphere. (^{40}K also decay by β^- emission to ^{40}Ca). The age of K-bearing minerals and rocks can thus be calculated from the relation

$$^{40}Ar = {}^{40}K\left(\frac{\lambda_K}{\lambda_K + \lambda_\beta}\right)[e^{\lambda t} - 1], \tag{13.12}$$

where ^{40}Ar is the amount of radiogenic argon extracted from the mineral, and λ, λ_K and λ_β are the total, K capture and negatron disintegration constants, respectively, of ^{40}K. $\lambda = \lambda_K + \lambda_\beta = 5.305 \times 10^{-10}$/yr corresponding to $T^{1/2} = 1.31 \times 10^9$ Yr, while $R = \lambda_K/\lambda = 0.112$. The ^{40}K is known from the potassium content of the mineral, measured by flame photometry, atomic absorption *et cetera*, while the ^{40}Ar content is determined by isotope dilution techniques after extraction and purification of Ar from the sample. The argon is gas extracted from the mineral, by fusing the latter in a furnace with induction or resistive heating, and is purified by passing the extracted gases into a purification train that leaves only noble gases. A known 'spike' of nearly pure ^{38}Ar is added to the purified gases, or admitted into the purification system, to equilibrate with the ^{40}Ar. The $^{40}Ar/^{38}Ar$ and $^{40}Ar/^{36}Ar$ ratios are determined in the mass spectrometer. The ^{40}Ar is calculated from the first ratio, whilst the second ratio enables a correction to be made for atmospheric argon (for which $^{40}Ar/^{36}Ar \sim 295$). The sensitivity of the method is vastly improved by operating the mass spectrometer in the 'static mode'. Here the mass spectrometer is evacuated to pressures $\sim 10^{-9}$–10^{-10} Torr by the use of ultra-high vacuum techniques, before isolating the system from the pump. The analysis gases are then introduced, raising the pressure to $\sim 10^{-7}$–10^{-8} Torr, and the isotope ratios measured.

The factor that governs the application of the method is the retention of radiogenic argon. Melting and crystallization of rocks and minerals results in complete loss of argon, and a resetting of the K–Ar clock. The same effect results from thermal metamorphic events which cause heating above the blocking temperature, which is rather low for biotites and feldspars. Hornblendes, however, show superior retentivity and the reliability of K–Ar ages for whole-rock basalts has also been shown in most cases. Some minerals, for example, pyroxenes, show excess argon, probably trapped during mineralization.

An important advance was the development of the ^{39}Ar–^{40}Ar technique

(Merrihue & Turner, 1966). In this method, the sample is irradiated in a reactor to produce ^{39}Ar from the reaction $^{39}K(n,p)^{39}Ar$. ^{39}Ar has a half-life of 269 yr and is released along with ^{40}Ar and other isotopes of argon during fusion of the sample. The argon is purified and $^{39}Ar/^{40}Ar$ ratios are determined with the mass spectrometer.

The production of ^{39}Ar is governed by

$$^{39}\mathrm{Ar} = {}^{39}\mathrm{K}\,\tau \int \phi(\varepsilon)\sigma(\varepsilon)d\varepsilon, \tag{13.13}$$

where τ = time of irradiation in the reactor and $\phi(\varepsilon)$ and $\sigma(\varepsilon)$ represent the neutron flux and cross section respectively. Hence

$$\frac{^{40}\mathrm{Ar}}{^{39}\mathrm{Ar}} = \frac{\lambda_K}{\lambda}\cdot\frac{^{40}\mathrm{K}}{^{39}\mathrm{K}}\cdot\frac{e^{\lambda t}-1}{\tau\int\phi(\varepsilon)\sigma(\varepsilon)d\varepsilon} = \frac{e^{\lambda t}-1}{J}, \tag{13.14}$$

The constant, J, is determined by irradiating a sample of known age along with the unknown sample. This method has the merits (a) that isotopic ratios only are measured and (b) that the amounts of K and Ar are determined from the same sample, thus eliminating inhomogeneities.

Another advantage of the K–Ar method is its use with the *incremental heating technique*, in which the temperature is slowly increased releasing the argon in a series of steps. At low temperatures, the argon is released from the surface and other low retentivity sites, where it accumulated during low grade metamorphic episodes, whilst at higher temperatures it is released from high retentivity sites. In this way both metamorphic and crystallization ages (plateaus) are observed (Dalrymple & Lanphere, 1971; Hussain & Schafer, 1973).

13.4 Other decay schemes

The Re–Os method based on the decay $^{187}Re \rightarrow {}^{187}Os + \beta^-$ has been applied to molybdenites (which contain 1% Re) (Herr & Merz, 1958), while the Lu–Hf method ($^{176}Lu \rightarrow {}^{176}Hf + \beta^-$) has also been suggested (Faure, 1977). A more promising method, however, is based on the Sm–Nd pair, $^{147}Sm \rightarrow {}^{143}Nd + {}^{4}He$. The great advantage of this method is that parent and daughter are geochemically coherent and, thus, are not fractionated in geological processes. Techniques have been developed to measure the isotopic ratios $^{143}Nd/^{144}Nd$ for the isochron method, and results have been reported for lunar and terrestrial rocks (Lugmair, Schenin & Marti, 1975).

13.5 Discussion of terrestrial and planetary rock ages determined by the mass spectrometric study of radiogenic isotopes

In this section, we summarize the major results that have been obtained by the application of this group of mass spectrometric methods.

Meteorites

The formation ages of meteorite samples have been determined by the Rb–Sr method, which has yielded excellent whole rock isochrons, despite the unfavourable Rb/Sr ratios. A large number of meteorites of all classes–carbonaceous, hypersothene, enstatite and bronzite chondrites, various achondrites, and silicate inclusions in iron meteorites have been studied (for example, Gopalan & Wetherill, 1969, 1970). Self-consistent data have also been obtained by the $^{207}Pb/^{206}Pb$ isochrons (Taksumoto, Knight & L'Allegre, 1976) and ^{39}Ar–^{40}Ar plateau ages (Podosek, 1971), whilst ^{147}Sm–^{143}Nd data (Lugmair, Schenin & Marti, 1975) are also available. The data reveal a strong clustering of ages at 4.57 BY (10^9 yr), indicating a contemporaneous formation of these objects, considered as fragments of parent bodies formed early in the solar system. This circumstantial evidence indicates the formation of the solar system at this time, as the small parent bodies cooled rapidly. The Sr-isotope data support this result as the lowest $^{87}Sr/^{86}Sr$ ratio has been observed for these bodies: that is, $0.698\,990 \pm 0.000\,047$ for achondrites (Papanastassiou & Wasserburg, 1969). This ratio is called BABI (basaltic achondrite best initially) – the figure is representative of the high accuracy of modern $^{87}Sr/^{86}Sr$ measurements. These is also evidence of collisional or shock metamorphism in some samples, for example, the Nakhla meteorite, which yields Rb–Sr and ^{39}Ar–^{40}Ar ages of about 1.4 BY, whilst some hypersthene chondrites yield 500 MY (10^6 yr) ages. The 'cosmic ray exposure ages' are also relevant to meteorites, as these indicate the time interval after break-up into metre-sized bodies during which they were exposed to cosmic ray protons. The latter produce a variety of spallation products, amongst which the $^{40}K/^{41}K$ and $^{21}Ne/^{22}Ne$ ratios are amenable to age measurement, by the decay of the respective parent radionuclides. The $^{40}K/^{41}K$ ratio was used by Voshage & Hintenberger (1967) who found exposure ages of about 10 MY for stony meteorites and about 500 MY for iron meteorites. The former result suggests that stony meteorites were expelled from their parent bodies only in recent times. The higher exposure age of the iron meteorites may be due to different mechanical strength or to a different point of origin.

The moon

A number of principal events in the history of the moon have been indicated by mass spectrometric dating techniques. Thus, (a) the early differentiation into the anorthositic crust above (and the magma source below) is represented by the Rb–Sr isochron age 4.55 ± 0.1 BY for the clasts from a mafic cumulate rock (Papanastassiou & Wasserburg, 1975), (b) similar ages are derived from plagioclases, troctolites, etc. from the so-called

ANT (anorthosite, norite, troctolite) rocks dominant in the highland, whilst (c) U–Pb diagrams obtained for crustal rocks also give concordant ages (Ringwood, 1979).

An important lunar event was the bombardment of the crust by giant asteroids resulting in the formation of maria (large basins). This event is indicated by the distribution of Rb–Sr ages of the highland rocks and ejecta, yielding values 3.85–4.05 BY (Kirsten, 1978). This age is also supported by Pb concordia curves. This was followed by the outpouring of basaltic magmas that filled the basins, an event that is indicated by the ^{39}Ar–^{40}Ar data and Rb–Sr mineral isochrons, which indicate a period of 'mare volcanism' from 3.1 to 3.9 BY (Ringwood, 1979). Cosmic ray exposure ages have also been measured for objects on the surface and yield information on transport processes in the regolith (top soil).

Terrestrial rock ages

Perhaps the major contribution of mass spectrometric dating techniques – especially the Rb–Sr whole-rock and mineral isochron ages and the K–Ar ages – has been the delineation of the major orogenic events in all continents of the globe, particularly in the Precambrian era, where a timescale has been established. Usually the dates cluster around certain values that are identified with the ages of 'orogenies' – tectonic events that affect large segments of continents and are associated with regional metamorphism and emplacement of granitic and other bodies of rocks. Examples are the Grenvellian (~1000 MY), Hudsonian (~1800 MY) and Kenoran (~2600 MY) orogenies of the Canadian Shield, and the Keewaneewan (~1100 MY) and Algonan (~2600 MY) orogenies of Minnesota (O'Nions & Pankhurst, 1978). Similar orogenic cycles have been delineated in South Africa (Clifford, Rooke & Allsopp, 1969; O'Nions & Pankhurst, 1978) and in Peninsular India (Crawford, 1969; Venkatasubramanian, Iyer & Pal, 1971; Venkatasubramanian, 1974). Recently, there has been a revival of interest in the study of early Archaen (> 2500 MY) crustal segments in the various continents. Thus 3700 MY old rocks have been found in Greenland (Amîtsoq gneisses, Isua conglomerates, etc.) and 3500 MY rocks in South Africa (komatiites, granite–greenstone belts of Rhodesia, etc.) (O'Nions & Pankhurst, 1978). The early origin of these rocks is also supported by the $^{87}Sr/^{86}Sr$ values, which are quite low (0.700–0.702), but greater than BABI value: the increase is due to Rb-enriched rocks of the early earth. However, even most of the younger Precambrian granites and gneisses tend to have low initial $^{87}Sr/^{86}Sr$ values, suggesting their 'juvenile' origin from the upper mantle. If the magma comprises recycled crustal material, the latter should have a very low Rb/Sr ratio.

The theory of continental drift received positive confirmation from mass spectrometric dating measurements. The theory implies that geological features, for example, truncated orogenies, should be shared by drifting continents that were joined at some time in the past. Thus, in Ghana, one has a belt of 2000 MY old rocks separated from younger (600 MY) rocks by a sharp line of discontinuity which could be traced to a particular location in Brazil (if South America had been joined to Africa) by a fit of continental boundaries. Age measurements in Brazil showed the same pattern as in Ghana – 2000 MY rocks separated from 600 MY rocks – confirming the continental drift hypothesis (Hurley & Rand, 1969).

The K–Ar method, as applied to young volcanic rocks, provided the vital evidence for geomagnetic polarity reversals. The remanent magnetization was measured in a sequence of volcanic rocks which were also dated by the K–Ar method, and the results clearly prove epochs of normal polarity, interspersed by events of reversed polarity, of the earth's magnetic field. The epochs in the sequence are designated Brunhes (normal: present to 0.7 MY), Matuyama (reversed: 0.7–2.4 MY), Gauss (normal: 2.4–3.3 MY) and Gilbert (reversed: 3.3–4.5 MY) (Cox, 1963). These phenomena are also related to the anomalous magnetic patterns (consisting of normal and reversed magnetization) discovered in the vicinity of mid-oceanic ridges, and are caused by sea-floor spreading, explained by plate tectonics as resulting from the motion of divergent oceanic plates.

The initial $^{87}Sr/^{86}Sr$ ratios of young volcanic rocks are consistent with their mantle origin, but the distinctly smaller ratios for oceanic dolerites, as compared to ridge-basalt, indicate an inhomogeneity of the upper mantle. This is rather surprising in view of the convective motions that are supposed to occur in the mantle (Armstrong, 1971).

Solar system

Mass spectrometry can also be employed for the measurement of the *formation time* of the solar system. This is based on 'extinct' radioactivities, that is, radioactive parents of short half-life which have now practically completely decayed into their daughter elements. The basis of this work was the discovery by Reynolds (1967) of excess ^{129}Xe in the rare gases extracted from meteorites. A fusion and purification system (similar to that used for argon dating – §13.3) was used with a static mass spectrometer. The excess ^{129}Xe was attributed to ^{129}I ($T_{1/2} = 17$ MY) that was incorporated into the meteorites when they were formed. The ratio $^{129}I/^{129}Xe$ permits one to calculate the time for the meteorites to cool to the blocking temperature for Xe. These intervals are small – t ~ 220 MY ($^{129}I/^{129}Xe \sim 10^{-4}$). Excess amounts of other Xe isotopes were found that were attributed to the extinct ^{244}Pu isotope. A large number of meteorite formation ages have been determined by this method, known as *xenonology* (Reynolds, 1967), and the values of t vary from 20–200 MY. Similar anomalies in oxygen isotopic ratios, $^{18}O/^{16}O$, were found by Clayton, Grossman & Mayeda (1973) in some meteorites. Since the mean 'element age', that is, the time of the 'big bang' for our galaxy, has been found by calculations based on nucleosynthesis models to be ~ 10 BY, this isotopic anomaly is an indication of a last minute 'spike', for example a supernova that introduced the anomalous isotopes and perhaps even triggered the formation of the solar system. An enrichment of ^{26}Mg has been observed in Ca-rich chondrules of the Allende meteorite which is attributed to the decay of ^{26}Al ($T_{1/2} = 0.74$ MY), again introduced shortly before the advent of condensation of the solar nebula (Wetherill, 1975).

13.6 Spontaneous fission, neutron-induced reactions and fossil reactors

Xenon isotopes are produced in the spontaneous fission of ^{238}U, leading to an excess of atmospheric levels. These xenon excesses have been ascertained and used for dating purposes (Teitsma, 1975). A more widely used technique, however, is the fission-track method (Fleischer, Buford & Walker, 1975) which depends upon the counting of microscopic fission tracks which have been made visible by suitable etching. This latter method has no mass spectrometric aspect.

The possibility also exists of abnormal isotopic distributions resulting from neutron-induced reactions. In these cases, the neutrons come primarily from spontaneous fission or (α, n) reactions on light elements. Inghram (1953) estimated the rate of production of neutrons as 5×10^{-2}/gs for pitchblende, and $\sim 10^{-7}$/gs for ordinary rock. Evidence of neutron-induced reactions was found in xenon extracted from old BiTe (Inghram & Reynolds, 1950). Here ^{129}Xe is enriched ~ 500% as a result of the $^{128}Te(n, \gamma)\,^{129}Te \rightarrow {}^{129}I + \beta^- \rightarrow {}^{129}Xe + 2\beta^-$ reactions. Neutron-induced reactions are also partly responsible for variations in nature of the $^{3}He/^{4}He$ ratio (Paneth, Reasbuck & Mayne, 1952).

Much interest was aroused by the discovery of the 'fossil' reactor in the ore bodies of Oklo district in Gabon (§9.7). This led to the discovery of a number of distinct zones in which the uranium, enriched to 50%, had much lower-than-normal $^{235}U/^{238}U$ ratios. Other isotopic anomalies were also found, for example, excesses over natural abundances of the neodymium isotopes 143, 144, 145, 146, 148 and 150. From the isotopic anomalies in Nd, Gd, Sn, Ru, Pd and U, it was possible to deduce certain of the reactor operating conditions. Thus, it appears that the richest 'hot spots' initiated the chain reaction with water and organic matter as the moderator in these 2000 MY deposits, in which the original $^{235}U/^{238}U$ ratio was higher than the present value. These regions were lens-shaped bodies of the order of a metre in size, and the duration of the reaction was ~ 0.15–1.5 MY with a neutron flux $\sim 10^{21}/cm^2$. Radiation damage in the surrounding rocks was quite extensive. The subject has been reviewed by M. Maurette (1976).

Non-radiogenic isotopes in geology

The other altered isotopic distributions found in nature are the result of small differences in the physical and chemical properties of the isotopes of an element. The circumstances under which these differences lead to isotopic fractionation have been the subject of extensive investigations, in which the mass spectrometer has established itself as the most convenient and accurate means of gauging the extent of this fractionation. The theoretical foundations of such studies in geochemistry were outlined in a classic paper by Urey (1947) in which he calculated isotopic fractionation factors for elements of geochemical interest, and suggested the determination of paleotemperatures from oxygen isotope ratios.

13.7 Effects leading to isotopic fractionation

The main phenomena leading to isotopic fractionation are physico–chemical effects, isotope exchange reactions and kinetic processes involving reaction rates. The physico–chemical effects form the basis of various methods for the separation of isotopes. These methods are listed below and briefly described. Isotope exchange reactions and kinetic effects are discussed thereafter.

Methods for isotope separation

Gaseous diffusion. The rate of diffusion of a gas through a porous membrane is inversely proportional to the square root of its molecular weight (Graham's Law). The isotope enrichment of a single stage can be greatly multiplied by the use of a multistage recycling arrangement.

Evaporation. Lighter isotopes evaporate more readily than heavier ones. Appreciable isotopic enrichment may occur if the evaporating molecules are prevented from condensing and if the evaporation proceeds at a slow rate.

Distillation. Slight differences in the vapour pressures of isotopes, causing heavier isotopes to have a higher boiling point, make possible a fractional distillation. Light and heavy water differ in boiling point by 1.4 °C.

Centrifuging. The molecules are subject to a pseudogravitational field which exerts a greater force on the heavier than on the lighter. The separation factor depends on the difference in the two masses and not, as in the diffusion method, on the square root of their ratio.

Thermal diffusion. In the presence of a temperature gradient, the heavier molecules of a gas tend to concentrate in either the hot or the cold region, depending on the nature of the intermolecular forces (Enskog, 1911; Chapman, 1917). The principle was first successfully employed for isotope separation by Clusius & Dickel (1938).

Electrolysis. The method stems from the observation that in the electrolysis of water the residual solution is enriched in deuterium. It has been much used for commercial production of heavy water.

As we shall see, certain of the above effects are of geologic significance.

Isotopic exchange reactions

Structurally similar molecules containing different isotopes of an element display slight differences in their vibrational frequencies, zero-point energies, etc., with the result that equilibrium processes in which they are involved, display isotopic fractionation (Urey & Grieff, 1935). In general, one is concerned with the overall ratio of the isotopes of an element in a particular compound as compared with a similar ratio in a second chemical compound. For an equilibrium process involving a redistribution of isotopes among molecules, namely,

$$aA_1 + bB_2 \leftrightarrows aA_2 + bB_1, \tag{13.15}$$

where A and B are molecules with different isotopic constitution, the equilibrium constant

$$K = \frac{(QA_2/QA_1)}{(QB_1/QB_2)}, \tag{13.16}$$

where Q's are the partition functions

$$Q = \sum g_i e^{-E_i/kT}, \tag{13.17}$$

the summation extending over the levels E_i of the respective molecules. The vibrational energies introduce the mass dependence, whilst the temperature dependence, clear from the above expression, is important for geologic purposes.

Kinetic effects

Because of the mass dependence of translational velocities of molecules, irreversible reactions show a preferential enrichment of the lighter isotopes in the products of the reaction. The basis for kinetic isotope effect calculations is the transition state theory.

13.8 Variations in the isotopic constitution of hydrogen

Normal hydrogen comprises ^1H (99.985%) and ^2D (0.015%). The large percentage mass difference between these two isotopes results in isotopic fractionations in nature which are appreciably greater than for any other element. For this reason, and because of the widespread distribution of hydrogen (as water) on the earth, there has been especial interest in studies of the variation in nature of the $^1H/^2D$ abundance ratio, which normally has a value ~ 6700.

Mass spectrometric determinations of the H/D ratio began with Bleakney & Gould (1933), were greatly refined in the Manhattan project (see Kirshenbaum, 1951) and have reached a high level of accuracy (0.1‰) in modern instruments utilizing double collectors and digital measurement techniques. The variations from a standard sample (SMOW – standard mean ocean water) are expressed as parts per mil as follows,

$$\delta D(‰) = \frac{(D/H)_{sample} - (D/H)_{standard}}{D_{standard}} \times 1000. \tag{13.18}$$

Thus, a positive sign for δD indicates that the sample is enriched in deuterium.

Isotopic fractionation of hydrogen occurs mainly because the vapour pressure varies inversely as the square root of the mass. Thus, $H_2^{16}O$ has a significantly higher vapour pressures than $D_2^{18}O$. During evaporation the vapour is enriched in H and ^{16}O, and the remaining liquid depleted in these isotopes. The fractionation factors are 1.085 and 1.010 for $HD^{16}O$ and $H_2^{16}O$ and $H_2^{16}O$ and $H_2^{18}O$, respectively. That is, in the former, water vapour is depleted in D by 85‰ whilst, in the latter, it is depleted in ^{18}O by 10‰. These values decrease with temperature, causing the following effects:

– a decrease in D/H (and $^{18}O/^{16}O$) from lower to higher altitudes due to decrease in temperature;

- a decrease in δD and $\delta^{18}O$ from oceanic coast to continental inland as rain or snow moves across the continent;
- seasonal variations in response to temperature changes.

During the condensation of a cloud, the continuous depletion of the vapour in heavy isotopes is described by the Rayleigh distillation equation:

$$R/R_0 = f^{\alpha - 1} \tag{13.19}$$

where R and R_0 are the final and initial ratios (D/H or $^{18}O/^{16}O$), f = fraction of remaining vapour and α = isotope fractionation factor for vapour–liquid. The concentrations of isotopes in precipitation and ground water are related linearly:

$$\delta D = 8(\delta\,^{18}O) + 10\% \text{ (meteoric water-line)}, \tag{13.20}$$

so that, at equilibrium, the isotopic fractionation is about eight times larger for D than for ^{18}O. The relationship was studied by Friedman, Redfield, Schoen & Harris (1964) who compared their δD results with $\delta\,^{18}O$ values obtained by others. This line of approach was put to good use by Craig & Boato (1955) who, working in Urey's laboratory, analysed several hundred fresh-water and thermal-water samples for both deuterium and ^{18}O content. Later work by Craig (1961) confirmed equation (13.20), the range of $\delta^{18}O$ values being 0 to − 60‰ and those of δD being + 10 to − 400‰. A constant D/^{18}O enrichment is taken as evidence that the water evaporated or condensed under equilibrium conditions. Deviations from the equilibrium line are observed for lakes and other water bodies where evaporation is intense and rapid. Dansgaard (1964) also demonstrated a linear relationship between the annual average air temperature and $\delta\,^{18}O$ (hence δD) values of the annual average precipitation.

The temperature dependence of the isotopic fractionation has been employed for obtaining paleoclimatic information on snow and ice deposited in the polar regions. Mass spectrometric δD studies on continuous ice cores recovered from continental ice sheets in Greenland and Antarctica have provided information on climatic variations over a time interval of ~ 100 000 yr, and data on the last ice age (70 000–12 000 yr ago) (Johnson, Dansgaard, Clansen & Langway, 1972).

Studies of the isotopic fractionation of D (and ^{18}O) form a valuable tool for hydrological studies. Ground water from the present-day arid zones (*e.g.* Middle East and North Africa) have revealed lower δD and $\delta^{18}O$ in the 'paleowaters' than in the recent ground waters, thus revealing colder and wetter conditions in the past. Isotopic composition also permits the study of the mixing of waters, for example infiltration of river water into ground water aquifers. δD and $\delta\,^{18}$O values are also correlated with salinity of the oceans (Craig & Gordon, 1965) and such correlations permit the study of the origin and circulation of the bottom water of the oceans of the world. The relationship between rainfall and run-off has also been investigated by $\delta\,^{18}$O studies (Fritz, Cherry, Weyer & Sklash, 1976), with the interesting result that after a heavy rainfall, the dominant contribution to the run-off is from older ground water. This result has been confirmed by tritium (radioactive) isotope studies. Also δD values can be used for this purpose.

Another application involves the origin of geothermal water and brines of various kinds on the basis of δD–$\delta\,^{18}$O relationships. Friedman, Redfield, Schoen & Harris (1964) found that the fumaroles of Yellowstone Park are greatly depleted (20–45‰) in deuterium, whilst δD and $\delta\,^{18}$O values for a number of steam and hot waters were measured by Craig (1963) and revealed the equilibrium of oxygen in the water with oxygen in the carbonate and silicate rocks. It was also shown that the contribution of 'juvenile' (that is, derived from the mantle rocks) water was small. A number of brines from various oil fields were examined by Clayton *et al.* (1966) and the δD *vs* δ^{18}O data indicated a larger contribution of connate water (from sedimentary rocks) than of sea water. The data also throw light on the origin of the brine pools in the geologically important Rift Valley of the Red Sea.

Finally, D/H ratios have been determined for various meteorites, including carbonaceous chrondrites which probably represent low-temperature accumulations of cosmic dust. Observed ratios are not significantly different from terrestrial values. Incidentally, the D/H value on the sun is very low ($\sim 10^{-6}$) because D is consumed preferentially in the nuclear reactions which fire the sun.

13.9 Variations in the isotopic constitution of carbon

The isotopic constitution of normal carbon is taken as ^{12}C(98.89%) and ^{13}C(1.11%) corresponding to a ^{12}C/^{13}C abundance ratio of 89.3. That the value of this ratio varies significantly in nature was first shown by Nier & Gulbranson (1939) and Murphey & Nier (1941) who found variations of $\sim$ 5%, and made the general observation that ^{13}C is concentrated in limestones whereas plant forms have a preference for ^{12}C. Subsequently Craig (1953) showed that terrestrial organic carbon and carbonate rocks constitute two well defined groups, the latter being richer in ^{13}C. Since then, much work has been done in this field: the earlier work has been summarized by Rankama (1954) and the later work by Faure (1977), Stahl (1979) and Hoefs (1980). Such studies are of great importance in petroleum geochemistry and in organic bio-geochemistry. The observations are made – usually with a double collector spectrometer – on CO_2 obtained by oxidation for graphites, or thermal decomposition or reaction with H_3PO_4 for carbonates, and the $\delta^{13}C$ values expressed with reference to a standard (see fig. 13.5). The usual standards are so-called 'PDB' (*Belemnitella Americana* from the Cretaceous Peedee formation, South Carolina: Hoefs, 1980) and Solenhoefen limestone (NBS-20) (see Commission of Atomic Weights, 1984, for sources of reference materials).

The main process responsible for fractionation of C isotopes in the biosphere is photosynthesis in which plant tissue is appreciably enriched in ^{12}C relative to the CO_2 that is absorbed. Mass spectrometric studies show that kinetic fractionation is involved in the various steps of the process, namely, diffusion of CO across the cell walls and dissolution in the cytoplasm, conversion into phosphoglyceric acid and subsequent synthesis of organic compounds, as discussed by Park & Epstein (1960) and Whelan, Sackett & Benedict (1973). Smith & Epstein (1971) distinguished two categories – terrestrial plants with $\delta^{13}C$ values of -24 to $-34‰$, and aquatic, desert and marsh plants with $\delta^{13}C$ values of -6 to $-20‰$ – thus revealing the environmental effects of photosynthetic fractionation. Organic matter in recent sediment also has $\delta^{13}C$ values of -20 to $-30‰$

Fig. 13.5. Variations in ^{13}C abundance for natural gases, crude oils and their source materials. (After Stahl, 1979.)

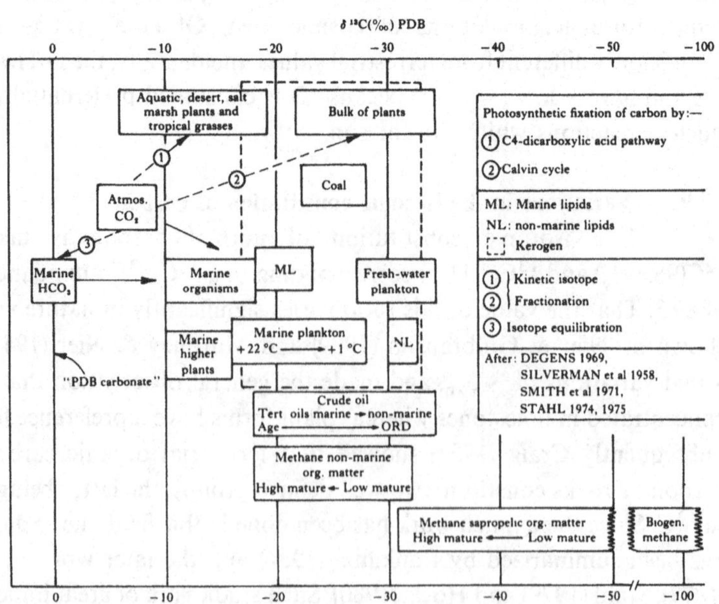

(Eckelmann, 1962; Schultz & Calder, 1976), indicating that it is derived from terrestrial plant debris. This work also showed that an age effect was involved, due to preferential destruction of proteins and carbohydrates, and enrichment in lipids, cellulose and lignin.

Carbon in coal has a $\delta\,^{13}C$ value of $\sim -25‰$ with little dependence on rank or age (Colombo, *et al.*, 1968). The enrichment in ^{12}C is consistent with the derivation of coal from biogenic material, for example from plants that had $\delta\,^{13}C$ values similar to that of their modern counterparts. Carbon isotopes also provide valuable information on the origin, maturation and correlation of sources of petroleum (Silverman, 1964, 1967). Terrestrial and marine plants and organisms deposited in sedimentary basins are transformed into humic acid complexes by bactericidal reaction, and these complexes then converted to kerogen, which is the source material of petroleum. Kerogen has $\delta\,^{13}C$ values of -17 to $-33‰$ and crude oils are depleted slightly further (-18 to $-34‰$). Thermocatalytic (natural) gases have $\delta\,^{13}C$ values of $-50‰$, whilst even lower values (to $-100‰$) appear to be characteristic of bactericidal gas formation. Hydrocarbon accumulations generally involve a 'maturation', that is, thermal transformation of kerogen buried in the host rocks. This process is accompanied by physico–chemical changes in the organic matter, for example reflectivity of a compound 'vitrianite' in the lignin–cellulose complex of plants. A plot of the $\delta\,^{13}C$ values of the methane-released versus the vitrianite-reflectance provides revealing insight into the organic source material and level of maturation. Thus the 'oil window' which corresponds to the depth where time–temperature parameters favour crude oils formation, also corresponds to $\delta\,^{13}C$ values of $-30‰$, in the studies of Stahl (1977). In addition, correlation studies between $\delta\,^{13}C$ values and kerogen, extract and oil, or distillation fractions, etc. have considerable diagnostic value. A study of carbon isotope abundances is thus a valuable tool for hydrocarbon exploration.

There is considerable evidence for the existence of life in sedimentary rocks of Precambrian age, for example the observation of morphologically preserved microscopic plants and the extraction of mixtures of organic compounds which normally form during the decomposition of organic matter. $^{12}C/^{13}C$ studies have contributed additional valuable evidence through large negative $\delta\,^{13}C$ values observed in sedimentary rocks as old as 3000 MY, such as the Onwerwacht series in South Africa. Early observations (Rankama, 1954) have been confirmed by isotope measurements of reduced carbon in Precambrian sedimentary rocks (Hoefs & Schidlowsky, 1967; McKirdy & Powell, 1974). The $\delta\,^{13}C$ values range from -15 to

$-40‰$, similar to the values in modern sediments. The scatter in values may be due to diagenesis of the rocks and formation of CH_4 by thermal decomposition of kerogen, thus leaving the residue enriched in ^{13}C. Carbon is also isotopically fractionated in equilibrium reactions involving CO_2 (gas), H_2O, and $CaCO_3$ (solid)

$$CO_2 + H_2O + CaCO_3 \rightleftharpoons Ca^{2+} + 2HCO^- . \tag{13.21}$$

For this reaction, $\delta\,^{13}C = 10‰$ at 25 °C, and also the temperature dependence of $\delta\,^{13}C$ is known. Marine carbonates yield $\delta\,^{13}C \sim 0‰$ and lacustrine carbonates negative $\delta\,^{13}C$ values due to addition of CO_2 derived from plant debris. Carbonates in carbonaceous chondrites, however, give large positive $\delta^{13}C$ values of $\sim +60‰$, which are attributed to inorganic synthesis of hydrocarbons by reactions between CO, H_2 and NH_3. Graphites also occur in igneous rocks, and some 'heavy' graphites with positive $\delta\,^{13}C$ can be produced inorganically by reduction of $CaCO_3$. The isotopic composition of carbon in hydrothermal deposits is variable as the $\delta\,^{13}C$ depends not only on the corresponding value for the mineralizing fluid but also on the pH, ionic strength and oxygen fugacity. These relationships were studied in detail for both ^{34}S and ^{13}C by Ohmoto (1972) and Rye & Ohmoto (1974) and are dealt with in §13.11 on sulphur isotope abundances.

13.10 Variations in the isotopic constitution of oxygen

Oxygen is probably the most important of the elements. It supports life, forms compounds with all other elements save the inert gases and is the most abundant constituent of the earth's crust. It exists in three stable isotopic forms, ^{16}O (99.758%), ^{17}O (0.038%) and ^{18}O (0.204%). Following Dole's discovery (1935, 1936) that atmospheric oxygen is richer in ^{18}O than fresh surface water, there have been many investigations of this and related effects. One of the outstanding results of these studies is the establishment of an oxygen isotope scale of paleotemperatures.

Oxygen isotope ratio measurements are carried out with a double collector mass spectrometer, using CO_2 as an inlet gas. Carbonates are treated with H_3PO_4, and silicates and oxides by fluorinating reagents (for example, BrF_5) that liberate O_2, which is converted into CO_2.

Urey suggested (1947) that paleotemperatures of ancient oceans could be estimated by measurement of oxygen isotope ratios. This involves the carbonate – water exchange represented by the equation

$$CaC\,^{16}O_3 + H_2\,^{18}O \rightleftharpoons CaC\,^{16}O_2\,^{18}O + H_2\,^{16}O \tag{13.22}$$

or

$$H_2\,^{18}O(liq) + C\,^{16}O_2(sol) \rightleftharpoons H_2\,^{16}O(liq) + C\,^{16}O\,^{18}O(sol), \quad (13.23)$$

in which oxygen exchanges between liquid water and the carbonate ion in solution. The equilibrium constant for this reaction is

$$K = \left(\frac{CaC\,^{16}O_2\,^{18}O}{CaC\,^{16}O_3}\right)\cdot\left(\frac{H_2\,^{16}O}{H_2\,^{18}O}\right) = \left(\frac{R_c}{R_w}\right), \quad (13.24)$$

where R_c and R_w denote the ($^{18}O/^{16}O$) ratios for carbonate and water. There is an appreciable temperature coefficient associated with this equilibrium constant, so that the determination of the $^{18}O/^{16}O$ ratios of suitable carbonates provides a means for calculating the temperatures of the sea waters in which they were formed. The relation between $\delta\,^{18}O$ values and temperature has been given

$$T = 16.9 - 4.2(\delta_c - \delta_w) + 0.13(\delta_c - \delta_w)^2, \quad (13.25)$$

where δ_c and δ_w are the $\delta\,^{18}O$ values for the calcite sample and water, respectively (Craig, 1965). Detailed studies of temperature variation in the upper 50 m of the ocean during the Pleistocene period have been made by Epstein & Mayeda (1953) and Emiliani and coworkers (1966) by $\delta\,^{18}O$ measurements on foraminifera. Fluctuations have been observed in temperature with a period $\sim 10^5$ yr and these are related to continental glaciation. Significant results are observed in the Mesozoic and Cenozoic seas, by the use of excellent materials made available through deep-sea sampling programmes and improved methods of biostratigraphy. The subject of one early paleotemperature determination was a Jurassic belemnite, an extinct animal that flourished 10^8 yr ago. Samples of successive layers of the creature's guard showed regular temperature variations. These are taken to be seasonal in character, indicating that the creature died at the age of four. The variation of $\delta\,^{18}O$ values with salinity has already been referred to.

Since oxygen is also an important constituent of rocks and minerals, oxygen isotope data are relevant to the study of their petrogenesis. There is good evidence that oxygen isotopes achieve equilibrium in rocks and minerals and that the difference in $\delta\,^{18}O$ values for a pair of minerals that have equilibrated oxygen with a common reservoir at temperature T is a function of T. Such relationships have been measured in hydrothermal experiments in which minerals are equilibrated with water of known isotopic composition, for example

$$\delta_c - \delta_w = 3.38 \times 10^6/T^2 - 3.40\,(\text{water-calcite}), \quad (13.26)$$

$$\delta_Q - \delta_w = 2.78 \times 10^6/T^2 - 3.40\,(\text{water-quartz}). \quad (13.27)$$

Thus, for the quartz–calcite pair, the difference in $\delta^{18}O$ values is $\delta_Q - \delta_c = 0.6 \times 10^6/T^2$, and similar relationships are known for other pairs of minerals. In this way, $\delta^{18}O$ values for mineral pairs yield temperatures of equilibrium with oxygen, and significant results have been obtained from the examination of plutonic and volcanic rocks. Thus, high equilibrium temperatures (~ 1000 °C) have been obtained for samples (rhyolite obsidians) from oceanic areas, indicating their origin by differentiation of basaltic or andesitic magmas, and not by crustal melting. Also, plutonic rocks have higher temperatures than their volcanic equivalents – as expected – due to slower cooling of the former. There is also a correlation between $\delta^{18}O$ values and chemical trends.

Oxygen isotope studies have elucidated several aspects of the interaction between igneous rocks and meteoric water (rocks stewed in their own juice). A study of batholiths revealed convective systems of meteoric–hydrothermal waters in their epizonal portions, which are characterized by low $\delta^{18}O$. In metamorphic rocks, the temperatures involved in metamorphism can be measured by $\delta^{18}O$ values of mineral pairs (for example, quartz–magnetite). These temperatures are found to increase with the grade of metamorphism, although 'isotopic reversals' occur in some complex polymetamorphic areas. Meteorites have also been studied, and the remarkable observation made (Clayton, Grossman & Mayeda, 1973) that certain high temperature phases of carbonaceous chondrites are anomalously enriched in ^{16}O. Detailed studies showed that this 'exotic' oxygen was probably introduced by reaction in a star (supernova) at the beginning of the solar system formation. This may suggest that a supernova triggered the formation of the solar system.

Figure 13.6 shows $\delta^{18}O$ values for rocks and minerals from various localities.

13.11 Variations in the isotopic constitution of sulphur

The isotopic constitution of S is ^{32}S (95.02%), ^{33}S (0.75%) and ^{34}S (4.21%); these values correspond to an abundance ratio $^{32}S/^{34}S = 22.6$. Thode and his coworkers (Thode, MacNamara & Collins, 1949; MacNamara & Thode, 1950, 1951; Thode, MacNamara & Fleming, 1953) and Trofimov (1949) first showed that the ^{33}S and ^{34}S abundances can vary in nature by as much as 4% and 8% respectively. Early work indicated that sulphates are enriched and sulphides depleted in ^{34}S, while the constitution of meteoritic sulphur is constant, corresponding to the overall average of sulphur of igneous origin. Actually, the $\delta^{34}S$ values reported so far (Nielsen, 1979) lie between −65‰ and +95‰ so that sulphur is the element in which fractionation processes operate most effectively in nature (see fig. 13.7). This is due to (a) the wide range of valence states available to sulphur (S^{2-} to S^{4+} and S^{6+}) under varying redox conditions and (b) the biologic sulphur cycle involving enzymatic complexing of the sulphate and the separation of the 'light' (sulphide) and 'heavy' (sulphate) sulphur by ground water. The isotopic composition of sulphur is thus a combination of both equilibrium and kinetic effects. Thode indicated that a significant exchange reaction involved in these processes is

$$H_2\,^{34}S + ^{32}SO^{2-}(\mathrm{sol}) \leftrightharpoons H_2\,^{32}S + ^{34}SO^{2-}(\mathrm{sol}), \tag{13.28}$$

for which the equilibrium constant is 1.074 at 25 °C. This exchange in which ^{34}S is concentrated in the sulphate can take place through the medium of the sulphur cycle in the sea, illustrated in fig. 13.8. Kinetic fractionation (due to differing strengths of the sulphur bond leading to a faster reaction with

Fig. 13.6. Natural variations in the isotopic abundance of ^{18}O. (Adapted from O'Neil, 1979.)

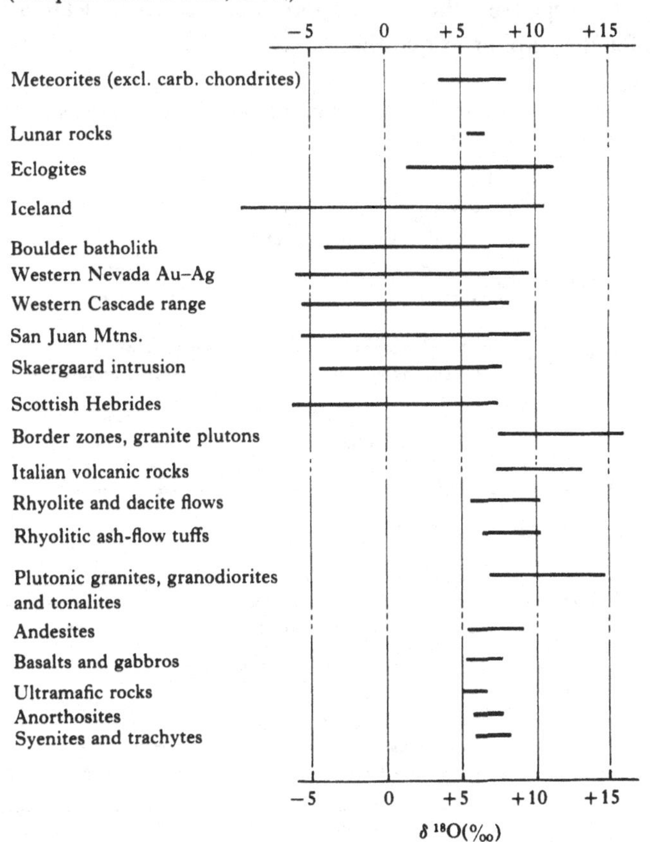

the lighter isotope) is also involved in bacterial sulphate reduction. These processes which cover both 'assimilatory reduction' (during uptake of sulphate in biosynthesis) and 'dissimilatory' or respiratory application, the latter involving anaerobic bacteria (for example, *desulfovibrio desulfuricans*), were discussed by MacNamara & Thode (1951) and later by Goldhaber & Kaplan (1974). Values of $\sim 25‰$ for the difference in $\delta\,^{34}S$ between H_2S and SO^{2-} have been observed in laboratory cultures while much greater values ($\sim 65‰$) are realized in natural samples (where equilibrium exchange reactions are also involved: Nielsen, 1979). The kinetic effects were also discussed by Rees (1973) and Schwarcz & Burnie (1973).

Sulphur isotope distributions in meteorites and magmatic rocks have been studied, and these yield $\delta^{34}S$ values of $+0.5$–$1.3‰$ for meteorites and an average of $\sim 1‰$ for terrestrial magmatic rocks. This clear-cut difference from biogenic sulphur led to the early generalization that a narrow range of $\delta^{34}S$ values around zero indicates a magmatic source, whilst large negative values with appreciable scatter indicate a biogenic source for the sulphur in question.

Fig. 13.7. Natural variations in the isotopic abundances of ^{34}S. (After Nielsen, 1979.)

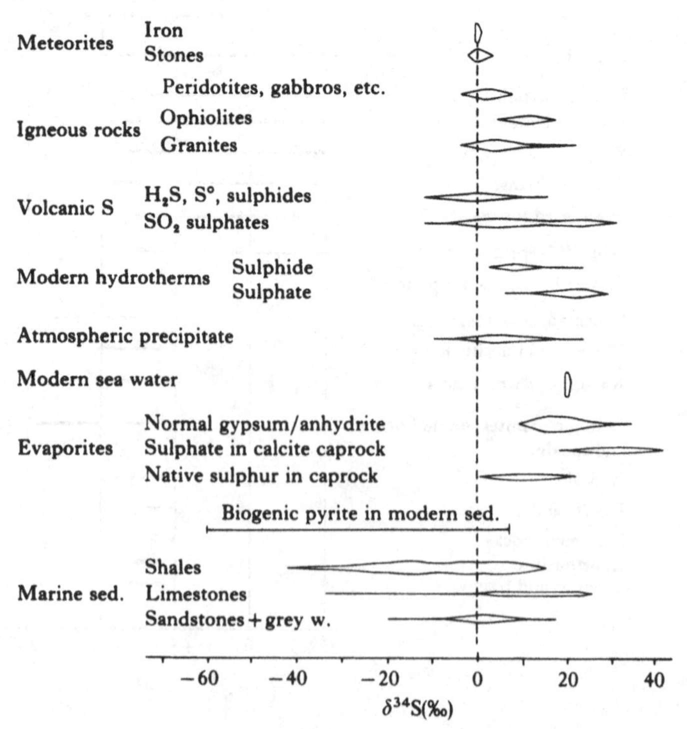

These principles were applied to the study of major sulphide ore deposits, but discrepancies were observed, especially for magmatic and hydrothermal ores. In such cases, the base metals had been transported by ore-forming solutions (the metals, for example as dissolved complexes in chloride solutions) and it was pointed out by Sakai (1968) and Ohmoto (1972) that the $\delta\,^{34}S$ values of the ore-forming fluid are considerably modified by the environmental conditions of ore deposition, that is, the pH values, oxygen fugacity (f_{O_2}) and temperature of the hydrothermal systems. Thus the isotopic composition of hydrothermal minerals is controlled by (a) the $\delta\,^{34}S$ value of the source, (b) the temperature that alters the proportions of sulphur-bearing species and (c) the proportions of oxidized and reduced sulphur species in solution. The last-named can be calculated in terms of the pH, f_{O_2} and T diagrams, which combine sulphur isotope contours and the stability fields of different sulphide minerals. Thus the trends of S-isotope fractionation (superimposed on such curves) yield information on the pH, f_{O_2} and T conditions that prevailed at the time of ore deposition.

Isotope exchange reactions among coexisting sulphide minerals and the $\delta\,^{34}S$ values of these coexisting minerals (for example, sphalerite and galena) can be used to determine their temperature of equilibrium (Kajiwara & Krouse, 1971).

The $\delta\,^{34}S$ values of petroleum vary from -8 to $+32‰$ and studies by Thode & Rees (1970) have shown the diagnostic value of such work. Also Thode & Monster (1965) found $\delta^{34}S$ enriched by $\sim 15‰$ with respect to the contemporaneous marine sulphate, and attributed this effect to bactericidal reaction. $\delta^{34}S$ in coals has also been studied. The study of marine sulphates of differing ages shows a continuous variation of $\delta^{34}S$ values in the oceans of the past.

Fig. 13.8. The sulphur cycle in the sea. (From Thode, Wanless & Wallouch, 1954.)

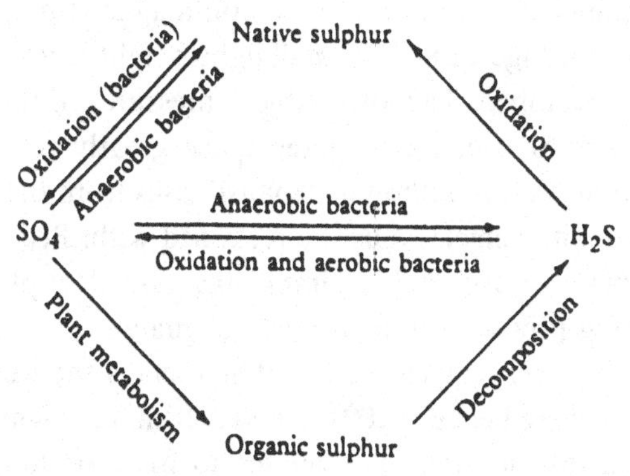

14

Mass spectrometry in upper atmosphere and space research

Since the early 1950's the mass spectrometer has enjoyed an important role in the investigation of the physico–chemical parameters of the upper atmosphere. In particular, for the ionosphere and the exosphere, it has provided the means of determining the atmospheric composition, including atomic and molecular components in both neutral and ionic states. This composition differs from that at lower levels (a) as a result of the absorption of solar radiation at UV, far UV, and X-ray wavelengths and (b) because of the much greater mean free paths. From mass spectrometric data the gas (total) densities and 'kinetic' temperatures may be computed. In addition, large-scale motion, 'thermal' and 'diffusive' equilibria (reflected in the relative intensities of the different constituents), as well as ion–molecule reactions involving the ionized components, can be studied. In more recent years, such studies have been extended to investigate the composition of atmospheres of other planets and the composition of the magnetosphere.

Apart from atmospheric studies, mass spectrometry has been applied to the investigation of the elemental and isotopic composition of meteorites and of lunar rocks (§13.5). In addition, it also played a key role in the Viking apparatus landed on Mars in the search for extra-terrestrial life.

14.1 Instrumentation

In general, small volume and low weight are prime requirements for rocket-borne instruments. On these counts, non-magnetic mass spectrometers offer some advantage, although small light-weight sector instruments have also been developed. The low ambient pressures in the upper atmosphere and the virtual vacuum in outer space greatly reduce the requirements for vacuum pumps, although exhaust gases from the rocket and degassing from satellite vehicles must be reckoned with. Because the spectrum of the upper atmosphere is simple (as also for planetary atmospheres) a resolving power ~ 50 is usually adequate.

The earliest instruments were carried on sounding rockets and were used to investigate the atmosphere between 100 km and 200 km in height. Later instruments have been (a) mounted in satellites to monitor long term variations in the atmosphere at high altitudes, including the magnetosphere, or (b) carried by balloons, to investigate further the region below 100 km.

The Bennett type rf energy-gain mass spectrometer (§6.4) was used for early studies (for example, OGO satellites of Goddard Space Flight Center: Townsend, 1952), while quadrupole and monopole spectrometers have been used extensively in rocket flights (Schaefer & Nichols 1961; Taylor, Brinton & Smith, 1962; Narcisi & Bailey, 1965). Other studies have used miniature single focusing sector instruments (see, for example, Nier, Hoffman, Johnson & Holmes, 1964; Hedin & Nier, 1966) or double focusing versions (Spencer & Reber, 1963). An important experiment on the Viking Mars probe employed a miniature Mattauch–Herzog instrument (Nier & Hayden, 1971; Hayden *et al.*, 1974). In these instruments ions are admitted directly to the analyser, whilst an electron bombardment source is provided for analysis of neutrals. A description of some representative instruments will illustrate the special features of the mass spectrometers used with rockets and satellites.

A diagram of the magnetic deflection instrument used by Hedin & Nier (1966) is shown in fig. 14.1. This was employed in rocket flight studies of the neutral composition of the atmosphere at altitudes ~100–200 km. As shown in the diagram, the atmosphere is sampled by an *open* source mounted on the side of the rocket. Ions are produced by electron bombardment, accelerated, and analysed in the sector field of a permanent magnet. Two collectors are employed, a Faraday cup for the relatively large ion currents at higher masses ($10 \leqslant M \leqslant 50$, namely, N_2^+, O_2^+, etc.) and an electron multiplier for the low ion currents at lower masses ($M \leqslant 10$, namely, He^+, N^{2+}, etc.). Mass spectra are obtained by discharging periodically the acceleration voltage across the condenser of an RC circuit whose time constant is ~2 s. The signal from the ion detectors is telemetered to the ground stations. A small sputter-ion pump (~1 l/s) evacuates the instrument prior to removal of the cap covering the ion source. At the desired height a cutter mechanism is activated and the cap is removed.

A miniature Mattauch–Herzog instrument, shown in fig. 14.2, was developed by Nier & Hayden (1971) for use in upper atmosphere rockets, whilst a modified version was employed for the Viking landing on Mars. The instrument is compact and comprises an electron bombardment source, an electrostatic analyser having $r_e = 4.27$ cm, a magnetic analyser with $r_m = 2.54$ cm (for the higher mass ions) and two collectors, for high and low mass ranges respectively. A sputter pump (~1 l/s) provides differential pumping between the source and analyser. The instrument has a high

Fig. 14.1. Single focusing magnetic mass spectrometer with open ion source. (After Hedin & Nier, 1966.)

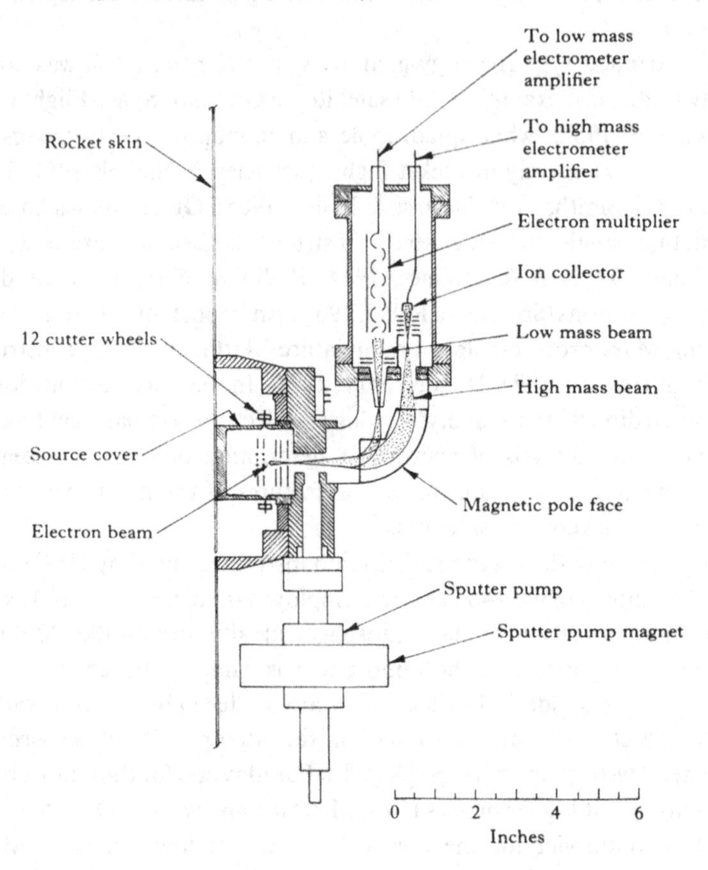

Fig. 14.2. Mattauch–Herzog instrument adapted for space-craft. (After Nier & Hayden, 1971.)

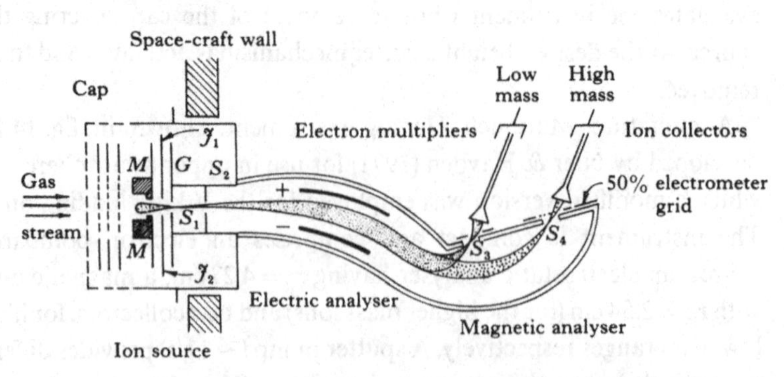

sensitivity – 10 μA/Torr for N_2 at 100 μA electron ionizing current.

The version which landed on Mars with the Viking space-craft included a gas chromatograph attachment. As described by Rushneck *et al.* (1978), the mass spectrometer is a small double focusing instrument with a 90° electrostatic analyser (radius 4.7 cm) followed by a 90° magnetic analyser (radius 3.8 cm). The magnet pole pieces form part of the vacuum housing whilst the analyser magnet also provides the field for the ion pump. Gas concentrations are measured using an atmospheric inlet coupled to an electron bombardment source.

Samples scooped from the Martian soil were heated, releasing vapours which were introduced into a gas chromotograph and eluted with H_2 at temperatures of 50–500 °C. A palladium-alloy separator removed the H_2-carrier gas, after which the effluents were analysed by the mass spectrometer. The first mass spectrum returned from Mars is shown in fig. 14.3.

A typical quadrupole instrument was described by Narcisi & Bailey (1965) (see fig. 14.4). In this case, the ions are accelerated into the entrance orifice, because the quadrupole structure is biased at − 128 V. Superimposed on this bias voltage are the normal analysing rf and dc voltages, which are scanned exponentially with the rf/dc ratio maintained constant. A zeolite adsorption pump is used as a vacuum pump whilst an electrometer is used to measure ion currents. Results obtained with a quadrupole filter mounted in a US Air Force research satellite have been summarized by Philbrick (1976). Similar work involving a monopole instrument on the ESRO satellite has been described by Fricke, Lane, Trinks & von Zahn (1975).

In relating the actual (ambient) densities of the atmospheric components to the measured ion currents certain experimental problems are encountered that are peculiar to these instruments. The first is that the satellite, or rocket velocities, exceed the average molecular velocities. As a result, the particle density in the vicinity of the ion source is a function of the velocity vector of the vehicle, being greatest when the vehicle 'looks' forward. In the

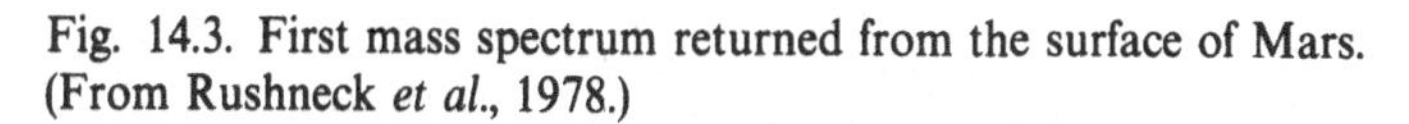

Fig. 14.3. First mass spectrum returned from the surface of Mars. (From Rushneck *et al.*, 1978.)

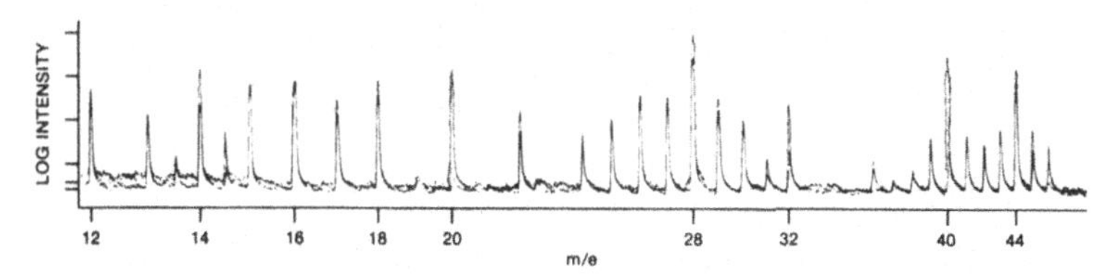

simplest case, if the source is represented by a flat plate, the actual particle density near the plate, n_s, is the sum of the density of outside particles moving toward the plate n_0, and the reflected density, n_r; that is, $n_s = n_0 + n_r$. Now n_0, as mentioned above, depends on the ambient density n_a, and the velocity vector $[V \cos\phi/\langle c\rangle]$, where V is the rocket speed, ϕ is the aspect angle between the normal to the plate and the velocity vector, and $\langle c\rangle$ is the most probable thermal speed of the molecules. The value of n_r depends on the 'accommodation coefficient', which describes the efficiency with which the particles are reflected by the plate material. Here n_r depends on the temperatures T_a and T_s, of the ambient region and the plate. This reflection effect can be corrected for, and the particle densities computed as a function of altitude, from the mass spectrometric data. 'Closed' ion sources, which have an orifice opening outside, bypass the effect of the accommodation coefficients, and enable the ambient temperature to be calculated from the ion current modulation resulting from the revolution of the spinning rocket (Hedin & Nier, 1966).

Fig. 14.4. Mounting of a quadrupole instrument in a rocket configuration. (After Narcisi & Bailey, 1965.)

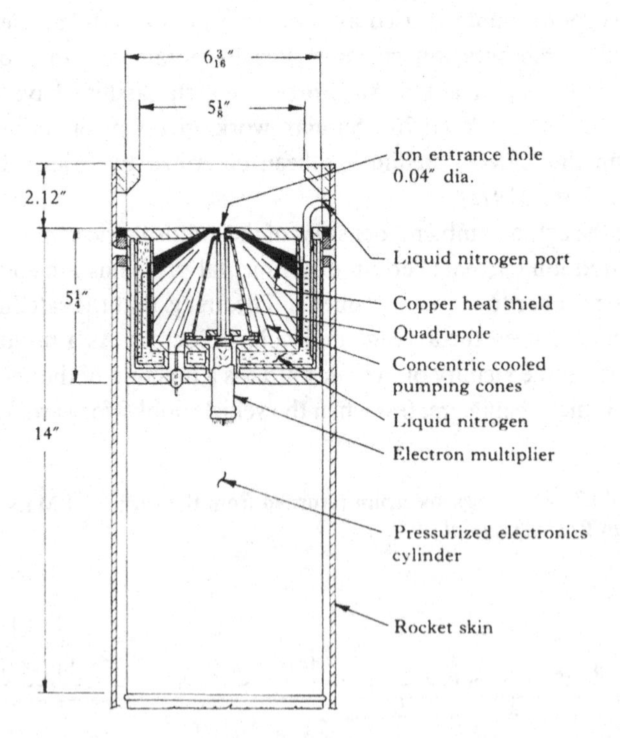

A second problem arises in determining the density of atomic oxygen, O, which is a major component of the atmosphere at higher altitudes ($>100\,\text{km}$). This highly reactive atom combines with the ion source surfaces, so that one must use wide-open sources. The ratios of the peaks at $M = 32$ and $M = 16$, under various source conditions, plus other criteria, have been used by Nier and coworkers (Lake & Nier, 1973) to correct for this effect.

Because of degassing of the satellite, or the contribution from rocket exhaust gases to the ambient gas that is being sampled by the ion source, the proper placement of the spectrometer inside the rocket or satellite is of great importance. Also, in view of the large number of other experiments in the payload, it is desirable that the bandwidth (number of channels) used for the transmission to earth of the mass spectral data be kept to a minimum. Programmed gain-switching of the output electrometers or logarithmic-gain electrometers are often used. In others cases, peak switching, instead of scanning of the complete spectrum (von Zahn, 1968), is used and only significant masses are measured.

14.2 Results

One of the most important results of mass spectrometric studies of the upper atmosphere is the determination of the *ionic* composition of the various layers of the ionosphere. These data have also elucidated the principal ionization mechanisms, arising from photoionization by solar reactions. The various regions of the ionosphere are shown in fig. 14.5, whilst the special features relating to the D (50–85 km), E (85–150 km) and F (above 150 km) regions are described briefly below.

D region. While the ions N_2^+, O_2^+ and NO^+ and OH^+ are found, a large fraction ($\sim 50\%$) of the ion current represents polymolecular clusters $(H_3O)^+(H_2O)_n$, at masses 37, 55, etc. A dominant contribution to ionization appears to result from Lyman H_α and other radiations of wavelengths $< 1800\,\text{Å}$ from the sun.

E and lower F regions. One of the main features is the preponderance of the NO^+ ion, along with N_2^+ and O_2^+. The molecules O_2 and N_2 are ionized mainly by the soft X-ray and far-UV radiations from the sun, whilst the NO^+ ions are produced according to the following reaction:

$$O_2^+ + N_2 \rightarrow NO^+ + NO. \tag{14.1}$$

NO^+ ions are also produced in the ion–atom interchange:

$$O^+ + N_2 \rightarrow NO^+ + N. \tag{14.2}$$

Metallic ions, Na^+, Mg^+ and Ca^+, are also found, which are probably caused by meteoric matter, and have been regarded as a possible cause of enhanced ionization (sporadic *E* layer).

Upper F region. In the *F* region, and higher *E* regions, while O_2^+, N_2^+ and NO^+ are present, the atomic constituents O^+ and N^+ increase with altitude. At still higher altitudes (~1000 km), the principal ions are H^+ and He^+.

The current status of atomic composition is summarized in recent COSPAR (Committee on Space Research) reports. The main features of the neutral atomic composition are (a) increase of atomic over molecular oxygen above 120 km and (b) preponderance of H and He above 1000 km. The explanation of the former in terms of ion–molecule reactions as well as the role of the latter in determining ionic compositions, have been discussed by Donahue (1968).

The production of atomic nitrogen and oxygen in ion–molecule reactions has been discussed by several authors (for example, Strobel, Hunten & McElroy, 1970; Strobel, 1971). Thus, atomic N is produced by

Fig. 14.5. Regions of the ionosphere. (After Kato, 1980.)

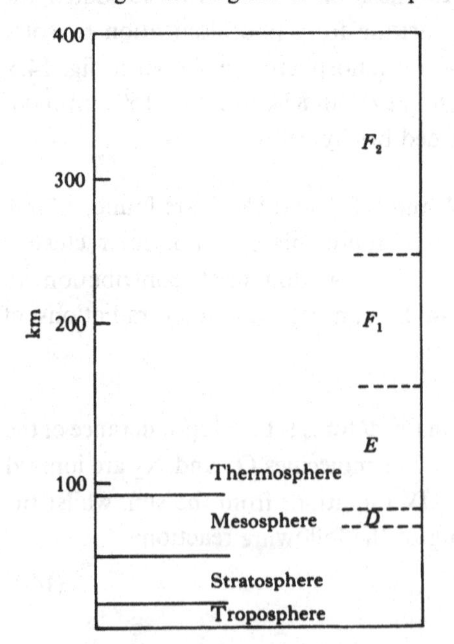

dissociative recombination of N_2^+ and NO^+ ions as well as by molecular photodissociation of N_2. An example of a recent study of N_2^+, NO^+, N^+ and other species with a rocket-borne mass spectrometer is that of Bibbo *et al.* (1979). Atomic O, which occurs in appreciable amounts above 150 km, is produced mainly by photodissociation. The effects of magnetic storms on the O/O_2 ratios in the thermosphere have been studied by Potter, Kayser & Nier (1978) and related to eddy-diffusion processes near the thermopause region. Similar variations in the topside ionosphere have been measured by Ivanov, Pakov & Pylof (1978).

Excited states of molecules are also indicated by mass spectrometric data on the basis of known ion–molecule reactions, for example, $^1S(O)$ and $^2D(N)$ atoms, and $^1\Delta_g(O_2)$, $^1\Sigma_g(O_2)$ and $^3\Sigma_u(O_2)$ molecules.

Mass spectrometric studies have also elaborated the diffusive equilibrium of atmospheric constituents. Generally the neutral particle number densities follow the Boltzmann law as a function of height, namely,

$$n_i = n_0\left(\frac{T(z_0)}{T(z)}\right)\exp\left[-\int_{z_0}^{z}\frac{dz}{H}\right], \tag{14.3}$$

with the scale height

$$H = \frac{RT(z)}{M(z)g(z)}, \tag{14.4}$$

where $T(z)$ is the temperature at height z, R is the gas constant, M is the molecular weight and g is the acceleration of gravity. At lower altitudes, the atmospheric components are thoroughly mixed due to convection and turbulence and the number density decreases according to the average molecular weight for air, $M = 29.0$ g/mol. At higher altitudes, the density of each component follows equation (14.3), with individual 'scale heights', H_i. This 'diffusive' equilibrium occurs above a certain height called the 'turbopause', and results in an increase in abundance of the lighter components (for example, H_2, He), and a decrease in the relative abundance of the heavier ones (for example, Ar). Moreover, the long lifetimes for ionized, dissociated and metastable states of the various constituents result in changes in composition of the upper atmosphere, which is a highly ionized plasma in the ionospheric regions (as discussed above).

Mass spectrometric measurements of $\{n(Ar)/n(N_2)\}$ or the parameter $\{n^2(N_2)/n(O)n(Ar)\}$, where the n refers to the number densities, have yielded a value ~ 120 km as the altitude for the onset of diffusive equilibrium (von

Zahn, 1968). Diurnal, latitudinal and seasonal changes are observed in the density and composition values and large scale motions (for example, equatorial electrojet in the tropics) have been inferred.

The density values can be obtained by calibration of the mass spectrometers for partial pressures of the various gases. Thus the atmospheric density, average molecular weight, etc., can be calculated and compared with 'model' values, and with satellite 'drag' measurements. For example, the data of Fricke *et al.* (1975) using the ESRO satellite, agree with the CIRA (COSPAR International Reference Atmosphere) 1972 model atmosphere within 20%.

Kinetic temperatures at various altitudes have been measured using the pressure (or density) modulations in the ion source of the mass spectrometer carried in a rotating rocket or satellite (see §14.1; Hedin & Nier, 1966).

Mass spectrometric measurements in the magnetosphere have also been made by Geiss *et al.* (1978) with a satellite-borne instrument. Ions H^+, He^+, O^+ of solar-wind origin, and the doubly charged species He^{2+} and O^{2+} are observed. The latter, especially, show an enhancement during magnetic storms.

The mass spectrometric apparatus and techniques which were developed for the study of the upper atmosphere of the earth have been applied more recently to the analysis of atmospheres of other planets, notably on the Viking probe to Mars (Nier *et al.*, 1976) and the Pioneer probe to Venus (Taylor *et al.*, 1979*a*, 1979*b*; von Zahn *et al.*, 1979; Niemann *et al.*, 1979; Hoffman *et al.*, 1979; Moroz, 1981). The partial pressures of the gases N_2, H_2O, and CO_2 are very low, so that Mars is very much depleted in the volatiles while the soil (on pyrolysis) yields larger amounts of chlorine and sulphur. In certain cases isotope ratios have been measured and found to differ from terrestrial values. For example, $[^{40}Ar/^{36}Ar]$ is ~ 10 times higher than for terrestrial samples. These results have an obvious bearing on our understanding of the origin and accretion of the planet. The Venusian atmosphere is composed of CO_2, N_2 and Ar while the water-vapour content is extremely low ($< 0.1\%$). The lunar atmosphere has an exceedingly low vapour pressure; in fact, even the higher-boiling-point 'volatiles' such as As, Sb, or Zn have been considerably depleted from the lunar surface rocks.

Analyses of the Martian soil for possible biogenic components (for example porphysins) were carried out *in situ* by the Viking probe, but yielded negative results (Rushneck *et al.*, 1978; Mazur *et al.*, 1978).

Finally, as this monograph goes to press, a coordinated research program is aimed at the study of Halley's Comet, which will make its next return in February, 1986. This program is motivated by the belief that comets are representative of the primordial material from which the solar system has evolved. Five spacecraft, carrying a variety of mass spectrometers, energy analysers, etc., are moving towards Halley's Comet to determine the gas composition close to the comet nucleus, the 'parent molecules', the interactions of the gas released with the solar wind, and the chemical composition of the dust particles released' (Kissel, 1985).

The methods and results relating to the extensive work on the geochronology and isotope geology of lunar samples are discussed in Chapter 13. These studies, while involving exotic sample material, were carried out in the laboratory and thus resemble more conventional work.

References

Chapter 1

Aston, F. W. (1919). *Philosophical Magazine*, **38**, 709.
Aston, F. W. (1920). *Philosophical Magazine*, **39**, 449.
Aston, F. W. (1923). *Philosophical Magazine*, **45**, 934.
Aston, F. W. (1927). *Proceedings of the Royal Society A*, **115**, 487.
Aston, F. W. (1942). *Mass Spectra and Isotopes*. London: Edward Arnold and Company.
Bainbridge, K. T. (1932*a*). *Physical Review*, **39**, 847.
Bainbridge, K. T. (1932*b*). *Physical Review*, **39**, 1021.
Bainbridge, K. T. (1933*a*). *Physical Review*, **43**, 103.
Bainbridge, K. T. (1933*b*). *Physical Review*, **43**, 1056.
Bainbridge, K. T. (1933*c*). *Journal of the Franklin Institute*, **215**, 509.
Bainbridge, K. T. (1933*d*). *Physical Review*, **44**, 123.
Classen, J. (1907). *Jahrbuch der Hamburgischen Wissenschaftlichen Anstalten.*
Costa, J. L. (1925). *Annales de Physique, Paris*, **4**, 425.
Dempster, A. J. (1918). *Physical Review*, **11**, 316.
Dempster, A. J. (1920). *Science*, **52**, 559.
Dempster, A. J. (1921). *Physical Review*, **18**, 415.
Dempster, A. J. (1922). *Physical Review*, **20**, 631.
Goldstein, E. (1886). *Sitzungsberichte der Königlich Preussischen Academie der Wissenschaften zu Berlin*, **39**, 691.
Herzog, R. (1934). *Zeitschrift für Physik*, **89**, 447.
Herzog, R. & Mattauch, J. H. E. (1934). *Annalen der Physik. Leipzig*, **19**, 345.
Kaufman, L. (1901). *Physikalische Zeitschrift*, **2**, 602.
Swann, W. F. G. (1930). *Journal of the Franklin Institute*, **210**, 751.
Thomson, J. J. (1897). *Philosophical Magazine V*, **44**, 293.
Thomson, J. J. (1912). *Philosophical Magazine VI*, **24**, 209, 668.
Thomson, J. J. (1913). *Rays of Positive Electricity*. London: Longmans Green and Co.
Wein W. (1898). *Verhandlungen der Physikalische Gesellschaft zu Berlin*, **17**, 1898.
Wien, W. (1902). *Annalen der Physik und Chemie, Leipzig*, **8**, 224.

Chapter 2

Alekseevskii, N. E., Prudkovskii, G. P., Kosourov, G. I. & Filimonov, S. I. (1955). *Doklady Akademi Nauk SSSR*, **100**, 229.
Aston, F. W. (1919). *Philosophical Magazine*, **38**, 709.
Bainbridge, K. T. (1947). *Solvay Report, Seventh Congress in Chemistry*, p. 5. Brussels: R. Stoops.
Bainbridge, K. T. (1949). *Physical Review*, **75**, 216A.

Bainbridge, K. T. (1953). *Experimental Nuclear Physics*, vol. 1, ed. Segré, E., p. 559. New York: Wiley.
Barber, N. F. (1933). *Proceedings of the Leeds Philosophical and Literary Society*, **2**, 427.
Barber, R. C. (1965). *Canadian Journal of Physics*, **43**, 716.
Barber, R. C., Bishop, R. L., Duckworth, H. E., Meredith, J. O., Southon, F. C. G., van Rookhuyzen, P. & Williams, P. (1971). *Review of Scientific Instruments*, **42**, 1.
Baril, M. & Kerwin, L. (1965). *Canadian Journal of Physics*, **43**, 1317.
Barnard, G. P. (1953). *Modern Mass Spectrometry*. London: The Institute of Physics.
Barnard, G. P. (1955). *Journal of Electronics*, **1**, 78.
Bartky, W. & Dempster, A. J. (1929). *Physical Review*, **33**, 1019.
Beiduk, F. M. & Konopinski, E. J. (1948). *Review of Scientific Instruments*, **19**, 594.
Bleakney, W. & Hipple, J. A. (1938). *Physical Review*, **53**, 520.
Bock, C. D. (1933). *Review of Scientific Instruments*, **4**, 575.
Brown, K. L., Belbeoch, R. & Bounin, P. (1964). *Review of Scientific Instruments*, **35**, 481.
Browne, C. P., Craig, D. S. & Williamson, R. M. (1951). *Review of Scientific Instruments*, **22**, 952.
Brüche, E. & Henneberg, W. (1935). DR Patent 651008.
Brüche, E. & Scherzer, O. (1934). *Geometrische Elektronenoptik*. Berlin: Julius Springer.
Camac, M. (1951). *Review of Scientific Instruments*, **22**, 197.
Cartan, L. (1937). *Journal de Physique et le Radium*, **8**, 453.
Coggeshall, N. D. (1946). *Physical Review*, **70**, 270.
Coggeshall, N. D. (1947). *Journal of Applied Physics*, **18**, 855.
Coggeshall, N. D. & Muskat, M. (1944). *Physical Review*, **66**, 187.
Cotte, M. (1938). *Annales de Physique, Paris*, **10**, 333.
Cross, W. G. (1951). *Review of Scientific Instruments*, **22**, 717.
Czok, U., Euler, K., Rauscher, M. & Wollnik, H. (1971). *Nuclear Instruments and Methods*, **92**, 365.
Dempster, A. J. (1918). *Physical Review*, **11**, 316.
Dietz, L. A. (1961). *Review of Scientific Instruments*, **32**, 859.
Enge, H. A. (1964). *Review of Scientific Instruments*, **35**, 278.
Enge, H. A. (1967). *Focusing of Charged Particles, vol. 2*, ed. Septier, A., p. 203. New York and London: Academic Press.
Ewald, H. & Liebl, H. (1955). *Zeitschrift für Naturforschung*, **10a**, 872.
Ewald, H. & Liebl, H. (1957*a*). *Zeitschrift für Naturforschung*, **12a**, 28.
Ewald, H. & Liebl, H. (1957*b*). *Zeitschrift für Naturforschung*, **12a**, 538.
Ewald, H. & Liebl, H. (1957*c*). *Zeitschrift für Naturforschung*, **14a**, 588.
Ewald, H., Liebl, H. & Sauermann, G. (1957). *Nuclear Masses and Their Determination*, ed. Hintenberger, H., p. 184. New York, London, Paris: Pergamon Press.
Ewald, H. & Sauermann, G. (1956). *Zeitschrift für Naturforschung*, **11a**, 173.
Ezoe, H. (1967). *Review of Scientific Instruments*, **38**, 390.
Ezoe, H. (1970). *Review of Scientific Instruments*, **41**, 952.
Fischer, D. (1952). *Zeitschrift für Physik*, **133**, 455.
Fujita, Y. & Matsuda, H. (1975). *Nuclear Instruments and Methods*, **123**, 495.
Fujita, Y., Matsuda, H. & Matsuo, T. (1977). *Nuclear Instruments and Methods*, **144**, 279.
Hachenberg, O. (1948). *Annalen der Physik, Leipzig*, **2**, 225.
Herb, R. G., Snowdon, S. C. & Sala, O. (1949). *Physical Review*, **75**, 246.
Herzog, R. (1934). *Zeitschrift für Physik*, **89**, 447.

Herzog, R. (1935). *Zeitschrift für Physik*, **97**, 596.
Herzog, R. (1951). *Acta Physica, Austria*, **4**, 431.
Herzog, R. (1953*a*). *Zeitschrift für Naturforschung*, **8a**, 191.
Herzog, R. (1953*b*). *Mass Spectroscopy in Physics Research*, p. 85. Washington: National Bureau of Standards Circular 522.
Herzog, R. F. K. (1955). *Zeitschrift für Naturforschung*, **10a**, 887.
Herzog, R. & Mattauch, J. H. E. (1934). *Annalen der Physik, Leipzig*, **19**, 345.
Hintenberger, H. (1948*a*). *Zeitschrift für Naturforschung*, **3a**, 125.
Hintenberger, H. (1948*b*). *Zeitschrift für Naturforschung*, **3a**, 669.
Hintenberger, H. (1949). *Review of Scientific Instruments*, **20**, 749.
Hinterberger, H. & König, L. A. (1959). *Advances in Mass Spectrometry*, vol. 1, ed. Waldron, J. D., p. 16. London, New York, Paris: Pergamon Press.
Hipple, J. A. & Sommer, H. (1953). *Mass Spectroscopy in Physics Research*, p. 123. Washington: National Bureau of Standards Circular 522.
Holmlid, L. (1975). *International Journal of Mass Spectrometry and Ion Physics*, **17**, 403.
Hübner, H. & Wollnik, H. (1970). *Nuclear Instruments and Methods*, **86**, 141.
Hughes, A. L. & McMillan, J. H. (1929). *Physical Review*, **34**, 291.
Hughes, A. L. & Rojansky, V. (1929). *Physical Review*, **34**, 284.
Ioanoviciu, D. (1973). *International Journal of Mass Spectrometry and Ion Physics*, **11**, 169.
Ioanoviciu, D. (1974). *International Journal of Mass Spectrometry and Ion Physics*, **15**, 89.
Ioanoviciu, D. & Cuna, C. (1974). *International Journal of Mass Spectrometry and Ion Physics*, **15**, 79.
Johnson, E. G. & Nier, A. O. (1953). *Physical Review*, **91**, 10.
Judd, D. L. (1949). *Review of Scientific Instruments*, **21**, 213.
Kerst, D. W. & Serber, R. (1941). *Physical Review*, **60**, 53.
Kerwin, L. (1949). *Review of Scientific Instruments*, **20**, 36.
Kerwin, L. (1963). *Mass Spectrometry*, ed. McDowell, C. A., p. 104. New York, San Francisco, Toronto and London: McGraw-Hill.
Kerwin, L. & Geoffrion, C. (1949). *Review of Scientific Instruments*, **20**, 381.
Kistemaker, J. & Zilvershoon, C. J. (1953). *Mass Spectroscopy in Physics Research*, p. 179. Washington: National Bureau of Standards Circular 522.
König, L. A. & Hintenberger, H. (1955). *Zeitschrift für Naturforschung*, **10a**, 877.
Kurie, F. D., Osoba, J. S. & Slack, L. S. (1948). *Review of Scientific Instruments*, **19**, 771.
Langer, L. M. & Cook, C. S. (1948). *Review of Scientific Instruments*, **19**, 257.
Lavatelli, L. S. (1946). PB-52433, US Department of Commerce, Office for Technical Services, MDDC Report 350.
Lee-Whiting, G. E. & Taylor, E. A. (1956). *Canadian Journal of Physics*, **35**, 1.
Matsuda, H. (1961). *Review of Scientific Instruments*, **32**, 850.
Matsuda, H. (1971). *Nuclear Instruments and Methods*, **91**, 637.
Matsuda, H. & Fujita, Y. (1975). *International Journal of Mass Spectrometry and Ion Physics*, **16**, 395.
Matsuda, H., Fukumoto, S. & Kuroda, Y. (1966). *Zeitschrift für Naturforschung*, **21a**, 25.
Matsuda, H. & Wollnik, H. (1970*a*). *Nuclear Instruments and Methods*, **77**, 40.
Matsuda, H. & Wollnik, H. (1970*b*). *Nuclear Instruments and Methods*, **77**, 283.
Matsuda, H. & Wollnik, H. (1972). *Nuclear Instruments and Methods*, **103**, 117.
Matsuo, T. (1975). *Nuclear Instruments and Methods*, **126**, 273.

Matsuo, T., Matsuda, H., Fujita, Y. & Wollnik, H. (1975). *Mass Spectroscopy (Japan)*, **24**, 19.

Matsuo, T., Matsuda, H. & Wollnik, H. (1972). *Nuclear Instruments and Methods*, **103**, 515.

Mattauch, J. H. E. (1936). *Physical Review*, **50**, 617.

Mattauch, J. H. E. & Herzog, R. (1934). *Zeitschrift für Physik*, **89**, 786.

Nier, A. O. (1940). *Review of Scientific Instruments*, **11**, 212.

Nier, A. O. (1960). *Review of Scientific Instruments*, **31**, 1127.

Nier, A. O. & Roberts, T. R. (1951). *Physical Review*, **81**, 507.

Paul, M. (1953). *Mass Spectroscopy in Physics Research*, p. 107. Washington: National Bureau of Standards Circular 522.

Penner, S. (1961). *Review of Scientific Instruments*, **32**, 150.

Ploch, W. & Walcher, W. (1950). *Zeitschrift für Physik*, **127**, 274.

Purcell, E. M. (1938). *Physical Review*, **54**, 818.

Reuterswärd, C. (1951). *Arkiv för Fysik*, **3**, 53.

Reuterswärd, C. (1952). *Arkiv för Fysik*, 4, 159.

Rogers, F. T. (1940). *Review of Scientific Instruments*, **11**, 19.

Rosenblum, E. S. (1949). *Review of Scientific Instruments*, **21**, 586.

Shull, F. B. & Dennison, D. M. (1947*a*). *Physical Review*, **71**, 681.

Shull, F. B. & Dennison, D. M. (1947*b*). *Physical Review*, **72**, 256.

Siegbahn, K. & Svartholm, N. (1946). *Nature*, **157**, 872; *Arkiv för Matematik, Astronomi och Fysik*, **33A**, no. 21.

Smith, L. G. (1967). *Proceedings of the Third International Conference on Atomic Masses*, ed. Barber, R. C., p. 811. Winnipeg: University of Manitoba Press.

Smythe, W. R. (1934). *Physical Review*, **45**, 299.

Smythe, W. R., Rumbaugh, L. H. & West, S. S. (1934). *Physical Review*, **45**, 724.

Stephens, W. E. (1934). *Physical Review*, **45**, 513.

Stevens, C. M. & Moreland, P. E. (1967). *Proceedings of the Third International Conference on Atomic Masses*, ed. Barber, R. C., p. 673. Winnipeg: University of Manitoba Press.

Stevens, C. M., Terandy, J., Lobell, G., Wolfe, J., Beyer, N. & Lewis, R. (1960). *Proceedings of the International Conference on Nuclidic Masses*, ed. Duckworth, H. E., p. 403. Toronto: University of Toronto Press.

Stevens, C. M., Terandy, J., Lobell, G., Wolfe, J., Lewis, R. & Beyer, N. (1963). *Advances in Mass Spectrometry*, vol. 2, ed. Elliott, R. M., p. 198. New York: Pergamon Press.

Svartholm, N. (1950). *Arkiv för Fysik*, **2**, 115.

Takeshita, I. (1966). *Zeitschrift für Naturforschung*, **21a**, 9.

Tasman, H. A. (1959). *Advances in Mass Spectrometry*, vol. 1, ed. Waldron, J. D., p. 36. London, New York, Paris: Pergamon Press.

Tasman, H. A. & Boerboom, A. J. H. (1959). *Zeitschrift für Naturforschung*, **14a**, 121.

Tasman, H. A., Boerboom, A. J. H. & Wachsmuth, H. (1959). *Zeitschrift für Naturforschung*, **14a**, 822.

Taya, S., Takiguchi, K., Kanomata, I. & Matsuda, H. (1978). *Nuclear Instruments and Methods*, **150**, 165.

Taya, S., Tsuyama, H., Kanomata, I., Noda, T. & Matsuda, H. (1978). *International Journal of Mass Spectrometry and Ion Physics*, **26**, 77.

Wachsmuth, H., Boerboom, A. J. H. & Tasman, H. A. (1959). *Zeitschrift für Naturforschung*, **14a**, 818.
Wien, W. (1902). *Annalen der Physik, Leipzig*, **8**, 224.
Wollnik, H. (1967*a*). *Focusing of Charged Particles*, vol. 2, ed. Septier, A., p. 163. New York, London: Academic Press.
Wollnik, H. (1967*b*). *Nuclear Instruments and Methods*, **52**, 250.
Wollnik, H. (1967*c*). *Nuclear Instruments and Methods*, **53**, 197.
Wollnik, H. (1968). *Nuclear Instruments and Methods*, **59**, 277.
Wollnik, H. (1972). *Nuclear Instruments and Methods*, **103**, 479.
Wollnik, H. & Ewald, H. (1965). *Nuclear Instruments and Methods*, **36**, 93.

Chapter 3

Ahearn, A. J. (1966). *Mass Spectrometric Analysis of Solids*, ed. Ahearn, A. J., p. 1. Amsterdam, London, New York: Elsevier.
Almén, O. & Nielsen, K. O. (1956). *Electromagnetically Enriched Isotopes and Mass Spectrometry*, ed. Smith, M. L., p. 23. London: Butterworths and Co. Ltd.
Almén O. & Nielsen, K. O. (1957). *Nuclear Instruments and Methods*, **1**, 302.
Anderson, C. A. (1969). *International Journal of Mass Spectrometry and Ion Physics*, **2**, 61.
Anderson, C. A. (1970). *International Journal of Mass Spectrometry and Ion Physics*, **3**, 413.
Aston, F. W. (1923). *Philosophical Magazine*, **47**, 385.
Aston, F. W. (1942). *Mass Spectra and Isotopes*, 2nd edn., pp. 2, 4, 7, 8, 34, 50, 87, 101, 111. London: Edward Arnold and Co.
Aston, F. W. & Thomson, J. J. (1921). *Nature*, **106**, 827.
Äystö, J., Rantala, V., Valli, K., Hillebrand, S., Kortelahti, M., Eskola, K. & Raunemaa, T. (1976). *Nuclear Instruments and Methods*, **139**, 325.
Backus, J. (1949). *Characteristics of Electrical Discharges in Magnetic Fields*, eds. Guthrie, A. & Wakerling, R. K., p. 345. New York: McGraw-Hill.
Bacon, F. M. (1978). *Review of Scientific Instruments*, **49**, 427.
Bacon, F. M., Bickes, R. W. & O'Hagan, J. B. (1978). *Review of Scientific Instruments*, **49**, 435.
Bainbridge, K. T. (1932). *Physical Review*, **39**, 847.
Ban, V. S. & Knox, B. E. (1969). *International Journal of Mass Spectrometry and Ion Physics*, **3**, 131.
Barber, M., Bordoli, R. S., Elliott, G. J., Sedgwick, R. D. & Tyler, A. N. (1982). *Analytical Chemistry*, **54**, 645A.
Barber, M., Bordoli, R. S., Sedgwick, R. D. & Tyler, A. N. (1981). *Biomedical Mass Spectrometry*, **8**, 492.
Barber, R. C., Bishop, R. L., Duckworth, H. E., Meredith, J. O., Southon, F.C.G., van Rookhuyzen, P. & Williams, P. (1971). *Review of Scientific Instruments*, **42**, 1.
Barnett, C. F., Stier, P. M. & Evans, G. E. (1953). *Review of Scientific Instruments*, **25**, 1112.
Barofsky, B. F. & Barofsky, E. (1974). *International Journal of Mass Spectrometry and Ion Physics*, **14**, 3.
Beckey, H. D. (1963). *Advances in Mass Spectrometry*, vol. 2, ed. Elliott, R. M., p.1. New York: Pergamon Press.
Beckey, H. D. (1971). *Field Ionization Mass Spectrometry*, Oxford: Pergamon Press.

Beckey, H. D. (1978). *Field Ionization and Field Desorption Mass Spectroscopy*, Oxford: Pergamon Press.

Beggs, D., Vestal, M. L., Fales, H. M. & Milne, G.W.A. (1971). *Review of Scientific Instruments*, **42**, 1578.

Bel'chenko, Y. I., Dimov, D. I. & Dudnikov, V. G. (1972). *Journal of Technical Physics (USSR)*, **43**, 1720.

Bennett, J. R. J. (1964). Thesis.

Bennett, J. R. J. (1972). *IEEE Transactions on Nuclear Science*, **NS19**, 48.

Benninghoven, A. (1975). *Surface Science*, **53**, 596.

Benninghoven, A. (1982). ed. *Proceedings of the International Conference on Ion Formation from Organic Solids*, Berlin: Springer-Verlag.

Benninghoven, A. & Sichtermann, W. (1978). *Analytical Chemistry*, **50**, 1180.

Bergström, I., Thulin, S., Svartholm, N. & Siegbahn, K. (1949). *Arkiv för Fysik*, **1**, 281.

Bernas, R. & Nier, A. O. (1948). *Review of Scientific Instruments*, **19**, 895.

Berthod, J. (1976). *Advances in Mass Spectrometry*, vol. 6, ed. West, A. R., p. 421. Barking: Applied Science Publishers Ltd.

Beynon, J. H., Fontaine, A. E. & Job, E. (1966). *Zeitschrift für Naturforschung*, **21a**, 776.

Billon, J. P. (1976). *Advances in Mass Spectrometry*, vol. 6, ed. West, A. R., p. 649. Barking: Applied Science Publishers Ltd.

Blaise, G. (1978). *Vide-Cauche Minces*, **33**, 1.

Bleakney, W. (1929). *Physical Review*, **34**, 157.

Bleakney, W. (1930). *Physical Review*, **36**, 1303.

Block, J. H. (1976). *Advances in Mass Spectrometry*, vol. 6, ed. West, A. R., p. 109. Barking: Applied Science Publishers.

Boehme, D. K., Goodings, J. M. & Ng. Chen Wai. (1977). *International Journal of Mass Spectrometry and Ion Physics*, **24**, 335.

Bohm, D. (1949). *Characteristics of Electrical Discharges in Magnetic Fields*, eds. Guthrie, A. & Wakerling, R. K., p. 87. New York: McGraw-Hill.

Bohr, N. (1948). *Kongelige Danske Videnskabernes Matematisk-fysiske Meddelelser, Selskab*, **18**, no. 8.

Briglia, D. D. & Rapp, D. (1965). *Journal of Chemical Physics*, **42**, 3201.

Brown, R., Craig, R. D. & Elliot, R. M. (1956). *Advances in Mass Spectrometry*, vol. 2, ed. Elliott, R. M., p. 141. New York: Pergamon Press.

Burhop, E. H., Massey, H. S. & Page, G. (1949). *Characteristics of Electrical Discharges in Magnetic Fields*, eds. Guthrie, A. & Wakerling, R. K., p. 107. New York: McGraw-Hill.

Cameron, A. E., Smith, D. H. & Walker, R. H. (1969). *Analytical Chemistry*, **41**, 525.

Castaing, R. & Slodzian, B. (1962). *Journal de Microscopie*, **1**, 395.

Čermák, V. & Herman, Z. (1962). *Collection of Czechoslovak Chemical Communications*, **27**, 406.

Chait, B. T. & Standing, K. G. (1981). *International Journal of Mass Spectrometry and Ion Physics*, **40**, 185.

Chakravarty, B., Venkatasubramanian, V. S. & Duckworth, H. E. (1962). *Advances in Mass Spectrometry*, vol. 2, ed. Elliott, R. M., p. 128. New York: Pergamon Press.

Cobic, P., Tosic, T. & Perovic, B. (1963). *Nuclear Instruments and Methods*, **24**, 358.

Coburn, J. W., Taglauer, E. & Kay, E. (1974). *Journal of Applied Physics*, **45**, 1779.

Cohen, M. J. & Karasek, F. W. (1970). *Journal of Chromatographic Science*, **8**, 330.

Colby, B. N. & Evans, C.N.E. (1976). *Advances in Mass Spectrometry*, vol. 6, ed. West,

A. R., p. 565. Barking: Applied Science Publishers.
Collins, L. E. & Gobbett, R. H. (1965). *Nuclear Instruments and Methods*, **35**, 277.
Conzemius, R. J. & Svec, H. J. (1972). *Trace Analysis by Mass Spectrometry*, ed. Ahearn, A. J., pp. 136–76. New York & London: Academic Press.
Craig, R. D. (1959). *Journal of Scientific Instruments*, **36**, 38.
Cross, R. H., Brown, H. L. & Anbar, M. (1976). *Review of Scientific Instruments*, **47**, 1270.
Davis, R. C., Morgan, O. B., Stewart, L. D. & Stirling, W. L. (1971). *Review of Scientific Instruments*, **43**, 278.
Davis, W. D. & Miller, H. C. (1969). *Journal of Applied Physics*, **49**, 2012.
Dawton, R.H.V.M. (1956). *Electromagnetically Enriched Isotopes and Mass Spectrometry*, ed. Smith, M. L., p. 37. London: Butterworths.
Dearnaley, G. (1973). *Ion Implantation*, eds. Dearnaley, G., Freeman, J. H., Nelson, R. S. & Stephen, J., p. 9. Amsterdam: North-Holland.
de Bièvre, P. (1978). *Advances in Mass Spectrometry*, vol. 7A, ed. Daly, N. R., p. 395. London: Heyden and Son Ltd.
Dempster, A. J. (1918). *Physical Review*, **11**, 316.
Dempster, A. J. (1935). *Nature*, **135**, 542.
Dempster, A. J. (1936). *Review of Scientific Instruments*, **7**, 46.
Dempster, A. J. (1946). *Bibliography of Scientific and Industrial Reports*, MDDC 370, US Department of Commerce.
Dück, P., Treu, W., Frohlich, H., Galster, W. & Voit, H. (1980). *Surface Science*, **95**, 603.
Ehlers, K. W. (1962). *Nuclear Instruments and Methods*, **18**, 571.
Ehlers, K. W. & Leung, K. N. (1980). *Review of Scientific Instruments*, **51**, 721.
Ewald, H. & Hintenberger, H. (1953). *Methoden und Anwendungen der Massenspektroskopie*, p. 23. Weinheim: Verlag Chemie GmbH.
Fabry, C. & Perot, A. (1900*a*). *Comptes Rendus, Académie des Sciences*, **130**, 406.
Fabry, C. & Perot, A. (1900*b*). *Journal de Physique*, **9**, 369.
Farrar, H. (1972). *Trace Analysis by Mass Spectrometry*, ed. Ahearn, A. J., p. 268. New York & London: Academic Press.
Field, F. H. (1972). *MTP International Review of Science: Physical Chemistry; Series 1, volume 5, Mass Spectrometry*, ed. Maccoll, A., p. 133. London: Butterworth.
Field, F. H., Lampe, F. W. & Franklin, J. L. (1957). *Journal of the American Chemical Society*, **79**, 2419.
Finkelstein, A. T. (1940). *Review of Scientific Instruments*, **11**, 94.
Fleming, W. H. & Thode, H. G. (1953 *a, b*). *Physical Review*, **90**, 857; **92**, 378.
Fox, R. E., Hickam, W. M., Kjeldaas, T. Jr & Grove, D. J. (1951). *Physical Review*, **84**, 859.
Franzen, J. (1963). *Zeitschrift für Naturforschung*, **18a**, 410.
Franzen, J. (1972). *Trace Analysis by Mass Spectrometry*, ed. Ahearn, A. J., chapter 1. New York: Academic Press.
Franzen, J. & Schuy, K. D. (1967). *Zeitschrift für Analytische Chemie*, **225**, 295.
Freeman, J. H. (1969). *Proceedings of the International Ion Source Conference*, p. 369. (INSTN–Saclay, France), also Atomic Energy Research Establishment Report R6138.
Frost, D. C., Mak, D. & McDowell, C. A. (1962). *Canadian Journal of Chemistry*, **40**, 1064.
Fuchs, G. (1972). *IEEE Transactions on Nuclear Science*, **NS19**, 160.
Gehrcke, E. & Reichenheim, O. (1906). *Berichte der Deutschen Physikalischen Gesellschaft*, **8**, 559.

Gehrcke, E. & Reichenheim, O. (1907). *Berichte der Deutschen Physikalischen Gesellschaft*, **9**, 76, 200, 376.
Gehrcke, E. & Reichenheim, O. (1908). *Berichte der Deutschen Physikalischen Gesellschaft*, **10**, 217.
Gierlich, H. H., Heindricks, A. & Beckey, H. D. (1974). *Review of Scientific Instruments*, **45**, 1208–11.
Goldstein, E. (1886). *Sitzungberichte der Königlich Preussischen Academie der Wissenschaften zu Berlin*, **39**, 691.
Golubev, V. P., Nalvaiko, V. L., Tokarev, G. M. & Tsepakhin, S. G. (1972). *Proceedings of the Proton Linear Accelerator Conference*. Los Alamos Report no. LA–5115, p. 356.
Gomer, R. & Inghram, M. G. (1955). *Journal of the American Chemical Society*, **77**, 500.
Gorman, J. G., Jones, E. J. & Hipple, J. A. (1951). *Analytical Chemistry*, **23**, 438.
Green, T. S. (1974). *Reports on Progress in Physics*, **37**, 1257.
Guyon, P. M. (1976). *Advances in Mass Spectrometry*, vol. 6, ed. West, A. R., p. 403. Barking: Applied Science Publishers.
Håkansson, P., Johansson, A., Kamensky, I., Sundquist, B., Fohlman, J. & Peterson, P. (1981). *IEEE Transactions on Nuclear Science*, **28**, 1776.
Hamilton-Gordon, W. & Osher, J. E. (1972). UCRL Report 74096, p. 1.
Hannay, N. B. (1954). *Review of Scientific Instruments*, **25**, 644.
Hannay, N. B. & Ahearn, A. J. (1954). *Analytical Chemistry*, **26**, 1056.
Harrison, W. W. & Magee, C. W. (1974). *Analytical Chemistry*, **46**, 461.
Hayden, J. L. & Nier, A. O. (1974). *International Journal of Mass Spectrometry and Ion Physics*, **15**, 37.
Herzog, R. F. & Viehböck, F. P. (1949). *Physical Review*, **76**, 855.
Hess, D. C. Wetherill, G. & Inghram, M. G. (1951). *Review of Scientific Instruments*, **22**, 838.
Hintenberger, H. (1966). *Advances in Mass Spectrometry*, vol. 3, ed. Mead, W. L., p. 517. London: Institute of Petroleum.
Hoegger, B. & Bommer, P. (1974). *International Journal of Mass Spectrometry & Ion Physics*, **13**, 35.
Honig, R. E. (1966). *Advances in Mass Spectrometry*, vol. 3, ed. Mead, W. L., p. 101. London: Institute of Petroleum.
Honig, R. E. & Woolston, J. R. (1963). *Applied Physics Letters*, **2**, 138.
Huber, B. A., Miller, T. M., Zeeman, H. D., Leon, R. J., Moseley, J. T. & Patterson, J. D. (1977). *Review of Scientific Instruments*, **48**, 1306.
Hunt, D. F., McEwen, C. M. & Harvey, T. H. (1975). *Analytical Chemistry*, **47**, 1730.
Hunt, D. F. & Sethi, P. (1978). *American Chemical Society Symposium Series*, **70**, 70.
Hurzeler, H., Inghram, M. G. & Morrison, J. O. (1958). *Journal of Chemical Physics*, **28**, 76.
Illgen, J., Kirchner, R. & Schulte, J. (1972). *IEEE Transactions on Nuclear Science*, **NS19**, 35.
Inghram, M. G. & Chupka, W. A. (1953). *Review of Scientific Instruments*, **24**, 518.
Inghram, M. G. & Gomer, R. (1954). *Journal of Chemical Physics*, **22**, 1279.
Inghram, M. G. & Gomer, R. (1955). *Zeitschrift für Naturforschung*, **10a**, 863.
Inghram, M. G. & Hayden, R. J. (1954). *Handbook on Mass Spectroscopy*, Nuclear Science Series, Report No. 14 (NRC–USA).
Inghram, M. G. & Reynolds, J. H. (1950). *Physical Review*, **78**, 822.

Isenor, N. R. (1964). *Canadian Journal of Physics*, **42**, 1413.
Isler, R. C. (1974). *Review of Scientific Instruments*, **45**, 308.
Jones, R. J. & Zucker, A. (1954). *Review of Scientific Instruments*, **25**, 362.
Kambara, H. & Kanomata, I. (1977). *International Journal of Mass Spectrometry and Ion Physics*, **24**, 453.
Kanno, H. (1971). *Bulletin of the Chemical Society of Japan*, **44**, 1808.
Kennett, T. J. & Thode, H. G. (1956). *Physical Review*, **103**, 323.
Kirchner, R. (1981). *Nuclear Instruments and Methods in Physics Research*, **186**, 283.
Kirchner, R. & Roeckl, E. (1976). *Nuclear Instruments and Methods*, **133**, 187.
Kistemaker, J., Rol, P. K., Shutten, J. & de Vries, C. (1956). *Electromagnetically Enriched Isotopes and Mass Spectrometry*, ed. Smith, M. L., p. 10. London: Butterworths.
Knox, B. E. (1972). *Trace Analysis by Mass Spectrometry*, ed. Ahearn, A. J., pp. 423–444. New York & London: Academic Press.
Kobayashi, M., Prelec, K. & Slyters, J. (1976). *Review of Scientific Instruments*, **47**, 1425.
Koch, J. (1942). *Electromagnetic Isotope Separation*, Thesis, University of Copenhagen.
Koch, J. (1953). *Mass Spectroscopy in Physics Research*, p. 165. Washington: National Bureau of Standards Circular 522.
Koch, J. & Bendt-Nielsen, B. (1944). *Matematisk-fysiske Meddelelser, Kongelige Danske Videnskabernes Selskab*, **21**, no. 8.
Kohno, I., Tonuma, T., Miyazawa, Y., Nakajima, S., Inoue, T., Shimomura, A. & Karasawa, T. (1968). *Nuclear Instruments and Methods*, **66**, 283.
Kunsman, C. H. (1925). *Science*, **62**, 269.
Kunsman, C. H. (1926). *Proceedings of the National Academy of Science*, Washington, **12**, 659.
Lamar, E. S., Samson, E. W. & Compton, K. T. (1935). *Physical Review*, **48**, 886.
Langmuir, I. (1929). *Physical Review*, **33**, 954.
Langmuir, I. & Kingdon, K. H. (1925). *Proceedings of the Royal Society*, **A**, **107**, 61.
Lawrence, G. P., Beauchamp, R. K. & McKibben, J. L. (1965). *Nuclear Instruments and Methods*, **32**, 357.
Lejejune, C. (1974). *Nuclear Instruments and Methods*, **116**, 417, 429.
Lindhard, J. & Scharff, M. (1961). *Physical Review*, **124**, 128.
Linlor, W. I. (1963). *Applied Physics Letters*, **3**, 210.
Lossing, F. P. & Tanaka, I. (1956). *Journal of Chemical Physics*, **25**, 1031.
Lyshede, J. M. (1941). *Matematisk-fysiske Meddelelser, Kongelige Danske Videnskabernes Selskab*, **18**, no. 13.
Macfarlane, R. D. (1982). *Accounts in Chemical Research*, **15**, 268.
Macfarlane, R. D. & Torgerson, D. F. (1976). *Science*, **191**, 920.
Macnamara, J. & Thode, H. G. (1950). *Physical Review*, **80**, 296.
Massey, H. S. W. (1976). *Negative Ions*. Cambridge University Press.
Marr, G. V. (1967). *Photoionization Processes in Gases*. New York: Academic Press.
Mather, R. E. & Todd, J. F. J. (1979). *International Journal of Mass Spectrometry and Ion Physics*, **30**, 1.
Matsuo, T., Matsuda, H. & Katakuse, I. (1979). *Analytical Chemistry*, **51**, 69.
Mattauch, J. H. E. (1937). *Naturwissenschaften*, **25**, 170.
Mattson, W. A., Bentz, B. L. & Harrison, W. W. (1976). *Analytical Chemistry*, **48**, 489.
Mazumdar, A. K., Wagner, H., Walcher, W. & Lund, T. (1976). *Nuclear Instruments and Methods*, **139**, 319.

McFarlin, W. A., Wilson, W. B., June, M. N. & Chapman, W. W. (1971). *IEEE Transactions on Nuclear Science*, **NS18**, 132.
McNeal, C. (1982). *Analytical Chemistry*, **54**, 43A.
McNeal, C. & Macfarlane, R. D. (1981). *Journal of the American Chemical Society*, **103**, 1609.
Melton, C. E. (1968). *International Journal of Mass Spectrometry and Ion Physics*, **1**, 353.
Middleton, R. (1977). *Nuclear Instruments and Methods*, **144**, 373.
Middleton, R. & Adams, C. T. (1974). *Nuclear Instruments and Methods*, **118**, 329.
Millikan, R. A., Sawyer, R. A. & Bowen, I. S. (1921). *Astrophysical Journal*, **53**, 150.
Moore, J. L. & Heald, E. F. (1978). *Advances in Mass Spectrometry*, vol. 7A, ed. Daly, N. R., p. 448. London: Heyden and Son Ltd.
Morgan, O. B., Kelley, G. G. & Davis. R. (1967). *Review of Scientific Instruments*, **38**, 467.
Morrison, G. & Slodzian, G. (1975). *Analytical Chemistry*, **47**, 933A.
Morrison, J. D., Hurzeler, H., Inghram, M. G. & Stanton, H. E. (1960). *Journal of Chemical Physics*, **33**, 821.
Müller, E. W. (1953). *Ergebnisse der Exakt Naturwissenschaften*, **27**, 290.
Müller, E. W., Panitz, J. A. & McLane, S. B. (1968). *Review of Scientific Instruments*, **39**, 83.
Munson, B. (1971). *Analytical Chemistry*, **43**, A28.
Munson, M. S. B. & Field, F. H. (1962). *Journal of the American Chemical Society*, **88**, 1621.
N'Guyen, L. & de Saint Simon, M. (1972). *International Journal of Mass Spectrometry and Ion Physics*, **9**, 299.
Nief, G. & Roth, E. (1974). *Advances in Mass Spectrometry*, vol. 5, ed. Quayle, A., p. 14. London: Institute of Petroleum.
Nielsen, K. O. (1957). *Nuclear Instruments and Methods*, **1**, 289.
Nier, A. O. C. (1940). *Review of Scientific Instruments*, **11**, 212.
Nier, A. O. C. (1947). *Review of Scientific Instruments*, **18**, 398.
Normand, C. E., Love, L. O., Bell, W. A. & Prater, W. K. (1956). *Electromagnetically Enriched Isotopes and Mass Spectrometry*, ed. Smith, M. L., p. 1. London: Butterworths.
Ormrod, J. H., MacDonald, J. R. & Duckworth, H. E. (1965). *Canadian Journal of Physics*, **43**, 275.
Palmer, G. H. (1959). *Advances in Mass Spectrometry*, ed. Waldron, J. D., p. 89. London, New York, Paris: Pergamon Press.
Parr, G. R. & Taylor, J. W. (1973). *Review of Scientific Instruments*, **44**, 1578.
Penning, F. M. (1937). *Physica*, **4**, 71.
Prelec, K. & Slyters, J. (1973). *Review of Scientific Instruments*, **44**, 1451.
Reddy, S. J., Rollgen, F. W., Maas, A. & Beckey, H. D. (1977). *International Journal of Mass Spectrometry and Ion Physics*, **17**, 147.
Rees, C. E. (1969). *International Journal of Mass Spectrometry and Ion Physics*, **3**, 71.
Reid, N. W. (1971). *International Journal of Mass Spectrometry and Ion Physics*, **6**, 1.
Reynolds, J. H. (1956). *Review of Scientific Instruments*, **27**, 928.
Reynolds, W. D. (1979). *Analytical Chemistry*, **51**, A283.
Robertson, A. J. B. & Viney, B. W. (1966). *Advances in Mass Spectrometry*, vol. 3, ed. Mead, W. L., p. 23. London: Institute of Petroleum.
Rollgen, F. W., Giessmann, E. & Reddy, S. J. (1976). *International Journal of Mass Spectrometry and Ion Physics*, **16**, 235.

Rose, P. H. & Galejs, A., 1965. See Dearnaley, Freeman, Nelson & Stephen, 1973, p. 255.
Rossman, K. J. R. & de Laeter, J. R. (1975). *International Journal of Mass Spectrometry and Ion Physics*, **16**, 385.
Sakudo, N., Togikuchi, M., Koike, H. & Kanomata, I. (1978). *Review of Scientific Instruments*, **49**, 940.
Schuy, K. D. & Hintenberger, H. (1963). *Zeitschrift für Analytische Chemie*, **197**, 98.
Schwegler, E. C. & White, F. A. (1968). *International Journal of Mass Spectrometry and Ion Physics*, **1**, 191.
Shields, W. R. (1966). US Department of Commerce, NBS Technical Note 277, *Analytical Mass Spectrometry Section: Instrumentation and Procedures for Isotopic Analysis.*
Sidenius, G. (1965). *Nuclear Instruments and Methods*, **38**, 19.
Sidenius, G. (1969). *Proceedings of the International Conference on Ion Sources*, **341**, (INSTN – Saclay, France).
Sidenius, G. (1978). *Nuclear Instruments and Methods*, **151**, 349.
Smith, D. H. & Christie, W. H. (1978). *International Journal of Mass Spectrometry and Ion Physics*, **26**, 61.
Stroud, P. T. (1969). *Journal of Physics, Part E*, **2**, 452.
Straus, H. A. (1941). *Physical Review*, **59**, 430.
Stüwer, D. (1976). *Advances in Mass Spectrometry*, vol. 6, ed. West, A. R., p. 655. Barking: Applied Science Publishers Ltd.
Sundquist, B. (1982). Proceedings of the Nordic Symposium on Ion Induced Desorption of Molecules from Bio-organic Solids, ed. *Nuclear Instruments and Methods*, **193**, no. 3.
Swingler, D. L. (1970). *Journal of Applied Physics*, **41**, 4096.
Tal'rose, V. L. & Lyubimova, A. K. (1952). *Doklady Akademii Nauk SSSR*, **86**, 909.
Terenin, A. & Popov, B. (1932). *Physikalische Zeitschrift der Sowjetunion*, **2**, 299.
Terra, G. U. F., Burnett, D. S. & Wasserburg, G. J. (1970). *Journal of Geophysical Research*, **75**, 2753.
Thode, H. G. & Graham, R. L. (1947). *Canadian Journal of Research*, **A**, **25**, 1.
Thomas, H. A. (1947). *Bulletion of the Texas Agricultural and Mechanical College Engineering Experimental Station*, no. 101.
Thonemann, P. C. (1953). *Progress in Nuclear Physics*, vol. 3. London: Pergamon Press.
Trt'yakov, Yu. P., Pasyuk, A. S., Kul'kina, L. P. & Kuznetsov, V. J. (1970). *Atomnaya Energiya*, **28**, 423.
Tuve, M. A., Dahl, O. & Hafstad, L. R. (1935). *Physical Review*, **48**, 241.
Van Voorhis, S. N., Kuper, J. B. H. & Harnwell, G. P. (1934). *Physical Review*, **45**, 492.
Vastola, F. J. & Pirone, A. J. (1966). *Advances in Mass Spectrometry*, vol. 4, ed. Kendrick, E., p. 107. London: Institute of Petroleum.
Venkatasubramanian, V. S. & Duckworth, H. E. (1963). *Canadian Journal of Physics*, **41**, 234.
Venkatasubramanian, V.S. & Swaminathan, S. (1966). *Proceedings of the Nuclear-Solid State Symposium*, **16B** (Bombay), p. 258.
Venkatasubramanian, V. S., Swaminathan, S. & Rajagopalan, P. T. (1977). *International Journal of Mass Spectrometry and Ion Physics*, **24**, 207.
von Ardenne, M. (1956). *Tabellen der Elektronenphysik, Ionenphysik und Vebermikroskopie*, p. 743. Berlin: VEB Deutcher Verlag der Wissenschaften.
Weissler, G. L., Samson, J. A. R., Ogawa, M. & Cook, G. R. (1959). *Journal of the Optical*

Society of America, **49**, 338.
Wetherill, G. (1953). *Physical Review*, **92**, 907.
Wilson, R. G. & Jamba, D. M. (1967). *Journal of Applied Physics*, **38**, 1976.
Woolston, J. R. & Honig, R. E. (1964). *Review of Scientific Instruments*, **35**, 69.
Yates, E. L. (1938). *Proceedings of the Royal Society*, **A**, **168**, 148.

Chapter 4

Allen, J. A. (1939). *Physical Review*, **55**, 966.
Allen, J. A. (1947). *Review of Scientific Instruments*, **18**, 739.
Allen, J. A. (1950). Preliminary Report, no. 10, Nuclear Energy Series (NRC–USA).
Aston, F. W. (1925). *See* Aston (1942), p. 87.
Aston, F. W. (1942). *Mass Spectra and Isotopes*. London: Edward Arnold Co.
Bainbridge, K. T. (1931). *Journal of the Franklin Institute*, **212**, 489.
Barnett, C. F., Evans, G. E. & Stier, P. H. (1954). *Review of Scientific Instruments*, **25**, 1112.
Bay, Z. (1938). *Nature*, **141**, 284, 1011.
Bay, Z. (1941). *Review of Scientific Instruments*, **12**, 127.
Becker, H., Dietz, E. & Gerhardt, U. (1972). *Review of Scientific Instruments*, **43**, 1587.
Berry, H. W. (1948). *Physical Review*, **74**, 848.
Brown, W. L., Buck, T. M., Gibson, W. M., Kerr, R. W. & Lie, H. P. (1970). *IEEE Transactions on Nuclear Science*, **17**, 329.
Carrico, J. P., Johnson, M. C. & Somer, T. A. (1973). *International Journal of Mass Spectrometry and Ion Physics*, **11**, 409.
Cohen, A. A. (1943). *Physical Review*, **63**, 219.
Daly, N. R., (1960). *Review of Scientific Instruments*, **31**, 264.
Daly, N. R., McCormick, A., Powell, R. E. & Hayes, R. (1973). *International Journal of Mass Spectrometry and Ion Physics*, **11**, 255.
Dietz, L. A. & Hanrahan, H. (1978). *Review of Scientific Instruments*, **49**, 1250.
Dietz, L. A. & Sheffield, C. (1973). *Review of Scientific Instruments*, **44**, 183.
Duclaux, J. & Jeantet, P. (1921). *Journal de Physique*, Paris, **2**, 154.
Evans, D. S. (1965). *Review of Scientific Instruments*, **36**, 375.
Evans, R. D. (1955). *The Atomic Nucleus*. New York: McGraw-Hill, chapter 28.
Goldstein, E. (1886). *Sitzungsberichte der Königlich Pruessischen Academie der Wissenschaften zu Berlin*, **39**, 691.
Goodrich, G. W. & Wiley, W. C. (1962). *Review of Scientific Instruments*, **33**, 761.
Gunn, R. (1932). *Physical Review*, **40**, 307.
Higatsberger, M. J. (1953). *Zietschrift für Naturforschung*, **8a**, 206.
Higatsberger, M. J., Demorest, H. L. & Nier, A. O. (1954). *Journal of Applied Physics*, **25**, 883.
Honig, R. E. (1972). *Trace Analysis by Mass Spectrometry*, ed. Ahearn, A. J., p. 101. New York: Academic Press.
Hull, A. W. (1932). *Physics*, **2**, 409.
Ihle, H. R. & Neubert, A. (1971). *International Journal of Mass Spectrometry and Ion Physics*, **7**, 189.
Inghram, M. G. & Hayden, R. J. (1954). *A Handbook of Mass Spectroscopy* (Nuclear Science Series Reprint No. 14, NRC–USA).

Inghram, M. G., Hayden, R. J. & Hess, D. C. (1953). *See* Inghram & Hayden (1954), p. 15.
Koenigsberger, J. & Kutchewski, J. (1910). *Physikalische Zeitschrift*, **11**, 666.
Le Caine, H. & Waghorne, J. H. (1941). *Canadian Journal of Physics*, **19A**, 21.
Lindhard, J. & Scharff, M. (1961). *Physical Review*, **124**, 128.
Metcalf, G. F. & Thomson, B. J. (1930). *Physical Review*, **36**, 1489.
Nguyen, L. -D. & Goby, G. (1978). *Advances in Mass Spectrometry*, vol. 7A, ed. Daly, N. R., p. 486. London: Heyden and Son.
Palevsky, H., Swank, R. K. & Grenchik, R. (1947). *Review of Scientific Instruments*, **18**, 298.
Raznikov, V. V., Dodonov, A. F. & Lanin, E. V. (1977). *International Journal of Mass Spectrometry and Ion Physics*, **25**, 295.
Richards, P. I. & Hays, E. E. (1950). *Review of Scientific Instruments*, **21**, 99.
Russell, R. D. & Ahearn, T. K. (1974). *Review of Scientific Instruments*, **45**, 1467.
Scherbatskoy, S. A., Gilmartin, T. H. & Swift, G. (1947). *Review of Scientific Instruments*, **18**, 415.
Schmidt, K. C. & Hendee, C. F. (1966). *IEEE Transactions on Nuclear Science*, **13**, 100.
Smith, L. G. (1951). *Review of Scientific Instruments*, **22**, 166.
Somer, T. A. & Graves, P. W. (1969). *IEEE Transactions on Nuclear Science*, **16**, 376.
Stickel, R. E., Kellert, F. G., Smith, K. A., Dunning, E. B. & Stebbings, R. F. (1980). *Review of Scientific Instruments*, **51**, 721.
Stoffels, J. J., Lagergren, C. R. & Hof, P. J. (1978). *International Journal of Mass Spectrometry and Ion Physics*, **28**, 159.
Thomas, D. G. A. & Finch, H. W. (1950). *Electrical Engineering*, **22**, 395.
Thomson, J. J. (1913). *Rays of Positive Electricity*. London: Longmans Green and Co.
Wiley, W. C. & Hendee, C. F. (1962). *IEEE Transactions on Nuclear Science*, **NS–9**, 103.
Zworykin, V. K., Morton, G. A. & Malter, L. (1936). *Proceedings of the Institute of Radio Engineers*, **24**, 351.

Chapter 5

Almén, O., Bruce, G. & Lundén, A. (1955). *See* Koch, (1958).
Almén, O., Bruce, G. & Lundén, A. (1958). *Nuclear Instruments and Methods*, **2**, 248.
Amiel, S. & Engler, G. (1976). *Electromagnetic Isotope Separators and Related Ion Accelerators. Nuclear Instruments and Methods*, **139**, 1.
Andersson, G. (1954). *Philosophical Magazine*, **45**, 621.
Andersson, G. & Holmen, G. (1973). *Proceedings of the Eighth International Conference on Mass Separators*, Gothenburg, Sweden: Chalmers University of Technology and University of Gothenburg.
Asada, T., Okuda, T., Ogata, K. & Yoshimoto, S. (1940). *Proceedings of the Physical and Mathematical Society of Japan*, **22**, 41.
Bainbridge, K. T. (1953). *Experimental Nuclear Physics*, vol. 1, ed. Segre, E., p. 559. New York: Wiley.
Bainbridge, K. T. & Ford, T. W. (1950). *See* Bainbridge (1953), p. 623.
Bainbridge, K. T. & Jordan, E. B. (1936). *Physical Review*, **50**, 282.
Bainbridge, K. T. & Moreland, P. E. (1960). *Proceedings of the International Conference on Nuclidic Masses*, ed. Duckworth, H. E., p. 460. Toronto: University of Toronto Press.
Barber, R. C., Bishop, R. L., Cambey, L. A., Duckworth, H. E., Macdougall, J. D., McLatchie, W., Ormrod, J. H. & van Rookhuyzen, P. (1964). *Nuclidic Masses*, ed. Johnson, W. H., p. 393. Vienna: Springer-Verlag.

Barber, R. C., Bishop, R. L., Duckworth, H. E., Meredith, J. O., Southon, F. C. G., van Rookhuyzen, P. & Williams, P. (1971). *Review of Scientific Instruments*, **42**, 1.

Barber, R. C., Meredith, J. O., Bishop, R. L., Duckworth, H. E., Kettner, M. E. & van Rookhuyzen, P. (1967). *Proceedings of the Third International Conference on Atomic Masses*, ed. Barber, R. C., p. 717. Winnipeg: University of Manitoba Press.

Becker, E. W. & Walcher, W. (1953). *Mass Spectroscopy in Physics Research*, p. 225. Washington: National Bureau of Standards Circular 522.

Beiduk, F. M. & Konopinski, E. J. (1948). *Review of Scientific Instruments*, **19**, 594.

Bennett, C. L. (1979). *American Scientist*, **67**, 450.

Bennett, C. L., Beukens, R. P., Clover, M. R., Gove, H. E., Liebert, R. B., Litherland, A. E., Purser, K. H. & Sondheim, W. E. (1979). *Science*, **198**, 508.

Bergstrom, I., Thulin, S., Svartholm, N. & Siegbahn, K. (1949). *Arkiv för Fysik*, **1**, no. 11.

Bernas, R. (1953). *Journal de Physique et le Radium*, **14**, 34.

Bernas, R. (1956). *See* Smith (1956), p. 262.

Bernas, R., Kaluszyner, L. & Duraux, J. (1954). *Journal de Physique et le Radium*, **15**, 273.

Bleakney, W. & Hipple, J. A. (1938). *Physical Review*, **53**, 521.

Blears, J. & Hill, R. W. (1948). *Review of Scientific Instruments*, **19**, 847.

Blears, J. & Mettrick, A. K. (1947). *Proceedings of the International Congress of Pure and Applied Chemistry*, **11**, 33.

Collins, T. L. & Bainbridge, K. T. (1957). *Nuclear Masses and Their Determination*, ed. Hintenberger, H., p. 213. New York, London, Paris: Pergamon Press.

Craig, R. D. & Errock, G. A. (1959). *Advances in Mass Spectrometry*, vol. 1, ed. Waldron, J. D., p. 66. London, New York, Paris: Pergamon Press.

Dawton, R. H. V. M. (1956). *See* Smith (1956), p. 208.

Dawton, R. H. V. M. & Smith, M. L. (1958). *Electromagnetic Isotope Separators and Applications of Electromagnetically Enriched Isotopes*, ed. Koch, J., p. 101. Amsterdam: North-Holland Publishing Company.

Dempster, A. J. (1918). *Physical Review*, **11**, 316.

Dempster, A. J. (1935). *Proceedings of the American Philosophical Society*, **75**, 755.

Duckworth, H. E. (1950). *Review of Scientific Instruments*, **21**, 54.

Dushman, S. (1962). *Scientific Foundation of Vacuum Technique*, 2nd edn, New York: Wiley.

Evans, S. & Graham, R. (1974). *Advances in Mass Spectrometry*, vol. 6, ed. West, A. R., p. 429. London: Applied Science Publishers.

Everling, F. (1957). *Nuclear Masses and Their Determination*, ed. Hintenberger, H., p. 253. New York, London, Paris: Pergamon Press.

Ewald, H. (1946). *Zeitschrift für Naturforschung*, **1**, 131.

Ewald, H. (1953). *Mass Spectroscopy in Physics Research*, p. 37. Washington: National Bureau of Standards Circular 522.

Ewald, H. (1960). *Proceedings of the International Conference on Nuclidic Masses*, ed. Duckworth, H. E., p. 491. Toronto: University of Toronto Press.

Ewald, H., Konecny, E. & Opowa, H. (1963). *Advances in Mass Spectrometry*, vol. 2, ed. Elliott, R. M., p. 189. New York: Pergamon Press.

Fenner, N. C., Jackson, M. C., Powell, R. E., Roberts, J. W. & Young, W. A. P. (1974). *International Journal of Mass Spectrometry and Ion Physics*, **14**, 245.

Fukumoto, S. & Matsuo, T. (1970). *Nuclear Instruments and Methods*, **83**, 58.

Fukumoto, S., Matsuo, T. & Matsuda, H. (1968). *Journal of the Physical Society of Japan*, **25**, 946.

Gall, R. N., Pliss, N. S. & Shcherbakov, A. P. (1980). *Advances in Mass Spectrometry*, vol. 8A, ed. Quayle, A., p. 1893. London: Heyden.
Gorman, J. G., Jones, E. J. & Hipple, J. A. (1951). *Analytical Chemistry*, **23**, 438.
Guthrie, A. & Wakerling, R. K. (1947). *Vacuum Equipment and Techniques*. New York: McGraw-Hill.
Hannay, N. B. (1954). *Review of Scientific Instruments*, **25**, 644.
Hintenberger, H. & König, L. A. (1959). *Advances in Mass Spectrometry*, vol. 1, ed. Waldrom, J. D., p. 16, London, New York, Paris: Pergamon Press.
Hipple, J. A. (1942). *Journal of Applied Physics*, **13**, 551.
Hipple, J. A. & Sommer, H. (1953). *Mass Spectroscopy in Physics Research*, p. 123. Washington: National Bureau of Standards Circular 522.
Inghram, M. G. & Hayden, R. J. (1954). *A Handbook on Mass Spectroscopy*, Nuclear Science Series, Report No. 14. Washington: National Academy of Science – National Research Council.
Inghram, M. G. & Hess, D. C. (1953). *See* Inghram & Hayden (1954), p. 23.
Inghram, M. G., Hess, D. C. & Hayden, R. J. (1953). *See* Inghram & Hayden (1954), p. 15.
Ioanoviciu, D. (1973). *International Journal of Mass Spectrometry and Ion Physics*, **12**, 115.
Johnson, E. G. & Nier, A. O. (1953). *Physical Review*, **91**, 10.
Katakuse, I., Nakabushi, H. & Ogata, K. (1976). *Atomic Masses and Fundamental Constants*, vol. 5, eds. Sanders, J. H. & Wapstra, A. H., p. 192. New York, London: Plenum Press.
Katakuse, I. & Ogata, K. (1972). *Atomic Masses and Fundamental Constants*, vol. 4, eds. Sanders, J. H. & Wapstra, A. H., p. 153. London, New York: Plenum Press.
Kayser, D. C., Halvorson, J. & Johnson, W. H. Jr (1976). *Atomic Masses and Fundamental Constants*, vol. 5, eds. Sanders, J. H. & Wapstra, A. H., p. 178. New York & London: Plenum Press.
Keim, C. P. (1955). *Third Annual Meeting, ASTM Committee E-14 on Mass Spectrometry.*
Kistemaker, J. & Zilverschoon, C. J. (1953). *Mass Spectroscopy in Physics Research*, p. 179. Washington: National Bureau of Standards Circular 522.
Koch, J. (1953). *Mass Spectroscopy in Physics Research*, p. 165. Washington: National Bureau of Standards Circular 522.
Koch, J. (1956). *See* Smith (1956), p. 214.
Koch, J. (1958). *Electromagnetic Isotope Separators and Applications of Electromagnetically Enriched Isotopes.* Amsterdam: North-Holland Press.
Koch, J. & Bendt-Nielsen, B. (1944). *Mathematisk-fysiske Meddelelser, Kongelige Danske Videnskabernes Selskab*, **21**, no. 8.
Kozier, K. S., Sharma, K. S., Barber, R. C., Barnard, J. W., Ellis, R. J., Derenchuk, V. P. & Duckworth, H. E. (1980). *Canadian Journal of Physics*, **58**, 1311.
Mariner, T. & Bleakney, W. (1949). *Review of Scientific Instruments*, **20**, 297.
Matsuda, H. (1976). *Atomic Masses and Fundamental Constants*, vol. 5, eds. Sanders, J. H. & Wapstra, A. H., p. 185. New York & London: Plenum Press.
Matsuda, H., Fukumoto, S., Kuroda, Y. & Nojiri, M. (1966). *Zeitschrift für Naturforschung*, **21a**, 25.
Matsuda, H., Fukumoto, S., Matsuo, T. & Nojiri, M. (1966). *Zeitschrift für Naturforschung*, **21a**, 1304.
Matsuda, H., Matsuo, T. & Takahashi, N. (1977). *International Journal of Mass Spectrometry and Ion Physics*, **25**, 229.

Mattauch, J. H. E. (1936). *Physical Review*, **50**, 617.

Mattauch, J. H. E. (1953). *Mass Spectroscopy in Physics Research*, p. 1. Washington: National Bureau of Standards Circular 522.

Mattauch, J. H. E. & Bieri, R. (1954). *Zeitschrift für Naturforschung*, **9a**, 303.

Mattauch, J. & Herzog, R. (1934). *Zeitschrift für Physik*, **89**, 786.

Monk, G. W., Graves, J. D. & Horton, J. L. (1947). *Review of Scientific Instruments*, **18**, 796.

Moreland, P. E., Rokop, D. J. & Stevens, C. M. (1970). *International Journal of Mass Spectrometry and Ion Physics*, **5**, 127.

Newman, E., Bell, W. A., Davis, W. C., Love, L. O., Prater, W. K., Spainhour, K. A., Tracy, J. G. & Veach, A. M. (1976). *Nuclear Instruments and Methods*, **139**, 87.

Nielsen, K. O. (1970). *Recent Developments in Mass Spectroscopy*, eds. Ogata, K. & Hayakawa, T., p. 506. Baltimore: University Park Press.

Nier, A. O. (1940). *Review of Scientific Instruments*, **11**, 212.

Nier, A. O. (1947). *Review of Scientific Instruments*, **18**, 398.

Nier, A. O. (1956). Private communication to Duckworth, H. E.

Nier, A. O. (1957). *Nuclear Masses and Their Determination*, ed. Hintenberger, H., p. 185. New York, London, Paris: Pergamon Press.

Nier, A. O. (1963). *Advances in Mass Spectrometry*, vol. 2, ed. Elliott, R. M., p. 214. New York: Pergamon Press.

Nier, A. O. & Roberts, T. R. (1951). *Physical Review*, **81**, 507.

Normand, C. E. (1956). *See* Smith (1956), p. 197.

Ogata, K. & Matsuda, H. (1953*a*). *Mass Spectroscopy in Physics Research*, p. 59. Washington: National Bureau of Standards Circular 522.

Ogata, K. & Matsuda, H. (1953*b*). *Physical Review*, **89**, 27.

Ogata, K. & Matsuda, H. (1955). *Zeitschrift für Naturforshung*, **10a**, 843.

Ogata, K. & Matsuda, H. (1957). *Nuclear Masses and Their Determination*, ed. Hintenberger, H., p. 202. New York & London: Pergamon Press.

Quisenberry, K. S., Scolman, T. T. & Nier, A. O. (1956). *Physical Review*, **102**, 1071.

Ravn, H. L., Kugler, E. & Sundell, S. (1981). *Proceedings of the Tenth International Conference on Electromagnetic Separators and Techniques Related to Their Applications, Nuclear Instruments and Methods*, **186.**

Redhead, P. A., Hobson, J. P. & Kernelsen, E. V. (1968). *The Physical Basis of Ultrahigh Vacuum*. London: Chapman and Hall, Ltd.

Ridley, R. G., Munro, R., Young, W. A. P., Hayes, R., Hardy, R. W. D. & Wilson, H. W. (1966). *Advances in Mass Spectrometry*, vol. 3, ed. Mead, W. L., p. 553. London: Applied Science Publishers.

Roberts, R. W. & Vanderslice, T. A. (1963). *Ultrahigh Vacuum and its Application*. Englewood, NJ: Prentice-Hall.

Robinson, N. W. (1968). *The Physical Principles of Ultra-High Vacuum Systems and Equipment*. London: Chapman and Hall, Ltd.

Robinson, C. F. & Hall, L. G. (1956). *Review of Scientific Instruments*, **27**, 504.

Roth, A. (1966). *Vacuum Sealing Techniques*, Oxford & New York: Pergamon Press.

Schmeing, H., Hardy, J. C., Hagberg, E., Perry, W. L., Wills, J. S., Camplan, J. & Rosenbaum, B. (1981). *Nuclear Instruments and Methods*, **186**, 47.

Schnitzer, R. & Anbar, M. (1976). *Proceedings of the 24th Annual Conference on Mass Spectroscopy, ASMS*, p. 361.

Schnitzer, R., Aberth, W. H. & Anbar, M. (1975). *Proceedings of the 23rd Annual Conference on Mass Spectroscopy, ASMS*, p. 479.

Schnitzer, R., Aberth, W. H., Brown, H. L. & Anbar, M. (1974). *Proceedings of the 22nd Annual Conference on Mass Spectroscopy, ASMS*, p. 64.
Shaw, A. E. & Rall, W. (1947). *Review of Scientific Instruments*, **18**, 278.
Sheffield, J. C. & White, F. A. (1958). *Applied Spectroscopy*, **12**, 12.
Smith, M. L. (1956). *Electromagnetically Enriched Isotopes and Mass Spectrometry*. London: Butterworths.
Stevens, C. M. (1963). *Advances in Mass Spectrometry*, vol. 2, ed. Elliott, R. M., p. 215. New York: Pergamon Press.
Stevens, C. M. & Moreland, P. E. (1967). *Proceedings of the Third International Conference on Atomic Masses*, ed. Barber, R. C., p. 673. Winnipeg: University of Manitoba Press.
Stevens, C. M., Terandy, J., Lobell, G., Wolfe, J., Beyer, N. & Lewis, R. (1960). *Proceedings of the International Conference on Nuclidic Masses*, ed. Duckworth, H. E., p. 403. Toronto: University of Toronto Press.
Stevens, C. M., Terandy, J., Lobell, G., Wolfe, J., Lewis, R. & Beyer, N. (1963). *Advances in Mass Spectrometry*, vol. 2, ed. Elliott, R. M., p. 198. New York: Pergamon Press.
Taya, S., Tsuyama, H., Kanomata, I., Noda, T. & Matsuda, H. (1978). *International Journal of Mass Spectrometry and Ion Physics*, **26**, 77.
Voorhies, H. G., Robinson, C. F., Hall, L. G., Brubaker, W. M. & Berry, C. E. (1959). *Advances in Mass Spectrometry*, ed. Waldron, J. D., p. 44. London: Pergamon Press.
Wagner, H. & Walcher, H. (1970). *International Conference on Electromagnetic Isotope Separators and the Techniques of Their Applications*, Marburg: Report BMBW–FB K70–28.
Walker, W. H. & Thode, H. G. (1953). *Physical Review*, **90**, 447.
White, F. A. & Collins, T. L. (1954). *Applied Spectroscopy*, **8**, 169.
White, F. A. & Forman, L. (1967). *Review of Scientific Instruments*, **38**, 355.
White, F. A., Sheffield, J. C. & Rourke, F. M. (1958). *Applied Spectroscopy*, **12**, 46.
Wilson, H. W. (1963). *Advances in Mass Spectrometry*, vol. 2, ed. Elliott, R. M., p. 206. New York: Pergamon Press.
Wilson, H. W., Munro, R., Hardy, R. W. D. & Daly, N. R. (1961). *Nuclear Instruments and Methods*, **13**, 269.

Chapter 6

Allermann, M., Kellerhals, H. & Wanczek, K. P. (1983). *International Journal of Mass Spectrometry and Ion Physics*, **46**, 139.
Alvarez, L. W. & Cornog, R. (1939). *Physical Review*, **56**, 379, 613.
Austin, W. E., Holme, A. E. & Leck, J. H. (1976). *Quadrupole Mass Spectrometry and its Applications*, ed. Dawson, P. H., p. 121. Amsterdam, Oxford, New York: Elsevier.
Baldeschwieler, J. D., Benz, H. & Llewellyn, P. M. (1968). *Advances in Mass Spectrometry*, vol. 4, ed. Kendrick, E., p. 113. London: Institute of Petroleum.
Bennett, W. H. (1948). *Physical Review*, **74**, 1222.
Bennett, W. H. (1950). *Journal of Applied Physics*, **21**, 143.
Bennett, W. H. (1953). *Mass Spectroscopy in Physics Research*, p. 111. Washington: National Bureau of Standards Circular 522.
Bloch, F. & Jeffries, C. D. (1950). *Physical Review*, **80**, 305.
Boyd, R. L. F. & Morris, D. (1955). *Proceedings of the Physics Society A*, **68**, 1.
Boyne, H. S. & Franklin, P. A. (1961). *Physical Review*, **123**, 242.

Brubaker, W. M. (1968). *Advances in Mass Spectrometry*, vol. 4, ed. Kendrick, E., p. 293. London: Institute of Petroleum.

Cameron, A. E. & Eggers, D. F. (1948). *Review of Scientific Instruments*, **19**, 605.

Chait, B. T. & Standing, K. G. (1981). *International Journal of Mass Spectrometry and Ion Physics*, **40**, 185.

Comisarow, M. B. (1978*a*). *Journal of Chemical Physics*, **69**, 4097.

Comisarow, M. B. (1978*b*). *Advances in Mass Spectrometry*, vol. 7B, ed. Daly, N. R., p. 1042. London: Heyden and Son.

Comisarow, M. B. & Marshall, A. G. (1976). *Journal of Chemical Physics*, **64**, 110.

Comisarow, M. B. & Marshall, A. G. (1980). *Advances in Mass Spectrometry*, vol. 8B, ed. Quayle, A., p. 1698. London: Heyden and Son.

Dawson, P. H. (1976*a*). *Quadrupole Mass Spectrometry and its Applications.* ed. Dawson, P. H., p. 9. Amsterdam, Oxford, New York: Elsevier.

Dawson, P. H. (1976*b*). *International Journal of Mass Spectrometry and Ion Physics*, **21**, 317.

Dawson, P. H. & Whetten, N. R. (1969). *Advances in Electronics and Electron Physics*, **27**, 60.

Dekleva, J. & Peterlin, A. (1955). *Review of Scientific Instruments*, **26**, 399.

Fischer, E. (1959). *Zeitschrift für Physik*, **156**, 26.

Fox, R. E., Hickam, W. M., Kjeldaas, T. & Grove, D. J. (1951). *Physical Review*, **84**, 859.

Glenn, W. E. (1952). *See* Inghram & Hayden (1954).

Goudsmit, S. A. (1948). *Physical Review*, **74**, 1537.

Halsted, J. B. (1974). *Advances in Mass Spectrometry*, vol. 6, ed. West, A. R., p. 901. Barking UK: Applied Science Publishers.

Hays, E. E., Richards, P. I. & Goudsmit, S. A. (1951). *Physical Review*, **84**, 824.

Herzog, R. F. (1976). *Quadrupole Mass Spectrometry*, ed. Dawson, P. H., p. 153. Amsterdam, Oxford, New York: Elsevier.

Hintenberger, H. & Mattauch, J. (1937). *Zeitschrift für Physik*, **106**, 279.

Hipple, J. A., Sommer, H. & Thomas, H. A. (1949). *Physical Review*, **76**, 1877.

Hipple, J. A., Sommer, H. & Thomas, H. A. (1950). *Physical Review*, **78**, 332.

Hipple, J. A. & Sommer, H. (1953). *Mass Spectroscopy in Physics Research*, National Bureau of Standards Circular 522, p. 123.

Inghram, M. G. & Hayden, R. J. (1954). *Handbook on Mass Spectroscopy*, Nuclear Series, Report No. 14 (NRC–USA).

Jeffries, C. D. (1951). *Physical Review*, **81**, 1040.

Katzenstein, H. S. & Friedland, S. S. (1955). *Review of Scientific Instruments*, **26**, 324.

Kerr, L. W. (1956). *Journal of Electronics*, **2**, 179.

Koets, E. (1976). *Atomic Masses and Fundamental Constants*, vol. 5, eds. Sanders J. H. & Wapstra, A. H., p. 164. New York and London: Plenum.

Koets, E., Kramer, J., Nonhebel, J. & Le Poole, J. B. (1980). *Atomic Masses and Fundamental Constants*, vol. 6, eds. Nolen, J. A. & Benenson, W., p. 275. New York & London: Plenum.

Macfarlane, R. D. & Torgerson, D. F. (1976*a*). *Science*, **191**, 920.

Macfarlane, R. D. & Torgerson, D. F. (1976*b*). *International Journal of Mass Spectrometry and Ion Physics*, **21**, 81.

Mamyrin, B. A., Alekseyenko, S. A., Aruyev, N. N., Ogurtsova, N. A. & Ioffe, A. F. (1975). *Atomic Masses and Fundamental Constants*, vol. 5, eds. Sanders, J. H. & Wapstra, A. H., p. 526. New York & London: Plenum.

Mamyrin, B. A., Karataeu, V. I., Smikk, D. J. & Zagulin, V. A. (1973). *Soviet Physics*

Journal of Experimental and Theoretical Physics, **31**, 1.
Mamyrin, B. A. & Smikk, D. J. (1979). *Soviet Physics Journal of Experimental and Theoretical Physics*, **49**, 762.
Mather, P. E., Lawson, G., Todd, J. & Baker, J. M. B. (1978). *International Journal of Mass Spectrometry and Ion Physics*, **28**, 347.
Matsuda, H. & Matsuo, T. (1977). *International Journal of Mass Spectrometry and Ion Physics*, **24**, 107.
Muller, R. A. (1977). *Science*, **196**, 489.
Oakey, N. S. & Macfarlane, R. D. (1967). *Nuclear Instruments and Methods*, **49**, 220.
Paul, W., Reinhard, H. P. & von Zahn, U. (1958). *Zeitschrift für Physik*, **152**, 143.
Paul, W. & Steinwedel, H. (1953). *Zeitschrift für Naturforschung*, **8a**, 448.
Peterlin, A. (1955). *Review of Scientific Instruments*, **26**, 398.
Petley, B. W. & Morris, K. (1968). *Journal of Physics E*, **1**, 417.
Redhead, P. A. (1952). *Canadian Journal of Physics*, **30**, 1.
Redhead, P. A. & Crowell, C. R. (1953). *Journal of Applied Physics*, **24**, 331.
Richards, J. A., Huey, R. M. & Hiller, J. (1973). *International Journal of Mass Spectrometry and Ion Physics*, **12**, 317.
Rothstein, S. M. (1978). *Advances in Mass Spectrometry*, vol. 7B, ed. Daly, N. R., p. 913. London: Heyden and Son.
Smith, L. G. (1951*a*). *Physical Review*, **81**, 295.
Smith, L. G. (1951*b*). *Review of Scientific Instruments*, **22**, 115.
Smith, L. G. (1952). *Physical Review*, **85**, 767.
Smith, L. G. (1953). *Mass Spectroscopy in Physics Research*, National Bureau of Standards Circular 522, p. 117.
Smith, L. G. (1958). *Physical Review*, **111**, 1606.
Smith, L. G. (1960). *Proceedings of the International Conference on Nuclidic Masses*, ed. Duckworth, H. E., p. 418. Toronto: University of Toronto Press.
Smith, L. G. (1967). *Proceedings of the Third International Conference on Atomic Masses*, ed. Barber, R. C., p. 811. Winnipeg: University of Manitoba Press.
Smith, L. G. (1971). *Physical Review*, **C4**, 22.
Smith, L. G. & Damm, C. C. (1956). *Review of Scientific Instruments*, **27**, 638.
Smythe, W. R. (1926). *Physical Review*, **28**, 1275.
Smythe, W. R. (1934). *Physical Review*, **45**, 299.
Smythe, W. R. & Mattauch, J. (1932). *Physical Review*, **40**, 429.
Sommer, H., Thomas, H. A. & Hipple, J. A. (1951). *Physical Review*, **82**, 697.
Stephens, W. E. (1946). *Physical Review*, **69**, 691.
Todd, J. F. J., Lawson, G. & Bonner, R. F. (1976). *Quadrupole Mass Spectrometry*, ed. Dawson, P. H., p. 222. Amsterdam, Oxford, New York: Elsevier.
Townsend, J. W. (1952). *Review of Scientific Instruments*, **23**, 538.
von Zahn, U. (1963). *Review of Scientific Instruments*, **34**, 1.
Wiechert, E. (1899). *Annalen der Physik* (Leipzig), **69**, 739.
Wilson, R. R. (1952). *See* Inghram & Hayden (1954).
Wiley, W. C. & McLaren, I. H. (1955). *Review of Scientific Instruments*, **26**, 1150.
Wolff, M. M. & Stephens, W. E. (1953). *Review of Scientific Instruments*, **24**, 616.

Chapter 7

Aston, F. W. (1942). *Mass Spectra and Isotopes*. London: Edward Arnold and Co.

Bainbridge, K. T. (1953). *Experimental Nuclear Physics*, vol. 1, ed. Segré E., New York: Wiley.

Bainbridge, K. T. (1947). *Solvay Report*, 7th Congress in Chemistry. Brussels: R. Stoops.

Bainbridge, K. T. & Ford, G. C. (1953). *Experimental Nuclear Physics*, vol. 1, ed. Segré, E., p. 623. New York: Wiley.

Bainbridge, K. T. & Nier, A. O. (1950). *Relative Isotopic Abundances of the Elements*, Nuclear Science Series, Preliminary Report no. 9, NRC–USA.

Bennett, C. L., Beukens, R. P., Clover, M. R., Gove, H. E., Liebert, R. B., Litherland, A. E., Purser, K. H. & Sondheim, W. E. (1977). *Science*, **198**, 508.

Bleakney, W. (1936). *American Physics Teacher*, **4**, 12.

Chait, E. M. & Hull, C. W. (1975). *American Laboratory*, **7**, 47.

Commission on Atomic Weights (1984*a*). *Pure and Applied Chemistry*, **56**, 653.

Commission on Atomic Weights (1984*b*). *Pure and Applied Chemistry*, **56**, 675.

de Bièvre, P. (1978). *Advances in Mass Spectrometry*, vol. 7A, ed. Daly, N. R., p. 395. London: Heyden and Son.

Duckworth, H. E. & Hogg, B. G. (1947). *Physical Review*, **71**, 212.

Eberhardt, A., Delwiche, R. & Geiss, J. (1964). *Zeitschrift für Naturforschung*, **19A**, 736.

Ewald, H. (1944). *Zeitschrift für Physik*, **122**, 487.

Garner, E. L., Machlan, L. A. & Shields, W. R. (1971). *National Bureau of Standards Special Publication*, pp. 260–77, Washington.

Halsted, R. E. & Nier, A. O. (1950). *Review of Scientific Instruments*, **21**, 1019.

Honig, R. E. (1945). *Journal of Applied Physics*, **16**, 646.

Inghram, M. G. (1948). *Advances in Electronics*, **1**, 219.

Inghram, M. G. (1954). *Annual Review of Nuclear Science*, **4**, 81.

Inghram, M. G. & Hayden, R. J. (1954). *A Handbook on Mass Spectroscopy*, Nuclear Science Series, Report no. 14 (NRC – USA).

Jackson, M. C. & Young, W. A. P. (1972). *Review of Scientific Instruments*, **44**, 32.

Kanno, H. (1971). *Bulletin of the Chemical Society of Japan*, **44**, 1808.

Litherland, A. E., Beukens, R. P., Kilius, L. R., Rucklidge, J. C., Gove, H. E., Elmore, D. & Purser, K. H. (1981). *Nuclear Instruments and Methods*, **186**, 463.

Mattauch, J. (1951). *Zeitschrift für Naturforschung*, **6a**, 391.

Mattauch, J. (1943). *Naturwissenschaften*, **31**, 487.

Mattauch, J. & Ewald, H. (1944). *Zeitschrift für Physik*, **122**, 314.

Mattauch, J. & Scheld, H. (1948). *Zeitschrift für Naturforschung*, **3a**, 105.

McKinney, C. R., McCrea, J. M., Epstein, S., Allen, H. A. & Urey, H. C. (1950). *Review of Scientific Instruments*, **21**, 724.

Moore, L. J., Heald, E.F. & Filliben, J. J. (1978). *Advances in Mass Spectrometry*, vol. 7A, ed. Daly, N. R., p. 448. London: Heyden and Son.

Muller, R. A. (1977). *Science*, **196**, 489.

Nelson, D. E., Korteling, R. G. & Stott, W. R. (1977). *Science*, **198**, 508.

Nguyen, L-D. & Goby, G. (1978). *Advances in Mass Spectrometry*, vol. 7A, ed. Daly, N. R., p. 486. London: Heyden and Son.

Nielsen, H. (1968). *Advances in Mass Spectrometry*, vol. 4, ed. Kendrick, E., p. 267. London: Institute of Petroleum.

Nier, A. O. (1950*a*). *Physical Review*, **77**, 789.

Nier, A. O. (1950*b*). *Physical Review*, **79**, 450.

Nier, A. O., Ney, E. P. & Inghram, M. G. (1947). *Review of Scientific Instruments*, **18**, 294.

Petruska, J. A., Thode, H. G. & Tomlinson, R. H. (1955). *Canadian Journal of Physics*, **33**, 693.

Purser, K. H., Williams, P., Litherland, A. E., Stein, J. D., Storms, H. A., Gove, H. E. & Stevens, C. M. (1981). *Nuclear Instruments and Methods*, **186**, 487.

Reynolds, R. J. (1950). *Physical Review*, **79**, 789.

Reynolds, R. J. (1956). *Review of Scientific Instruments*, **27**, 928.

Straus, H. A. (1941). *Physical Review*, **59**, 430.

Swann, W. F. G. (1931). *Journal of the Franklin Institute*, **212**, 439.

White, F. A., Collins, T. L. & Rourke, F. M. (1955). *Physical Review*, **97**, 566.

White, F. A., Collins, T. L. & Rourke, F. M. (1956). *Physical Review*, **101**, 1786.

Chapter 8

Aston F. W. (1942). *Mass Spectra and Isotopes*, 2nd edn. London: Edward Arnold and Company.

Audi, G., Epherre, M., Thibault, C., Klapisch, R., Huber, G., Touchard, F. & Wollnik, H. (1980). *Atomic Masses and Fundamental Constants*, vol. 6, eds. Nolen, J. A. & Benenson, W., p. 281. New York & London: Plenum.

Bainbridge, K. T. & Dewdney, J. W. (1967). *Proceedings of the Third International Conference on Atomic Masses*, ed. Barber, R. C., p. 758. Winnipeg: University of Manitoba Press.

Barber, R.C., Barnard, J. W., Haque, S. S., Kozier, K. S., Meredith, J. O., Sharma, K. S., Southon, F. C. G., Williams, P. & Duckworth, H. E. (1976). *Atomic Masses and Fundamental Constants*, vol. 5, eds. Sanders, J. H. & Wapstra, A. H., p. 170. New York & London: Plenum Press.

Barber, R. C., Bishop, R. L., Cambey, L. A., Duckworth, H. E., Macdougall, J. D., McLatchie, W., Ormrod, J. H. & van Rookhuyzen, P. (1964). *Proceedings of the Second International Conference on Nuclidic Masses*, ed. Johnson, W. H. Jr, p. 393. Vienna: Springer-Verlag.

Barber, R. C., Bishop, R. L., Duckworth, H. E., Meredith, J. O., Southon, F. C. G., van Rookhuyzen, P. & Williams, P. (1971). *Review of Scientific Instruments*, **42**, 1.

Barber, R. C., Meredith, J. O., Southon, F. C. G., Williams, P., Barnard, J. W., Sharma, K. S. & Duckworth, H. E. (1973). *Physical Review Letters*, **31**, 728.

Barber, R. C., McLatchie, W., Bishop, R. L., van Rookhuyzen, P. & Duckworth, H. E. (1963). *Canadian Journal of Physics*, **41**, 1482.

Bearden, J. A. & Thomsen, J. S. (1957). *Il Nuovo Cimento Supplement*, **5**, 267.

Benson, J. L. & Johnson, W. H. Jr (1966). *Physical Review*, **141**, 1112.

Bergkvist, K.-E. (1973). *Nuclear Physics*, **B39**, 317.

Bernas, R. (1970). *Recent Developments in Mass Spectrometry*, eds. Ogata, K. & Hayakawa, T., p. 535. Tokyo: University of Tokyo Press.

Blair, J. M., Halverson, J., Johnson, W. H. Jr & Smith, R. (1980). *Atomic Masses and Fundamental Constants*, vol. 6, eds. Nolen, J. A. & Benenson, W., p. 267. New York & London: Plenum.

Bleakney, W. (1936). *American Physics Teacher*, **4**, 12.

Cohen, E. R. & Wapstra, A. H. (1983). *Nuclear Instruments and Methods*, **211**, 153.

Collins, T. L., Nier, A. O. & Johnson, W. H. Jr (1952). *Physical Review*, **86**, 408.

Dempster, A. J. (1938). *Physical Review*, **53**, 64, 869.

Demirkhanov, R. A., Dorokhov, V. V. & Dzkuya, M. I. (1972). *Atomic Masses and

Fundamental Constants, vol. 4, eds. Sanders, J. H. & Wapstra, A. H., London, New York: Plenum Press.
Duckworth, H. E. (1958). *Mass Spectroscopy*. Cambridge University Press.
Duckworth, H. E. & Johnson, H. A. (1950). *Physical Review*, **78**, 179.
Duckworth, H. E., Kegley, C. L., Olson, J. M. & Stanford, G. S. (1951). *Physical Review*, **83**, 1114.
Duckworth, H. E. & Preston, R. S. (1951). *Physical Review*, **82**, 468.
Duckworth, H. E., Woodcock, K. S. & Preston, R. S. (1950). *Physical Review*, **79**, 198.
Elsasser, W. (1934). *Journal de Physique, Paris*, **5**, 389, 625.
Epherre, M., Audi, G., Thibault, C., Klapisch, R., Huber, G., Touchard, F. & Wollnik, H. (1979). *Physical Review*, **C19**, 1504.
Epherre, M., Audi, G., Thibault, C. & Klapisch, R. (1980). *Atomic Masses and Fundamental Constants*, vol. 6, eds. Nolen, J. A. & Benenson, W., p. 299. New York & London: Plenum.
Everling, F. (1957). *Nuclear Masses and Their Determination*, ed. Hintenberger, H., p. 253. New York, London, Paris: Pergamon Press.
Ewald, H. (1953). *Mass Spectroscopy in Physics Research*, p. 37. Washington: US National Bureau of Standards Circular 522.
Feenberg, E. (1947). *Reviews of Modern Physics*, **19**, 239.
Feenberg, E. & Hammack, K. C. (1949). *Physical Review*, **75**, 1877.
Giese, C. F. & Collins, T. L. (1954). *Physical Review*, **96**, 833A.
Greenwood, R. C., Helmer, R. G., Gehrke, R. J. & Chrien, R. E. (1980). *Atomic Masses and Fundamental Constants*, vol. 6, eds. Nolen, J. A. & Benenson, W., p. 219. New York & London: Plenum Press.
Hansen, P. G. & Nielsen, O. B. (1981). *Proceedings, 4th International Conference on Nuclei Far From Stability*. ed. Geneva: CERN Report 81–09.
Haxel, O., Jensen, J. H. D. & Suess, H. E. (1949). *Physical Review*, **75**, 1766.
Helmer, R. G., Greenwood, R. C. & Gehrke, R. J. (1976). *Atomic Masses and Fundamental Constants*, vol. 5, eds. Sanders, J. H. & Wapstra, A. H., p. 30. New York & London: Plenum.
Helmer, R. G., van Assche, P. H. M. & van der Leun, C. (1979). *Atomic Data and Nuclear Data Tables*, **24**, 39.
Hogg, B. G. & Duckworth, H. E. (1954). *Canadian Journal of Physics*, **32**, 65.
Kayser, D. C., Britten, R. A. & Johnson, W. H. (1972). *Atomic Masses and Fundamental Constants*, vol. 4, eds. Sanders, J. H. & Wapstra, A. H., p. 172. New York & London: Plenum Press.
Kayser, D. C., Halverson, J. & Johnson, W. H. Jr (1976). *Atomic Masses and Fundamental Constants*, vol. 5, eds. Sanders, J. H. & Wapstra, A. H., p. 178. New York & London: Plenum Press.
Kerr, D. P. & Bainbridge, K. T. (1970). *Recent Developments in Mass Spectroscopy*, eds. Ogata, K. & Hayakawa, T., p. 490. Tokyo: University of Tokyo Press.
Klapisch, R. (1976). *Proceedings, 3rd International Conference on Nuclei Far From Stability*, ed. Geneva: CERN Report 76–13.
Klepper, O. (1984). ed. *Atomic Masses and Fundamental Constants*, vol. 7. T. H. D. Darmstadt.
Koets, E. (1976). *Atomic Masses and Fundamental Constants*, vol. 5, eds. Sanders, J. H. & Wapstra, A. H., p. 164. New York & London: Plenum Press.
Koets, E., Kramer, J., Nonhebel, J. & LePoole, J. B. (1980). *Atomic Masses and*

Fundamental Constants, vol. 6, eds. Nolen, J. A. & Benenson, W., p. 275. New York & London: Plenum Press.

Kohman, T. P., Mattauch, J. H. E. & Wapstra, A. H. (1958). *Science*, **127**, 1431.

Kozier, K. S. (1979). Ph.D. Thesis, University of Manitoba, Winnipeg, Manitoba, Canada, Unpublished.

Kozier, K. S., Sharma, K. S., Barber, R. C., Barnard, J. W., Ellis, R. J., Derenchuk, V. P. & Duckworth, H. E. (1979). *Canadian Journal of Physics*, **57**, 266.

Kozier, K. S., Sharma, K. S., Barber, R. C., Barnard, J. W., Ellis, R. J., Derenchuk, V. P. & Duckworth, H. E. (1980). *Canadian Journal of Physics*, **58**, 1311.

Margenau, H. (1934). *Physical Review*, **46**, 613.

Matsuda, H., Fukumoto, S. & Matsuo, T. (1967). *Proceedings of the Third International Conference on Atomic Masses*, ed. Barber, R. C., p. 733. Winnipeg: University of Manitoba Press.

Mattauch, J. (1960), *Proceedings of the International Conference on Atomic Masses*, ed. Duckworth, H. E., p. 3. Toronto: University of Toronto Press.

Mayer, M. G. (1948). *Physical Review*, **74**, 235.

Mayer, M. G. (1949). *Physical Review*, **75**, 1969.

Mayer, M. G. (1950). *Physical Review*, **78**, 17.

Meredith, J. O., Southon, F. C. G., Barber, R. C., Williams, P. & Duckworth, H. E. (1972). *International Journal of Mass Spectrometry and Ion Physics*, **10**, 359.

Moreland, P. E. & Bainbridge, K. T. (1964). *Proceedings of the Second International Conference on Nuclidic Masses*, ed. Johnson, W. H. Jr, p. 423. Vienna: Springer Verlag.

Nakabushi, H., Katakuse, I. & Ogata, K. (1970). *Recent Developments in Mass Spectroscopy*, eds. Ogata, K. & Hayakawa, T., p. 482. Tokyo: University of Tokyo Press.

Nier, A. O. (1950). *Physical Review*, **77**, 789.

Nier, A. O. (1953). *Mass Spectroscopy in Physics Research*, p. 29. Washington: US National Bureau of Standards Circular 522.

Nier, A. O. (1957). *Nuclear Masses and Their Determination*, ed. Hintenberger, H., p. 185. New York, London, Paris: Pergamon Press.

Nolen, J. A. & Benenson, W. (1980). eds. *Atomic Masses and Fundamental Constants*, vol. 6, ed. New York & London: Plenum.

Nordheim, L. W. (1949). *Physical Review*, **75**, 1894.

Ogata, K., Matsumoto, S., Nakabushi, H. & Katakuse, I. (1967). *Proceedings of the Third International Conference on Atomic Masses*, ed. Barber, R. C., p. 748. Winnipeg: University of Manitoba Press.

Ogle, W., Wahlborn, S., Piepenbring, R. & Fredriksson, S. (1972). *Reviews of Modern Physics*, **43**, 424.

Petit-Clerc, Y. & Carette, J. D. (1968). *Vacuum*, **18**, 7.

Petit-Clerc, Y. & Carette, J. D. (1970). *Abstract of the 38th ACFAS Congress*, Quebec, Canada.

Quisenberry, K. S., Scolman, T. T. & Nier, A. O. (1956). *Physical Review*, **102**, 1071.

Rudstam, G. (1976). *Nuclear Instruments and Methods*, **139**, 239.

Rytz, A. Greenberg, B. & Gorman, D. J. (1972). *Atomic Masses and Fundamental Constants*, vol. 4, eds. Sanders, J. H. & Wapstra, A. H., p. 1. London & New York: Plenum.

Sanders, J. H. & Wapstra, A. H. (1972). *Atomic Masses and Fundamental Constants*, vol. 4, ed. London, New York: Plenum.

Sanders, J. H. & Wapstra, A. H. (1976). *Atomic Masses and Fundamental Constants*, vol. 5, ed. New York, London: Plenum.
Sharma, K. S., Kozier, K. S., Barnard, J. W., Barber, R. C., Haque, S. S. & Duckworth, H. E. (1977). *Canadian Journal of Physics*, **55**, 506.
Smith, L. G. (1960). *Proceedings of the International Conference on Nuclidic Masses*, ed. Duckworth, H. E., p. 418. Toronto: University of Toronto Press.
Smith, L. G. (1972). *Atomic Masses and Fundamental Constants*, vol. 4, eds. Sanders, J. H. & Wapstra, A. H., p. 164, London & New York: Plenum Press.
Smith, L. G. & Damm, C. C. (1953a). *Physical Review*, **91**, 481A.
Smith, L. G. & Damm, C. C. (1953b). *Physical Review*, **90**, 324.
Smith, L. G. & Damm, C. C. (1956). *Review of Scientific Instruments*, **727**, 638.
Smith, L. G. & Wapstra, A. H. (1975). *Physical Review*, **11C**, 1392.
Southon, F. C. G. (1973). Ph.D. Thesis, University of Manitoba, Winnipeg, Manitoba, Canada, Unpublished.
Southon, F. C. G., Meredith, J. O., Barber, R. C. & Duckworth, H. E. (1977). *Canadian Journal of Physics*, **55**, 383.
Stevens, C. M. & Moreland, P. E. (1967). *Proceedings of the Third International Conference on Atomic Masses*, ed. Barber, R. C., p. 673. Winnipeg: University of Manitoba Press.
Stevens, C. M. & Moreland, P. E. (1970). *Recent Developments in Mass Spectroscopy*, eds. Ogata, K. & Hayakawa, T., p. 1296. Tokyo: University of Tokyo Press.
Stewart, R. L. (1934). *Physical Review*, **45**, 488.
Swann, W. F. G. (1931). *Journal of the Franklin Institute*, **212**, 439.
Tal'rose (1985). *Tenth International Mass Spectrometry Conference*, Swansea.
Taylor, B. N., Parker, W. H. & Langenberg, D. N. (1969). *The Fundamental Constants and Quantum Electrodynamics*. New York & London: Academic Press.
Thibault, C., Epherre, M., Audi, G., Klapisch, R., Huber, G., Touchard, F. & Wollnik, H. (1980). *Atomic Masses and Fundamental Constants*, vol, 6, eds. Nolen, J. A. & Benenson, W., p. 291. New York & London: Plenum.
Thibault, C., Klapisch, R., Rigaud, C., Poskanzer, A. M., Prieels, R., Lessard, L. & Reisdorf, W. (1975). *Physical Review*, **C12**, 644.
Wapstra, A. H. & Gove, N. B. (1971). *Nuclear Data Tables*, **9**, 267.
Wapstra, A. H. & Bos, K. (1977*a*). *Atomic Data and Nuclear Data Tables*, **19**, 177.
Wapstra, A. H. & Bos, K. (1977*b*). *Atomic Data and Nuclear Data Tables*, **20**, 1.
Wichers, E. (1956). *Journal of the American Chemical Society*, **78**, 3235.
Williams, P. & Duckworth, H. E. (1972). *Scientific Progress*, **60**, 319.

Chapter 9

Armbruster, P., Asghar, M., Bocquet, J. P., Decker, R., Ewald, H., Greif, J., Moll, E., Pfeiffer, B., Schrader, H., Schussler, F., Siegert, G. & Wollnik, H. (1976). *Nuclear Instruments and Methods*, **139**, 213.
Arrol, W. J., Chackett, K. F. & Epstein, S. (1947). *Proceedings of the Conference on Nuclear Chemistry* (McMaster University), p. 108.
Bodu, R. Bouziques, H. Morin, N. & Pfiffelmann, J.-P. (1972). *Comptes Rendus de l'Académie des Sciences*, **275D**, 1731.
Bogdanov, D. D., Demyanov, A. V., Karnankhov, V. A., Petrov, L. A., Plohocki, A., Subbotin, V. G. & Voboril, J. (1976). *Proceedings of the Third International Conference on Nuclei Far from Stability*, CERN Report 76–13, p. 299. Geneva.

Bohr, N. & Wheeler, J. A. (1939). *Physical Review*, **56**, 426.
Borg, S., Bergstrom, I., Holm, G. B., Rydberg, B., DeGeer, L.-E., Rudstam, G., Grapengiesser, B., Lund, E. & Westgaard, L. (1971). *Nuclear Instruments and Methods*, **91**, 1.
Burkard, K. H. & Roeckl, E. (1976). *Nuclear Instruments and Methods*, **133**, 187.
Cohen, B. L., Cohen, A. F. & Coley, C. D. (1956). *Physical Review*, **104**, 1046.
Crocker, I. H., Werner, R. D. & Cherrin, W. (1968). *Advances in Mass Spectrometry*, vol. 4, ed. Kendrick, E., p. 955. London: Institute of Petroleum.
Crouch, E. A. C. (1977). *Atomic Data and Nuclear Data Tables*, **19**, 419.
de Bièvre, P. (1978). *Advances in Mass Spectrometry*, vol. 7A, ed. Daly, N. R., p. 395. London: Heyden and Son.
Dempster, A. J. (1947). *Physical Review*, **71**, 829.
Dempster, A. J. & Wilkins, T. R. (1938). *Physical Review*, **54**, 315.
Devillers, C., Lecompte, T., Lucas, M. & Hagemann, R. (1978). *Advances in Mass Spectrometry*, vol. 7A, ed. Daly, N. R., p. 553. London: Heyden and Son.
Ellis, R. J., Hall, B. J., Dyck, G. R., Lander, C. A., Sharma, K. S., Barber, R. C. & Duckworth, H. E. (1984). *Physics Letters*, **136B**, 146.
Ewald, H., Konecny, E., Opower, H. & Rosler, H. (1964). *Zeitschrift für Naturforschung*, **19a**, 194.
Flammersfeld, A. & Mattauch, J. (1943). *Naturwissenschaften*, **31**, 66.
Fleming, E. H., Ghiorso, A. & Cunningham, B. B. (1951). *Physical Review*, **82**, 967.
Geiger, J. S., Hogg, B. G., Duckworth, H. E. & Dewdney, J. W. (1953). *Physical Review*, **89**, 621.
Gill, R. L., Stelts, M. L., Chrien, R. E., Manzella, V. & Lion, H. (1981). *Nuclear Instruments and Methods*, **186**, 243.
Greth, W. E., Gangadharan, S. & Wolke, R. I. (1970). *Journal of Inorganic and Nuclear Chemistry*, **32**, 2113.
Hahn, O., Strassmann, F. & Walling, E. (1937). *Naturwissenschaften*, **25**, 189.
Hahn, O. & Strassmann, F. (1939). *Naturwissenschaften*, **27**, 11, 89.
Halsted, R. E. (1952). *Physical Review*, **88**, 666.
Hansen, P. G. (1979). *Annual Reviews of Nuclear and Particle Physics*, **29**, 69.
Hansen, P. G. & Nielsen, O. B. (1981). *Proceedings of the 4th International Conference on Nuclei Far From Stability*, CERN Report 81–09: Geneva.
Haxton, W. C., Stephenson, G. J. & Strottman, D. (1918). *Physical Review Letters*, **47**, 153.
Hayden, R. J. & Inghram, M. G. (1953). *Mass Spectroscopy in Physics Research*, p. 85. Washington: US National Bureau of Standards Circular 522.
Hayden, R. J., Reynolds, J. H. & Inghram, M. G. (1949). *Physical Review*, **75**, 1500.
Hess, D. C. & Inghram, M. G. (1949). *Physical Review*, **76**, 1717.
Hintenberger, H., Herr, W. & Voshage, H. (1954). *Physical Review*, **95**, 1690.
Inghram, M. G., Hayden, R. J. & Hess, D. C. (1947). *Physical Review*, **72**, 349, 967.
Johnson, W. H. Jr & Nier, A. O. (1957). *Physical Review*, **105**, 1014.
Kingdon, K. H., Pollock, H. C., Booth, E. T. & Dunning, J. R. (1940). *Physical Review*, **57**, 749.
Kirsten, T., Schaeffer, O., Norton, E. & Stoenner, R. W. (1968). *Physical Review Letters*, **20**, 1300.
Klapisch, R. (1976). *Proceedings of the 3rd International Conference on Nuclei Far From Stability*, CERN Report 76–13: Geneva.

Klapisch, R., Gradsztajn, E., Yion, F., Epherre, M. & Bernas, R. (1966). *Advances in Mass Spectrometry*, vol. 3, ed. Mead, W. L., p. 547. London: The Institute of Petroleum.

Kofoed-Hansen, O. & Nielsen, K. O. (1951). *Physical Review*, **82**, 96.

Lapp, R. E., Van Horn, J. R. & Dempster, A. J. (1947). *Physical Review*, **71**, 745.

Leland, W. T. (1949). *Physical Review*, **76**, 1722.

McLatchie, W., Barber, R. C., Duckworth, H. E. & van Rookhuyzen, P. (1964). *Physics Letters*, **10**, 330.

McConnell, J. R. & Talbert, W. L. Jr. (1975). *Nuclear Instruments and Methods*, **128**, 227.

McMullen, C. C., Fritze, K. & Tomlinson, R. H. (1966). *Canadian Journal of Physics*, **44**, 3033.

Martell, E. A. & Libby, W. F. (1950). *Physical Review*, **80**, 977.

Mattauch, J. (1934). *Zeitschrift für Physik*, **91**, 361.

Mattauch, J. (1937). *Naturwissenchaften*, **25**, 189.

Mattauch, J. & Lichtblau, H. (1939). *Zeitschrift für Physik*, **111**, 514.

Meitner, L. (1926). *Naturwissenschaften*, **14**, 719.

Mulholland, G. I. & Kohman, T. P. (1952). *Physical Review*, **87**, 681.

Neuilly, M., Bussac, J., Frejacques, C., Nief, G., Vendryes, G. & Yvon, J. (1972). *Comptes Rendus de l'Académie des Sciences*, **275D**, 1847.

Nier, A. O. (1935). *Physical Review*, **48**, 283.

Nier, A. O. (1939). *Physical Review*, **55**, 150.

Nier, A. O., Booth, E. T., Dunning, J. R. & Grosse, A. V. (1940). *Physical Review*, **57**, 546, 748.

Ogata, K., Okano, J. & Takaoka, N. (1966). *Advances in Mass Spectrometry*, vol. 3, ed. Mead, W. L., p. 603. London: Institute of Petroleum.

Petruska, J. A., Melaika, E. A. & Tomlinson, R. H. (1955). *Canadian Journal of Physics*, **33**, 640.

Petruska, J. A., Thode, H. G. & Tomlinson, R. H. (1955). *Canadian Journal of Physics*, **33**, 693.

Plutonium Project Report. (1946). *Journal of the American Chemical Society*, **68**, 2411.

Pringle, R. W., Standil, S. & Roulston, K. (1950). *Physical Review*, **78**, 303.

Pringle, R. W., Standil, S., Taylor, H. W. & Fryer, G. (1951). *Physical Review*, **84**, 1066.

Rasmussen, J. O., Reynolds, F. L., Thompson, S. G. & Ghiorso, A. (1950). *Physical Review*, **80**, 475.

Ravn, H. L. (1979). *Physics Reports*, **54**, 201.

Reynolds, J. H. (1950). *Physical Review*, **79**, 789.

Rider, B. F., Peterson, J. P. Jr & Ruiz, C. P. (1963). *Nuclear Science and Engineering*, **15**, 284.

Rosman, K. J. R., De Laeter, J. R., Boldeman, J. W. & Thode, H. G. (1983). *Canadian Journal of Physics*, **61**, 1490.

Roth, E., Lucas, M., Lecompte, T., Devillers, C. & Hagemann, R. (1975). *Le Phenoménè d' Oklo*, p. 489, International Atomic Energy Agency, Vienna.

Sayag, G.-J. (1951). *Comptes Rendus Hebdomadaires des Séances de l'Académie des Sciences*, Paris, **232**, 2091.

Sharma, K. S., Ellis, R. J., Derenchuk, V. P., Barber, R. C. & Duckworth, H. E. (1980). *Physics Letters*, **91B**, 211.

Simpson, J. J, Campbell, J. L. & Mahm, H. (1982). Private communication to the authors.

Smythe, W. R. & Hemmendinger, A. (1937). *Physical Review*, **51**, 178.

Talbert, W. L. Jr, Wohn, F. K., Landin, A. R., Pacer, J. C., Gill, R. L., Cullison, M. A., Sheppard, G. A., Burke, K. A., Malaby, K. L. & Voigt, A. F. (1976). *Nuclear Instruments and Methods*, **139**, 257.

Thode, H. G. & Graham, R. L. (1947). *Canadian Journal of Research A*, **25**, 1.

Wanless, R. K. & Thode, H. G. (1953). *Canadian Journal of Physics*, **31**, 517.

Watt, D. E. & Glover, R. N. (1962). *Philosophical Magazine*, **7**, 105.

Weaver, B. (1950). *Physical Review*, **80**, 301.

White, F. A., Collins, T. L. & Rourke, F. M. (1955). *Physical Review*, **97**, 566.

White, F. A., Sheffield, J. C. & Rourke, F. M. (1958). *Applied Spectroscopy*, **12**, 46.

Chapter 10

Baldeschwieler, J. D., Benz, H. & Llewellyn, P. M. (1968). *Advances in Mass Spectrometry*, vol. 4, ed. Kendrick, E., p. 113. London: Institute of Petroleum.

Baldeschwieler, J. D. & Woodgate, S. S. (1971). *Accounts of Chemical Research*, **4**, 114.

Barnard, G. P. (1953). *Modern Mass Spectrometry*. London: The Institute of Physics.

Bleakney, W. (1929). *Physical Review*, **34**, 157.

Bleakney, W. (1930*a*). *Physical Review*, **35**, 1180.

Bleakney, W. (1930*b*). *Physical Review*, **36**, 1303.

Bourne, A. J. & Danby, C. J. (1968). *Journal of Scientific Instruments, Series 2*, **1**, 155.

Brehm, B. & von Puttkamer, E. (1968). *Advances in Mass Spectrometry*, vol. 4, ed. Kendrick, E., p. 591. London: Institute of Petroleum.

Čermák, V. & Herman, Z. (1961). *Nucleonics*, **19**, 106.

Champion, R. L., Doverspike, L. D. & Bailey, T. L. (1966). *Journal of Chemical Physics*, **45**, 4377.

Davidson, W. R., Powers, T., Sue, T. & Sue, D. H. (1977). *International Journal of Mass Spectrometry and Ion Physics*, **24**, 83.

Dempster, A. J. (1916). *Philosophical Magazine*, **31**, 438.

Dibeler, V. H. (1947). *Journal of Research of the National Bureau of Standards*, **38**, 329.

Ditchburn, R. W. & Arnot, F. L. (1929). *Proceedings of the Royal Society A*, **123**, 516.

Dromey, R. G., Morrison, J. D. & Traeger, J. C. (1971). *International Journal of Mass Spectrometry and Ion Physics*, **6**, 57.

Drowart, J. (1985). *Tenth International Mass Spectrometry Conference*, Swansea.

Drowart, J. & Goldfinger, P. (1967). *Angewandte Chemie*, **79**, 589.

Durup, J. (1974). *Advances in Mass Spectrometry*, vol. 6, ed. West, A. R., p. 691. London: Institute of Petroleum.

Eland, J. H. D. (1980). *Advances in Mass Spectrometry*, vol. 8A, ed. Quayle, A., p. 17. London: Institute of Petroleum.

Eyring, H. (1955). *Annual Review of Nuclear Science*, **5**, 241.

Fand, R. & McMahon, C. (1978). *International Journal of Mass Spectrometry and Ion Physics*, **27**, 163.

Fano, U. & Cooper, J. W. (1965). *Physical Review*, **137A**, 1364.

Fehsenfeld, F. C., Ferguson, E. E. & Schmeltekopf, A. L. (1966). *Journal of Chemical Physics*, **44**, 3022.

Ferguson, E. E. (1975). *Annual Reviews of Physical Chemistry*, **26**, 17.

Fox, R. E., Hickam, W. M., Kjeldaas, T. & Grove, D. J. (1951). *Physical Review*, **84**, 859.

Fox, R. E., Hickam, W. M., Grove, D. J. & Kjeldaas, T. (1955). *Review of Scientific Instruments*, **26**, 1101.

Fox, R. E. & Hipple, J. A. (1948). *Review of Scientific Instruments*, **19**, 462.
Franck, J. & Hertz, G. (1913). *Verhandlungen der Physkalische Gesellschaft*, **15**, 373.
Franklin, J. L. (1972). ed. *Ion–Molecule Reactions*. New York: Plenum Press.
Franklin, J. L. & Harland. P. W. (1974). *Annual Reviews of Physical Chemistry*, **25**, 485.
Freiser, B. S. (1978). *International Journal of Mass Spectrometry and Ion Physics*, **26**, 39.
Fristrom, R. M. (1975). *International Journal of Mass Spectrometry and Ion Physics*, **16**, 15.
Futrell, J. H. & Miller, C. D. (1966). *Review of Scientific Instruments*, **37**, 1521.
Futrell, J. H. & Tiernan, O. (1972). *See* Franklin (1972), p. 487.
Geltmann, S. (1956). *Physical Review*, **102**, 171.
Giese, C. F. & Maier, W. B. (1963). *Journal of Chemical Physics*, **39**, 739.
Goldfinger, P. (1965). *Mass Spectrometry*, ed. Reed, R. I., p. 265. London: Academic Press.
Grimley, R. T. (1967). *Characterization of High Temperature Vapours*, ed. Margrave, J. L. New York: Wiley.
Gross, M. L. & Wilkins, C. L. (1971). *Analytical Chemistry*, **43**, 65A.
Hagstrum, H. D. (1951). *Reviews of Modern Physics*, **23**, 185.
Hagstrum, H. D. & Tate, J. T. (1941). *Physical Review*, **59**, 354.
Hanson, E. E. (1937). *Physical Review*, **51**, 86.
Hasted, J. B. (1974). *Advances in Mass Spectrometry*, vol. 6, ed. West, A. R., p. 901. London: Institute of Petroleum.
Herman, Z. & Wolfgang, R. (1972). *See* Franklin (1972).
Herod, A. A. & Harrison, A. G. (1970). *International Journal of Mass Spectrometry and Ion Physics*, **4**, 415.
Herzberg, G. (1950). *Spectra of Diatomic Molecules*. Van Nostrand.
Hipple, J. A., Fox, R. E. & Condon, E. U. (1946). *Physical Review*, **69**, 347.
Hipple, J. A. & Stevenson, D. P. (1943). *Physical Review*, **63**, 121.
Hogness, T. R. & Lunn, E. G. (1925). *Physical Review*, **26**, 44.
Honig, R. E. (1954). *Journal of Chemical Physics*, **22**, 126, 1610.
Inghram, M. G. (1953). *Mass Spectroscopy in Physics Research*, p. 151. Washington: National Bureau of Standards Circular 522.
Jennings, K. R. (1978). *Advances in Mass Spectrometry*, vol. 7A, ed. Daly, N. R., p. 209. London: Institute of Petroleum.
Johnstone, R. A. W. & McMaster, B. N. (1974). *Advances in Mass Spectrometry*, vol. 6, ed. West, A. R., p. 451. Barking: Applied Science Publishers.
Kallmann, H. & Bredig, M. (1925). *Zeitschrift für Physik*, **34**, 736.
Katzenstein, H. S. & Friedland, S. S. (1955). *Review of Scientific Instruments*, **26**, 324.
Lenard, P. (1902). *Annalen der Physik* (Leipzig), **8**, 149.
Lindholm, E. (1954). *Zeitschrift für Naturforschung*, **9a**, 535.
Lozier, W. W. (1931). *Physical Review*, **36**, 1285.
Lozier, W. W. (1933). *Physical Review*, **44**, 575.
Lozier, W. W. (1934). *Physical Review*, **46**, 268.
McDaniel, E. W., Čermák, V., Dalgarno, A., Ferguson, E. E. & Friedman, L. (1970). *Ion–Molecule Reactions*. New York: Wiley-Interscience.
McDowell, C. A. (1954). *Applied Mass Spectrometry*, p. 129. London: Institute of Petroleum.

Maeda, K., Semchuk, G. P. & Lossing, F. P. (1968). *International Journal of Mass Spectrometry and Ion Physics*, **1**, 395.
Mann, M. M., Hustrulid, A. & Tate, J. T. (1940). *Physical Review*, **58**, 340.
Mariner, T. & Bleakney, W. (1947). *Physical Review*, **72**, 807.
Marchant, P., Pacquet, C. & Marmet, P. (1969). *Physical Review*, **180**, 123.
Marmet, P. & Morrison, J. D. (1962). *Journal of Chemical Physics*, **36**, 1238.
Marr, G. V. (1967). *Photoionization Processes in Gases*, chapter 6. New York: Academic Press.
Marshall, A. G. (1985). *Tenth International Mass Spectrometry Conference*, Swansea.
Massev. H. S. W. & Burhop, E. H. S. (1972). *Contemporary Physics*, **13**, 135, 375
Mattauch, J. & Lichtblau, H. (1939). *Physikalische Zeitschrift*, **40**, 16.
Morrison, J. D. (1953*a*). *Review of Scientific Instruments*, **25**, 291.
Morrison, J. D. (1953*b*). *Journal of Chemical Physics*, **21**, 1767.
Morrison, J. D. (1954). *Journal of Chemical Physics*, **22**, 1219.
Morrison, J. D. & Traeger, J. C. (1971). *International Journal of Mass Spectrometry and Ion Physics*, **7**, 391.
Mulliken, R. S. (1932). *Review of Modern Physics*, **4**, 1.
Niwa, Y., Nishimura, T., Nozoye, H. & Tsuchiya, T. (1979). *International Journal of Mass Spectrometry and Ion Physics*, **30**, 63.
Robertson, A. J. B. (1955). *Mass Spectrometry*, p. 42. London: Methuen.
Rosenstock, H. M. (1968). *Advances in Mass Spectrometry*, vol. 4, ed. Kendrick, E., p. 523. London: Institute of Petroleum.
Rosenstock, H. M. (1976). *International Journal of Mass Spectrometry and Ion Physics*, **20**, 139.
Rosenstock, H. M., Wallstein, M. B., Wahrhaftig, A. L. & Eyring, H. (1952). *Proceedings of the National Academy of Sciences of the United States of America*, **38**, 667.
Simpson, J. A. & Kuyatt, C. E. (1963). *Review of Scientific Instruments*, **34**, 265.
Smyth, H. D. (1922). *Proceedings of the Royal Society A*, **102**, 283.
Smyth, H. D. (1925). *Physical Review*, **25**, 452.
Stevenson, D. P. (1942). *Journal of Chemical Physics*, **10**, 291.
Stevenson, D. P. (1943). *Journal of the American Chemical Society*, **65**, 209.
Stevenson, D. P. (1950). *Journal of Chemical Physics*, **18**, 1347.
Stevenson, D. P. (1951). *Discussions of the Faraday Society*, **10**, 35.
Stevenson, D. P. (1963). *Mass Spectrometry*, ed. McDowell, C. A., p. 589. New York: McGraw-Hill.
Stockbauer, R. (1977). *International Journal of Mass Spectrometry and Ion Physics*, **25**, 89, 401.
Studniarz, S. A. & Franklin, J. L. (1968). *Journal of Chemical Physics*, **49**, 2652.
Tal'rose, V. L. & Frankevich, E. L. (1960). See Tal'rose, V. L. (1962). *Pure and Applied Chemistry*, **5**, 455.
Tal'rose, V. L. (1966). *Advances in Mass Spectrometry*, vol. 3, ed. Mead, W. L., p. 211. London: Institute of Petroleum.
Tate, J. T. & Smith, P. T. (1932). *Physical Review*, **39**, 270.
Thomas, R., Barassin, A. & Burke, R. R. (1978). *International Journal of Mass Spectrometry and Ion Physics*, **28**, 275.
Turner, B. R., Fineman, M. A. & Stebbings, R. F. (1965). *Journal of Chemical Physics*, **42**, 4088.
Vasile, M. J. & Smolinsky, G. (1974). *Advances in Mass Spectrometry*, vol. 6, ed. West,

A. R., p. 743. London: Institute of Petroleum.
Waldron, J. D. & Wood, K. (1952). *Mass Spectrometry*, p. 16. London: Institute of Petroleum.
Wigner, E. P. (1948). *Physical Review*, **73**, 1002.
Winters, R. E. & Collins, J. E. (1966). *Journal of Chemical Physics*, **45**, 1931.
Worley, R. E. (1943). *Physical Review*, **64**, 207.
Worley, R. E. (1953). *Physical Review*, **89**, 863.
Worley, R. E. & Jenkins, F. A. (1938). *Physical Review*, **54**, 305.

Chapter 11

Arpino, P. J. & Guiochon, G. (1979). *Analytical Chemistry*, **51**, 682A.
Beckey, H. D., Knoppel, H., Metzinger, G. & Schulze, P. (1966). *Advances in Mass Spectrometry*, vol. 3, ed. Mead, W. L., p. 35. London: Institute of Petroleum.
Benninghoven, A., Jaspers, D. & Sichtermann, W. (1978). *Advances in Mass Spectrometry*, vol. 7B, ed. Daly, N. R., p. 1433. London: Institute of Petroleum.
Beynon, J. H. (1959). *Advances in Mass Spectrometry*, ed. Waldron, J. D., p. 328. London, New York, Paris, Los Angeles: Pergamon.
Beynon, J. H. (1968). *Advances in Mass Spectrometry*, vol. 4, ed. Kendrick, E., p. 123. London: The Institute of Petroleum.
Beynon, J. H. (1978). *Pure and Applied Chemistry*, **50**, 65.
Beynon, J. H. & Williams, A. E. (1963). *Mass and Abundance Tables for Use in Mass Spectrometry*. Amsterdam, London, New York: Elsevier.
Biemann, K. (1972). *Biochemical Applications of Mass Spectrometry*, ed. Walker, G. R., New York: Wiley.
Blakley, C. R., McAdams, M. J. & Vestal, M. L. (1980). *Advances in Mass Spectrometry*, vol. 8A, ed. Quayle, A., p. 1616. London: Institute of Petroleum.
Block, J. (1968). *Advances in Mass Spectrometry*, vol. 4, ed. Kendrick, E., p. 791. London: Institute of Petroleum.
Bowie, J. D. (1975). *MTP International Review of Science, Physical Chemistry Series 2, vol. 5, Mass Spectrometry*, ed. Maccoll, A., p. 89. London: Butterworths.
Budzikiewicz, H., Djerassi, C. & Williams, D. H. (1967). *Mass Spectrometry of Organic Compounds*, San Francisco: Holden–Day.
Burlingame, A. L., Shackleton, C., Howe, I. & Chishov, O. S. (1978). *Analytical Chemistry*, **50**, 346R.
Cooks, R. G., Louris, J. N., Wright, L. G., Schoen, A. E., Dobberstein, P. and Pesch, R. (1985). *Tenth International Mass Spectrometry Conference*, Swansea.
Daly, N. R. (1978*a*). ed. *Advances in Mass Spectrometry*, vol. 7B, section 10, p. 975 *et seq.* London: Institute of Petroleum.
Daly, N. R. (1978*b*). ed. *Advances in Mass Spectrometry*, vol. 7B, section 13, p.1483 *et seq.* London: Institute of Petroleum.
Derrick, P. J. (1978). *Advances in Mass Spectrometry*, vol. 7A, ed. Daly, N. R., p. 143. London: Institute of Petroleum.
Fales, H. M., Milne, G. W. A., Winkler, H. U., Beckey, H. D., Damico, J. N. & Barrow, R. (1975). *Analytical Chemistry*, **47**, 207.
Fenselau, C. (1977). *Analytical Chemistry*, **49**, 563A.
Field, F. H. (1968). *Advances in Mass Spectrometry*, vol. 4, ed. Kendrick, E., p. 645. London: Institute of Petroleum.

Games, D. E. (1979). *Mass Spectrometry Specialist Periodical Reports*, vol. 5, ed. Johnstone, R. A. W., p. 285. London: Burlington House, The Chemical Society.
Gierlich, H. H. & Röllgen, F. W. (1979). *International Journal of Mass Spectrometry and Ion Physics*, **29**, 363.
Gudzinowicz, B., Gudzinowicz, M. J. & Martin, H. F. (1976). *Fundamentals of Integrated GC–MS*. New York, London: Marcel Dekker.
Henneberg, D. (1980). *Advances in Mass Spectrometry*, vol. 8B., ed. Quayle, A., p. 1511. London: Institute of Petroleum.
Holmes, J. C. (1975). *MTP International Review of Science, Physical Chemistry, Series 2, vol. 5, Mass Spectrometry*, ed. Maccoll, A., p. 207. London: Butterworths.
Holmes, J. C. & Morrell, F. A. (1957). *Applied Spectroscopy*, **11**, 86.
Horman, I. (1979). *Mass Spectrometry Specialist Periodical Reports*, vol. 5, p. 211. London: Burlington House, The Chemical Society.
Johnstone, R. A. W. (1972). *Mass Spectrometry for Organic Chemists*. Cambridge University Press.
Johnstone, R. A. W. (1981). *Mass Spectrometry Specialist Periodical Reports*, vol. 6. London: Royal Society of Chemistry.
Krahmer, U. I. & McCloskey, J. A. (1978). *Advances in Mass Spectrometry*, vol. 7B, ed. Daly, N. R., p. 1483. London: Institute of Petroleum.
Macfarlane, R. D. (1982). *Accounts of Chemical Research*, **15**, 268.
Macfarlane, R. D., McNeal, C. J. & Hunt, J. E. (1980). *Advances in Mass Spectrometry*, vol. 8A, ed. Quayle, A., p. 349. London: Institute of Petroleum.
McCormick, A. (1977). *Mass Spectrometry Specialist Periodical Reports*, vol. 4, ed. Johnstone, R. A. W., p. 85. London: Burlington House, The Chemical Society.
McCormick, A, (1979). *Mass Spectrometry Specialist Periodical Reports*, vol. 5, ed. Johnstone, R. A. W., p. 121. London: Burlington House, The Chemical Society.
McCormick, A. (1981). *Mass Spectrometry Specialist Periodical Reports*, vol. 6, ed. Johnstone, R. A. W., p. 153. London: Burlington House, The Chemical Society.
McLafferty, F. W. (1980). *Interpretation of Mass Spectra*, 3rd edn. Mill Valley, CA: University Science Books.
McLafferty, F. W. (1983). ed. *Tandem Mass Spectrometry*. New York: Wiley.
McLafferty, F. W. (1985). *Tenth International Mass Spectrometry Conference*, Swansea.
McLafferty, F. W, Todd, P. J., McGilvery, D. C., Baldwin, M. A., Bockhoff, F. M., Wendel, G. J., Wixom, M. R. & Niemi, T. E. (1980). *Advances in Mass Spectrometry*, vol. 8B, ed. Quayle, A., p. 1589. London: Institute of Petroleum.
McLafferty, F. W., Wachs, T., Koppel, C., Dymerski, P. P. & Bockhoff, F. M. (1978). *Advances in Mass Spectrometry*, vol. 7B, ed. Daly, N. R., p. 1231. London: Institute of Petroleum.
McLafferty, F. W. & Venkataraghavan, R. (1979) *Journal of Chromatographic Science*, **17**, 24.
Melera, A. (1980). *Advances in Mass Spectrometry*, vol. 8A, ed. Quayle, A., p. 1616. London: Institute of Petroleum.
Mellon, F. A. (1975). *Mass Spectrometry Specialist Periodical Reports*, vol. 3, ed. Johnstone, R. A. W., p. 117. London: Burlington House, The Chemical Society.
Mellon, F. A. (1977). *Mass Spectrometry Specialist Periodical Reports*, vol. 4, ed. Johnstone, R. A. W., p. 59. London: Burlington House, The Chemical Society.
Mellon, F. A. (1979). *Mass Spectrometry Specialist Periodical Reports*, vol. 5, ed.

Johnstone, R. A. W., p. 100. London: Burlington House, The Chemical Society.
Milne, G. W. A., Heller, S. R., Heller, R. S. & Martinsen, D. P. (1980). *Advances in Mass Spectrometry*, vol. 8B, ed. Quayle, A., p. 1578. London: Institute of Petroleum.
Puzo, G. & Prome, J. C. (1978). *Advances in Mass Spectrometry*, vol. 7B, ed. Daly, N. R., p. 1596. London: Institute of Petroleum.
Schrader, S. R. (1974). *Introductory Mass Spectroscopy*. Boston: Allyn and Bacon.
Schulten, H. R. (1978). *Advances in Mass Spectrometry*, vol. 7A, ed. Daly, N. R., p. 83. London: Institute of Petroleum.
Stenhagen, E., Abrahamson, S. & McLafferty, F. W. (1969). *Atlas of Mass Spectral Data*. New York, London, Sydney, Toronto: Wiley.
Stenhagen, E., Abrahamson, S. & McLafferty, F. W. (1974). *Registry of Mass Spectral Data*. New York, London, Sydney, Toronto: Wiley.
Tunnicliff, D. D., Wadsworth, P. A. & Schissler, D. O. (1965). *Analytical Chemistry*, **37**, 543.
Ward, S. D. (1971). *Mass Spectrometry Specialist Periodical Reports*, vol. 1, ed. Johnstone, R. A. W., p. 253. London: Burlington House, The Chemical Society.
Ward, S. D. (1973). *Mass Spectrometry Specialist Periodical Reports*, vol. 2, ed. Johnstone, R. A. W., p. 264. London: Burlington House, The Chemical Society.
Wilson, J. M. (1971). *Mass Spectrometry Specialist Periodical Reports*, vol. 1, ed. Johnstone, R. A. W., p. 1. London: Burlington House, The Chemical Society.
Wilson, J. M. (1973). *Mass Spectrometry Specialist Periodical Reports*, vol. 2, ed. Johnstone, R. A. W., p. 1. London: Burlington House, The Chemical Society.
Wilson, J. M. (1975). *Mass Spectrometry Specialist Periodical Reports*, vol. 3, ed. Johnstone, R. A. W., p. 86. London: Burlington House, The Chemical Society.
Wilson, J. M. (1977). *Mass Spectrometry Specialist Periodical Reports*, vol. 4, ed. Johnstone, R. A. W., p. 102. London: Burlington House, The Chemical Society.
Wood, G. W. (1982). *Tetrahedron*, **38**, 1125.

Chapter 12

Anderson, C. A. (1972). *Science*, **175**, 863.
Anderson, C. A. & Hinthorne, J. R. (1973). *Analytical Chemistry*, **45**, 1421.
Benninghoven, A. & Loebach, E. (1971). *Review of Scientific Instruments*, **42**, 49.
Benninghoven, A., Plog, C. & Trietz, N. (1974). *International Journal of Mass Spectrometry and Ion Physics*, **13**, 415.
Bingham, R. A. & Salter, P. L. (1976). *International Journal of Mass Spectrometry and Ion Physics*, **21**, 133.
Bohr, N. (1948). *Kongelige Danske Videnskab Selskab Matematisk-Fysiske Meddelelser*, **18**, 8.
Carter, G. & Grant, W. A. (1977). *Ion Implantation in Semiconductor and Other Materials*. New York: Wiley.
Castaing, R. & Hennequin, J. F. (1971). *Advances in Mass Spectrometry*, vol. 5, ed. Quayle, A., p. 419. London: Institute of Petroleum.
Castaing, R. & Slodzian, G. (1962). *Journal de Microscopie (Paris)*, **1**, 395.
Chakravarty, B., Venkatasubramanian, V. S. & Duckworth, H. E. (1963). *Advances in Mass Spectrometry*, vol. 2, ed. Elliott, R. M., p. 128. New York: The Macmillan Company.
Chu, W. K., Mayer, J. W. & Nicolet, M. A. (1978). *Backscattering Spectrometry*. New York: Academic Press.

Conzemius, R. J. & Svec, H. J. (1972). *Trace Analysis by Mass Spectrometry*, ed. Ahearn, A. J., chapter 5. New York: Academic Press.

Dearnaley, G., Freeman, J. H., Nelson, R. S. & Stephen, J. (1973). *Ion Implantation.* Amsterdam: North Holland.

de Grieve, F., Figaret, P. & Laty, P. (1979). *International Journal of Mass Spectrometry and Ion Physics*, **29**, 351.

Donnelly, J. P. (1981). *Nuclear Instruments and Methods*, **182/183**, 553.

Evans, C. A. (1971). *Advances in Mass Spectrometry*, vol. 5, ed. Quayle, A., p. 436. London: Institute of Petroleum.

Evans, C. A. & Pensler, J. P. (1970). *Analytical Chemistry*, **42**, 1060.

Farrar, H. (1972). *Trace Analysis by Mass Spectrometry*, ed. Ahearn, A. J., chapter 8. New York: Academic Press.

Freeman, J. H. (1967). *Proceedings of the International Conference on Applications of Ion Beams to Semiconductor Technology*, ed. Glotin, P. M. Grenoble: CEN.

Furstenau, N., Hillenkamp, F. & Nitsche, R. (1979). *International Journal of Mass Spectrometry and Ion Physics*, **31**, 85.

Gemell, D. S. (1974). *Reviews of Modern Physics*, **46**, 129.

Hirvonen, J. K. (1980). *Ion Implantation, Treatise on Materials Science and Technology*, **18**, New York & London: Academic Press.

Honig, R. E. (1966) *Advances in Mass Spectrometry*, vol. 3, ed. Mead, W. L., p. 126. London: Institute of Petroleum.

Honig, R. E. (1976). *Advances in Mass Spectrometry*, vol. 6, ed. West, A. R., p. 337. London: Institute of Petroleum.

Hutcheon, I. D., Steele, J. M., Smith, J. V. & Clayton, R. N. (1978). *Proceedings of the Ninth Lunar Conference*, **1**, 1345.

James, J. A. & Williams, J. L. (1959). *Advances in Mass Spectrometry*, ed. Waldron, J. D., p. 157. London: Pergamon Press.

Jaworski, J. F. & Morrison, G. H. (1974). *Analytical Chemistry*, **46**, 2080.

Kiko, J., Muller, R. W. & Kirsten, E. (1979). *International Journal of Mass Spectrometry and Ion Physics*, **11**, 23.

Kloppel, K. D. & Seidl, D. (1979). *International Journal of Mass Spectrometry and Ion Physics*, **31**, 151.

Kovalev, I. D., Maksimov, G. A., Suchkov, A. I. & Larin, N. V. (1978). *International Journal of Mass Spectrometry and Ion Physics*, **27**, 101.

Liebl, H. (1967). *Journal of Applied Physics*, **38**, 5277.

Lindhard, J. (1954). *Kongelige Danske Videnskabernes Selskab Matematisk-Fysiske Meddelelser*, **28**, no. 8.

Morobito, J. M. & Lewis, R. K. (1973). *Analytical Chemistry*, **45**, 863.

Morrison, G. H. & Slodzian, G. (1975). *Analytical Chemistry*, **47**, 933A.

Nielsen, K. O. (1957). *Nuclear Instruments and Methods*, **1**, 289.

Pawel, R. E., Pensler, J. P. & Evans, C. A., Jr (1972). *Journal of the Electrochemical Society*, **119**, 24.

Pickar, K. A. (1975). *Applied Solid State Science*, vol. 5, ed. Wolfe, R., p. 152. New York: Academic Press.

Ryssel, H. & Glawischnig, H. (1982). *Ion Implantation Techniques.* Berlin: Springer-Verlag.

Schilling, J. H. & Büger, P. A. (1978). *International Journal of Mass Spectrometry and Ion Physics*, **26**, 163.

Slodzian, G. (1975). *Surface Science*, **48**, 161.
Slodzian, G. & Havette, A. (1976). *Advances in Mass Spectrometry*, vol. 6, ed. West, A. R., p. 483. London: Institute of Petroleum.
Stüwer, J. (1976). *International Journal of Mass Spectrometry and Ion Physics*, **20**, 387.
Taylor, S. R. (1971). *Geochimica et Cosmochimica Acta*, **35**, 1187.
Townsend, P. D., Kelly, J. C. & Hartley, N. E. W. (1976). *Ion Implantation, Sputtering and Their Applications.* New York: Academic Press.
Van Hoye, E., Gijbels, R. & Adams, F. (1980). *Advances in Mass Spectrometry*, vol. 8, ed. Quayle, A., p. 357. London: Heyden and Son Ltd.
Wilson, R. G. (1967). *Proceedings of the International Conference on Applications of Ion Beams*, p. 105, Grenoble: CEN.
Wilson, R. G. & Brewer, G. R. (1979). *Ion Beams – with Applications to Ion Implantation.* Malabar, Florida: Robert E. Krieger.
Wolicki, E. A. (1979). *IEEE Transactions on Nuclear Science*, **26**, 1800.
Woolston, J. B. (1972). *Trace Analysis by Mass Spectrometry*, ed. Ahearn, A. J., chapter 7. New York: Academic Press.
Westmore, J. B., Ens, W. & Standing K. G. (1982). *Biomedical Mass Spectrometry*, **9**, 119.

Chapter 13

Ahrens, L. H. (1949). *Bulletin of the Geological Society of America*, **60**, 217.
Aldrich, L. T. & Nier, A. O. (1948). *Physical Review*, **74**, 1225.
Armstrong, R. L. (1971). *Earth and Planetary Science Letters*, **12**, 137.
Aston, F. W. (1929). *Nature, London*, **123**, 313.
Aston, F. W. (1933). *Proceedings of the Royal Society A*, **140**, 534.
Bleakney, W. & Gould, A. J. (1933). *Physical Review*, **44**, 265.
Catanzaro, E. J. (1968). *Radiometric Dating for Geologists*, ed. Hamilton, E. I. & Farquhar, R. M., New York: Interscience.
Chapman, S. (1917). *Philosophical Transactions of the Royal Society of London, Ser. A*, **217**, 184.
Clayton, R. N., Grossman, L. & Mayeda, T. (1973). *Science*, **182**, 485.
Clayton, R. N., Friedman, I., Graf, D. L., Mayeda, T. K., Meents, W. F. & Shimp, N. F. (1966). *Journal of Geophysical Research*, **71**, 3869.
Clifford, T. N., Rooke, J. M. & Allsopp, H. L. (1969). *Geochimica et Cosmochimica Acta*, **33**, 973.
Clusius, K. & Dickel, G. (1938). *Naturwissenschaften*, **26**, 546.
Colombo, U., Gazzarrini, F., Gonfiantini, R., Kneuper, G., Teichmueller, R. (1968). *Zeitschrift für Angewandte Geologie*, **14**, 257.
Commission on Atomic Weights (1984). *Pure and Applied Chemistry*, **56**, 675.
Cox, A. (1963). *Science*, **163**, 237.
Craig, H. (1953). *Geochimica et Cosmochimica Acta*, **3**, 53.
Craig, H. (1961). *Science*, **133**, 1833.
Craig, H. (1963). *Nuclear Geology on Geothermal Areas*, Conference, Spoleto, Italy, ed. Tongiorgi, E., p. 53. Pisa: Consiglio Nazionale delle Ricerche.
Craig, H. (1965). *Stable Isotopes in Oceanographic Studies and Paleotemperatures*, Conference, Spoleto, Italy, p. 24. Pisa: Consiglio Nazionale delle Ricerche.
Craig, H. & Boato, C. (1955). *Annual Review of Physical Chemistry*, **6**, 403.
Craig, H. & Gordon, L. I. (1965). *Stable Isotopes in Oceanographic Studies and*

Paleotemperatures, Conference, Spoleto, Italy, p. 122. Pisa: Consiglio Nazionale delle Ricerche.

Crawford, A. R. (1969). *Geological Society of India*, **10**. 117.

Cumming, G. L. & Richards, J. R. (1975). *Earth and Planetary Science Letters*, **28**, 155.

Dalrymple, G. B. & Lanphere, M. A. (1971). *Earth and Planetary Science Letters*, **12**, 300.

Dansgaard, W. (1964). *Tellus*, **16**, 436.

Doe, B. R. (1967). *Journal of Petrology*, **8**, 51.

Doe, B. R. (1970). *Lead Isotopes*. Berlin, New York: Springer-Verlag.

Dole, M. (1935). *Journal of the American Chemical Society*, **57**, 2731.

Dole, M. (1936). *Journal of Chemical Physics*, **4**, 268.

Eckelmann, W. R. (1962). *American Association of Petroleum Geologists Bulletin*, **46**, 699.

Emiliani, C. (1966). *Science*, **154**, 851.

Enskog, D. (1911). *Physikalische Zeitschrift*, **12**, 56, 533.

Epstein, S. & Mayeda, E. (1953). *Geochimica et Cosmochimica Acta*, **4**, 213.

Faure, G. (1977). *Principles of Isotope Geology*. New York: Wiley.

Fleischer, R. L., Buford, P. & Walker, R. M. (1975). *Nuclear Tracks in Solids: Principles and Applications*. Berkeley: University of California Press.

Friedman, I., Redfield, A. C., Schoen, B. & Harris, J. (1964). *Reviews of Geophysics*, **2**, 177.

Fritz, P., Cherry, J. A., Weyer, K. U. & Sklash, M. (1976). *Interpretation of Environmental Isotope and Hydrochemical Data in Groundwater Hydrology*, p. 111. Vienna: International Atomic Energy Agency.

Gebauer, D. & Grünenfelder, M. (1979). *Lectures in Isotope Geology*. eds. Jäger, E. & Hunzicker, J. C., p. 105. Berlin, Heidelberg, New York: Springer-Verlag.

Goldhaber, M. B. & Kaplan, I. R. (1974). 'The sedimentary S-cycle', in *The Sea*, vol. 4, ed. Goldberg, E. D. New York: Wiley.

Gopalan, K. & Wetherill, G. W. (1969). *Journal of Geophysical Research*, **73**, 7133

Gopalan, K. & Wetherill, G. W. (1970). *Journal of Geophysical Research*, **75**, 3547.

Hann, O. & Walling, E. (1938). *Zietschrift für Anorganische und Allgemeine Chemie*, **236**, 78.

Hart, S. R. (1964). *Journal of Geology*, **72**, 493.

Herr, W. & Merz, E. (1958). *Zeitschrift für Naturforschung*, **13a**, 231.

Hoefs, J. (1980). *Stables Isotope Geochemistry*, 2nd edn. Berlin, Heidelberg, New York: Springer-Verlag.

Hoefs, J. & Schidlowsky, M. (1967). *Science*, **155**, 1096.

Holmes, A. (1946). *Nature*, **157**, 680.

Houtermans, G. (1951). *Naturwissenschaften*, **38**, 132.

Hurley, P. M. & Rand, J. R. (1969). *Science*, **164**, 1229.

Hussain, L. & Schafer, G. A. (1973). *Science*, **130**, 1358.

Inghram, M. G. (1953). *Proceedings of the Conference on Application of Nuclear Processes to Geological Problems*, p. 35. Geneva, Wisconsin, U.S.A.

Inghram, M. G. & Reynolds, J. H. (1950). *Physical Review*, **78**, 822.

Johnson, S. J., Dansgaard, W., Clansen, H. B. & Langway, C. C. (1972). *Nature*, **235**, 429.

Kajiwara, Y. & Krouse, H. R. (1971). *Canadian Journal of Earth Sciences*, **8**, 1397.

Kanaseiwitch, E. R. (1968). *Radiometric Dating for Geologists*, eds. Hamilton, E. I. & Farquhar, R. M. New York: Interscience.

Kirshenbaum, I. (1951). *Physical Properties and Analysis of Heavy Water*. New York: McGraw-Hill.
Kirsten, T. (1978). *The Origin of the Solar System*, ed. Dermott, S. F., p. 312. Chichester, New York, Brisbane, Toronto: John Wiley and Sons.
Krogh, T. E. (1973). *Geochemica et Cosmochimica Acta*, **37**, 485.
Libby, W. F. (1955). *Radiocarbon Dating*. Chicago: University of Chicago Press.
Litherland, A. E., Beukens, R. P., Kilius, L. R., Rucklidge, J. C., Gove, H. E., Elmore, D. & Purser, K. H. (1981). *Nuclear Instruments and Methods*, **186**, 463.
Lugmair, G. W., Schenin, M. D. & Marti, K. (1975). *Proceedings of the Lunar Society Sixth Conference*, **2**, 1419.
MacNamara, J. & Thode, H. G. (1950). *Physical Review*, **78**, 307.
MacNamara, J. & Thode, H. G. (1951). *Research*, **4**, 582.
Maurette, M. (1976). *Annual Review of Nuclear Science*, **26**, 351.
McKirdy, D. M. & Powell, T. G. (1974). *Geology*, **2**, 591.
Merrihue, C. & Turner, G. (1966). *Journal of Geophysical Research*, **77**, 2852.
Moorbath, S. (1969). *Journal of Geology*, **5**, 154.
Moorbath, S. & Pankhurst, R. J. (1976). *Nature*, **262**, 124.
Muller, R. A. (1977). *Science*, **196**, 489.
Murphey, B. F. & Nier, A. O. (1941). *Physical Review*, **59**, 771.
Murthy, V. R. & Patterson, C. (1962). *Journal of Geophysical Research*, **67**, 116.
Nielsen, H. (1979). *Lectures in Isotope Geology*, eds. Jäger, E. & Hunzicker, J. C., p. 283. Berlin, Heidelberg, New York: Springer-Verlag.
Nier, A. O. (1935). *Physical Review*, **48**, 283.
Nier, A. O. (1938). *Journal of the American Chemical Society*, **60**, 1571.
Nier, A. O. (1939*a*). *Physical Review*, **55**, 150.
Nier, A. O. (1939*b*). *Physical Review*, **55**, 153.
Nier, A. O. & Gulbranson, E. A. (1939). *Journal of the American Chemical Society*, **61**, 697.
Nier, A. O., Thompson, R. W. & Murphey, B. F. (1941). *Physical Review*, **60**, 112.
Ohmoto, H. (1972). *Economic Geology*, **67**, 551.
O'Neil, J. R. (1979). *Lectures in Isotope Geology*, eds. Jäger, E. & Hunzicker, J. C., p. 235. Berlin, Heidelberg, New York: Springer-Verlag.
O'Nions, R. K. & Pankhurst, R. J. (1978). *Earth and Planetary Science Letters*, **38**, 211.
Paneth, F. A., Reasbuck, P. & Mayne, K. I. (1952). *Geochimica et Cosmochimica Acta*, **2**, 300.
Papanastassiou, D. A. & Wasserburg, G. J. (1969). *Earth and Planetary Science Letters*, **5**, 361.
Papanastassiou, D. A. & Wasserburg, G. J. (1975). *Proceedings of the Sixth Lunar Science Conference*, **2**, 1467.
Park, R. & Epstein, S. (1960). *Geochimica et Cosmochimica Acta*, **21**, 110.
Patterson, C. (1955). *Geochimica et Cosmochimica Acta*, **7**, 151.
Patterson, C. (1956). *Geochimica et Cosmochimica Acta*, **10**, 230.
Podosek, F. (1971). *Geochimica et Cosmochimica Acta*, **35**, 157.
Rankama, K. (1954). *Isotope Geology*. Elmsford, New York: Pergamon Press.
Rees, C. E. (1973). *Geochimica et Cosmochimica Acta*, **35**, 625.
Reynolds, J. H. (1967). *Annual Review of Nuclear Science*, **25**, 283.

Richards, T. W. & Lembert, M. E. (1914). *Journal of the American Chemical Society*, **36**, 1329.
Ringwood, A. E. (1979). *Origin of the Earth and Moon*. Berlin, New York: Springer-Verlag.
Russell, R. D. (1972). *Reviews of Geophysical Space Research*, **10**, 529.
Russell, R. D. & Farquhar, R. M. (1960). *Lead Isotopes in Geology*. New York: Interscience.
Rye, R. O. & Ohmoto, H. (1974). *Economic Geology*, **69**, 826.
Sakai, H. (1968). *Geochemical Journal*, **2**, 29.
Schultz, D. A. & Calder, E. A. (1976). *Geochimica et Cosmochimica Acta*, **48**, 381.
Schwarcz, H. P. & Burnie, S. W. (1973). *Mineralium Deposita*, **8**, 264.
Silverman, S. R. (1964). *Isotopic and Cosmic Geochemistry*, eds. Craig, H., Miller, S. L. & Wasserburg, G. J. p. 92. Amsterdam: North Holland.
Silverman, S. R. (1967). *Journal of the Oil Chemistry Society*, **44**, 691.
Smith, B. N. & Epstein, S. (1971). *Plant Physiology*, **47**, 380.
Soddy, F. (1913). *Annual Reports – Chemical Society*, p. 269.
Stacey, J. S. & Kramers, J. D. (1975). *Earth and Planetary Science Letters*, **26**, 207.
Stahl, W. J. (1977). *Chemical Geology*, **20**, 121.
Stahl. W. J. (1979). *Lectures on Isotope Geology*, eds. Jäger, E. & Hunzicker, H., p. 274. Berlin, Heidelberg, New York: Springer-Verlag.
Stanton, R. L. & Russell, R. D. (1959). *Economic Geology*, **54**, 588.
Taksumoto, M., Knight, R. & L'Allegre, C. (1976). *Science*, **180**, 1279.
Teitsma, A. (1975). *Earth and Planetary Science Letters*, **36**, 263.
Thode, H. G., MacNamara, J. & Collins, C. B. (1949). *Canadian Journal of Research*, **B27**, 361.
Thode, H. G., MacNamara, J. & Fleming, W. H. (1953). *Geochimica et Cosmochimica Acta*, **3**, 235.
Thode, H. G. & Monster, J. (1965). *Memoir – American Association of Petroleum Geologists*, **4**, 367.
Thode, H. G. & Rees, R. E. (1970). *Endeavour*, **29**, 24.
Thode, H. G., Wanless, R. K. & Wallouch, R. (1954). *Geochimica et Cosmochimica Acta*, **5**, 286.
Tilton, G. R. (1960). *Journal of Geophysical Research*, **71**, 6095.
Trofimov, A. (1949). *Doklady Akademii Nauk SSSR*, **66**, 181.
Urey, H. C. (1947). *Journal of the American Chemical Society*, **69**, 562.
Urey, H. C. & Grieff, L. J. (1935). *Journal of the American Chemical Society*, **57**, 321.
Venkatasubramanian, V. S. (1974). *Geological Society of India*, **15**, 463.
Venkatasubramanian, V. S., Iyer, S. S. & Pal, S. (1971). *American Journal of Science*, **270**, 43.
Venkatasubramanian, V. S., Jayaram, S., Subramanian, V. (1981). *Journal of the Geological Society of India*, **23**, 219.
von Grosse, A. (1932). *Physical Review*, **42**, 565.
von Grosse, A. (1934). *Journal of Physical Chemistry*, **38**, 487.
von Weizsäcker, C. F. (1937). *Physikalische Zeitschrift*, **38**, 623.
Voshage, H. & Hintenberger, H. (1967). *Zeitschrift für Naturforschung*, **22a**, 477.
Wasserburg, G. J. (1963). *Journal of Geophysical Research*, **68**, 4823.
Wetherill, G. W. (1975). *Annual Review of Nuclear Science*, **25**, 283.

Wetherill, G. W., Tilton, G. O., Davis, G. R. & Aldrich, L. T. (1956). *Geochimica et Cosmochimica Acta*, **9**, 292.

Whelan, T., Sackett, W. M. & Benedict, C. R. (1973). *Plant Physiology*, **51**, 1051.

Wood, A. (1904). *Cambridge Philosophical Society Proceedings*, **12**, 477.

Wood, A. (1905). *Philosophical Magazine*, **9**, 550.

Chapter 14

Bibbo, G., Carver, J. H., Davis, L. A., Horton, B. H. & Lean, J. L. (1979). *Space Research*, **19**, 255.

Donahue, T. M. (1968). *Science*, **159**, 489.

Fricke, K. H., Lane, V., Trinks, H. & von Zahn, U. (1975). *Space Research*, **16**, 273.

Geiss, J., Balsiger, H., Eberhardt, P., Walker, H. P., Weber, L., Young, D. T. & Rosenbauer, H. (1978). *Space Science Reviews*, **22**, 537.

Hayden, J. L., Nier, A. O., French, J. B., Reid, N. M. & Duckett, R. J. (1974). *International Journal of Mass Spectrometry and Ion Physics*, **15**, 37.

Hedin, A. E. & Nier, A. O. (1966). *Journal of Geophysical Research*, **71**, 4121.

Hoffman, J. H., Hodges, R. R., McElroy, M. B., Donahue, T. M. & Kolpin, M. (1979). *Science*, **203**, 802.

Ivanov, G. V., Pakov, L. A. & Pylof, Y. P. (1978). *Space Research*, **19**, 287.

Kissel, J. (1985). *Tenth International Mass Spectrometry Conference*, Swansea.

Lake, L. & Nier, A. O. (1973). *Journal of Geophysical Research*, **78**, 1645.

Mazur, P., Barghoorn, E. S., Halvorson, H. D., Jukes, T. H., Kaplan, I. R. & Margulis, L. (1978). *Space Science Reviews*, **22**, 3.

Moroz, V. I. (1981). *Space Science Reviews*, **29**, 3.

Narcisi, R. S. & Bailey, A. D. (1965). *Journal of Geophysical Research*, **70**, 3687.

Niemann, H. B., Hartle, R. E., Kasprzak, W. T., Spencer, N. W., Hunter, D. M. & Carignan, G. N. (1979). *Science*, **203**, 770.

Nier, A. O., Hanson, W. B., Seiff, A., McElroy, M. B., Spencer, N. W., Duckett, R. J., Knight, T. C. D. & Cook, W. S. (1976). *Science*, **193**, 786.

Nier, A. O. & Hayden, J. L. (1971). *International Journal of Mass Spectrometry and Ion Physics*, **6**, 339.

Nier, A. O., Hoffman, J. H., Johnson, C. Y. & Holmes, J. C. (1964). *Journal of Geophysical Research*, **69**, 979.

Philbrick, C. R. (1976). *Space Research*, **17**, 289.

Potter, W. G., Kayser, B. E. & Nier, A. O. (1978). *Space Research*, **19**, 257.

Rushneck, R. D., Diaz, A. V., Howarth, D. W., Rampacek, J., Olson, K. W., Dencker, W. D., Smith, P., McDavid, L., Tomassian, A., Harris, M., Bulota, K., Biemann, K., LaFleur, A. L., Biller, J. E. & Owen, T. (1978). *Review of Scientific Instruments*, **49**, 817.

Schaefer, E. J. & Nichols, M. H. (1961). *American Rocket Society Journal*, **31**, 1773.

Spencer, N. W. & Reber, C. A. (1963). *Space Research*, **3**, 1151.

Strobel, D. F. (1971). *Journal of Geophysical Research*, **76**, 8384.

Strobel, D. F., Hunten, D. M. & McElroy, M. B. (1970). *Journal of Geophysical Research*, **75**, 4307.

Taylor, H. A., Brinton, H. C., Bauer, S. J., Hartle, R. E., Cloutier, P. A., Michel, F. C., Daniell, R. E., Donahue, T. M. & Maehl, R. C. (1979*a*). *Science*, **203**, 755.

Taylor, H. A., Brinton, H. C., Bauer, S. J., Hartle, R. E., Donahue, T. M., Cloutier, P. A., Michel, F. C., Daniell, R. E. & Blackwell, B. H. (1979*b*) *Science*, **203**, 752.

Taylor, H. A., Brinton, H. C. & Smith, C. R. (1962). *Proceedings of the 8th Annual Aero-Space Symposium*, p. 1. Washington: Instrument Society of America.

Townsend, J. W. (1952). *Review of Scientific Instruments*, **23**, 538.

von Zahn, U. (1968). *Advances in Mass Spectrometry*, vol. 4, ed. Kendrick, E., p. 869. London: Institute of Petroleum.

von Zahn, U., Krankowsky, D., Mauersberger, K., Nier, A. O. & Hunten, D. M. (1979). *Science*, **203**, 768.

Index